SRA NUMBER WORLDS™

A Prevention/Intervention Math Program

Teacher Edition
Level A

Sharon Griffin

 Building Blocks Douglas H. Clements

Julie Sarama

SRA
Columbus, OH

The **McGraw-Hill** Companies

Author
Sharon Griffin
Associate Professor of Education and
Adjunct Associate Professor of Psychology
Clark University
Worcester, Massachusetts

Building Blocks Authors

Douglas H. Clements
Professor of Early Childhood
and Mathematics Education
University at Buffalo
State University of New York, New York

Julie Sarama
Associate Professor of Mathematics Education
University at Buffalo
State University of New York, New York

Contributing Writers
Sherry Booth, *Math Curriculum Developer,* Raleigh, North Carolina
Elizabeth Jimenez, *English Language Learner Consultant,* Pomona, California

Program Reviewers

Jean Delwiche
Almaden Country School
San Jose, California

Cheryl Glorioso
Santa Ana Unified School District
Santa Ana, California

Sharon LaPoint
School District of Indian River County
Vero Beach, Florida

Leigh Lidrbauch
Pasadena Independent School District
Pasadena, Texas

Dave Maresh
Morongo Unified School District
Yucca Valley, California

Mary Mayberry
Mon Valley Education Consortium, AIU 3
Clairton, Pennsylvania

Lauren Parente
Mountain Lakes School District
Mountain Lakes, New Jersey

Juan Regalado
Houston Independent School District
Houston, Texas

M. Kate Thiry
Dublin City School District
Dublin, Ohio

Susan C. Vohrer
Baltimore County Public Schools
Baltimore, Maryland

SRAonline.com

Send all inquiries to:
SRA/McGraw-Hill
8787 Orion Place
Columbus, OH 43240-4027

ISBN 0-07-605335-0

3 4 5 6 7 8 9 BCM 12 11 10 09 08 07 06

Acknowledgments
The research conducted to establish reliability and validity and to create developmental norms was made possible by generous grants from the James S. McDonnell Foundation.

This curriculum was supported in part by the National Science Foundation under Grant No. ESI-9730804, "Building Blocks-Foundations for Mathematical Thinking, Pre-Kindergarten to Grade 2: Research-based Materials Development" to Douglas H. Clements and Julie Sarama. The curriculum was also based partly upon work supported in part by the Institute of Educational Sciences (U.S. Dept. of Education, under the Interagency Educational Research Initiative, or IERI, a collaboration of the IES, NSF, and NICHHD) under Grant No. R305K05157, "Scaling Up TRIAD: Teaching Early Mathematics for Understanding with Trajectories and Technologies" and by the IERI through a National Science Foundation NSF Grant No. REC-0228440, "Scaling Up the Implementation of a Pre-Kindergarten Mathematics Curricula: Teaching for Understanding with Trajectories and Technologies." Any opinions, findings, and conclusions or recommendations expressed in this material are those of the authors and do not necessarily reflect the views of the funding agencies.

Photo Credit
C12 ©PhotoDisc/Getty Images, Inc.

Build a solid foundation with

SRA NUMBER WORLDS™

A Prevention/Intervention Math Program

Make a world of difference in your students' **math skills** with a versatile program that has **proven results** in the classroom.

Number Worlds is an intensive intervention program that focuses on students who are one or more grade levels behind in elementary mathematics. It provides all the tools teachers need to assess students' abilities, individualize instruction, build foundational skills and concepts, and make learning fun. And only *Number Worlds* includes a prevention program for Grades Pre-K–1. It's a unique course full of activities that build foundational math skills and prepare younger children to understand more complex concepts later.

☑ Targeted instruction
Through intense identification and development of core concepts, *Number Worlds* creates competency that quickly puts students on-level with their peer groups. The program provides hands-on activities proven effective with even the lowest-level students, including:
- Computer activities
- Discussion activities
- Paper-and-pencil activities

☑ Precise assessment
Number Worlds' easy-to-use assessment component pinpoints the exact unit in which students should begin the curriculum. Weekly and unit tests monitor progress with open response and/or multiple-choice questions to identify when students are ready to return to the main math curriculum.

☑ Flexibility for teachers and students
Number Worlds' lessons are flexible for use in many settings:
- Resource room
- After school
- Summer school

Teacher's aides and parents can use *Number Worlds* after class or at home.

☑ Comprehensive, fully-integrated program
Complete *Number Worlds* program kits are available at every level and include:
- Teacher Edition
- Student workbooks (Levels C–H)
- Student worksheet blackline masters
- Manipulatives (Levels A–F)
- Software for assessment, placement, professional development, and activities

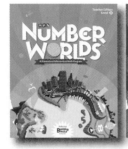

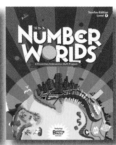

Number Worlds
Teacher Editions

Number Worlds
Manipulatives and Games

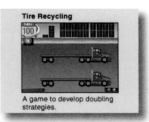

Building Blocks
Screen Capture

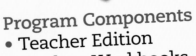

Program Components
- Teacher Edition
- Student Workbooks
- Assessment
- Manipulatives
- Games
- Technology featuring **Building Blocks** software

Number Worlds Results

Number Worlds has been developed and refined since the mid-1980s and has been the only such program to show proven results through years of rigorous field testing. These tests show how students who began at a disadvantage surpassed the performance of students who began on-level with their peers, simply with the help of the **Number Worlds** program.

One of the tests was a longitudinal study conducted to measure the progress of three groups of children from the beginning of Kindergarten to the end of Grade 2. The treatment and control groups both tested one to two years behind normative measures in mathematical knowledge, while the normative group was on track. The treatment group received the **Number Worlds** program while the other two groups used a variety of other mathematical programs during the entire course of the study.

The chart below shows the progress of each group in mean developmental mathematics-level scores as measured by the **Number Worlds** test. The treatment group using the **Number Worlds** program met and exceeded normative mean developmental-level scores by the end of Grade 2. Meanwhile, the control group continued to fall behind their peers.

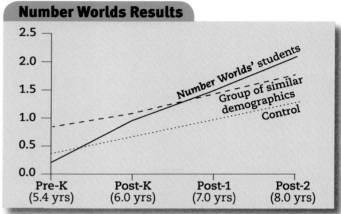

Longitudinal study showing mean developmental scores in mathematical knowledge during Grades K–2

Building Blocks Software Results

The *Building Blocks* software, incorporated into the **Number Worlds** program, is the result of National Science Foundation-funded research. *Building Blocks* includes research-based computer tools with activities and a management system that guides children through research-based learning trajectories.

The program is designed to
- Build upon young children's experiences with mathematics with activities that integrate ways to explore and represent mathematics
- Involve children in "doing mathematics"
- Establish a solid foundation
- Develop a strong conceptual framework
- Emphasize the development of children's mathematical thinking and reasoning abilities
- Develop learning in line with state and national standards

In research studies, *Building Blocks* software was shown to increase young children's knowledge of multiple essential mathematical concepts and skills. One study tested *Building Blocks* against a comparable preschool math program and a no-treatment control group. All classrooms were randomly assigned, the "gold standard" of scientific evaluation. *Building Blocks* children significantly outperformed both the comparison group and control group of children. Results indicate strong positive effects with achievement gains near or exceeding those recorded for individual tutoring.

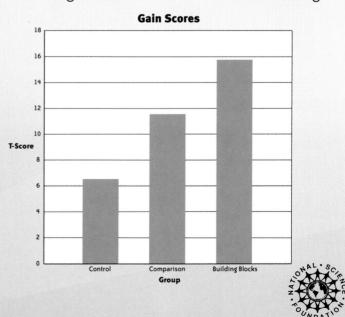

Get students back on track with confidence

Number Worlds is the only program that includes a prevention instruction section for students in **Grades Pre-K–1**. This unique 30-week course of daily instruction improves students' grasp of the world of math so they can move forward with the head start they need.

For students in **Grades 2–6** who are one or more grade levels behind in math, *Number Worlds* intervention program is an invaluable tool. It builds on students' current level of understanding with six 4-week intensive units per grade.

With each daily lesson, teachers have the opportunity to reach the program's knowledge objectives through problem-solving activities, small-group interaction, and discussion. By asking good questions in the classroom, teachers can encourage learning while defining areas that require extra work, ultimately helping to bring students to their appropriate grade level.

Fundamental Concepts Levels A–C

Level A	Level B	Level C
Children acquire well-developed counting and quality schemas.	Children develop a well-consolidated central conceptual structure for single-digit numbers.	Children link their central conceptual structure of number to the formal symbol system.

Core Content Topics Levels D–H

Level D	Level E	Level F	Level G	Level H
Number Sense	Number Sense	Number Sense	Number Sense	Number Sense
Number Patterns and Relationships (Algebra)	Number Patterns and Relationships (Algebra)	Number Patterns and Relationships (Algebra)	Number Patterns and Relationships (Algebra)	Number Patterns and Relationships (Algebra)
Addition	Addition	Addition & Subtraction	Multiplication	Fractions, Decimals & Percents
Subtraction	Subtraction	Multiplication	Division	Multiplication & Division
Geometry & Measurement	Geometry & Measurement	Geometry & Measurement	Geometry & Measurement	Geometry & Measurement
Data Analysis & Applications	Data Analysis & Applications	Data Analysis & Applications	Data Analysis & Applications	Data Analysis & Applications

Number Worlds Lesson Planner provides a wide array of helpful information before lessons even begin. Background overviews, activity ideas, and tips prepare teachers to the fullest extent.

Weekly Planners map out an entire week of lessons, complete with pacing options, goals, and the resources necessary to get the most out of every class period.

Math Background gives teachers math **context for the lesson**.

Get **insight into your students' capabilities** and how their minds work.

Manipulative-rich lessons are proven to help students turn **abstract concepts into concrete understanding**.

Building Blocks' **research-based software** gives teachers over 150 activity choices that go hand-in-hand with the lessons.

Define **key vocabulary in English or Spanish** to improve students' understanding of concepts.

Math at Home extends learning to provide the extra practice students need and encourage support at home.

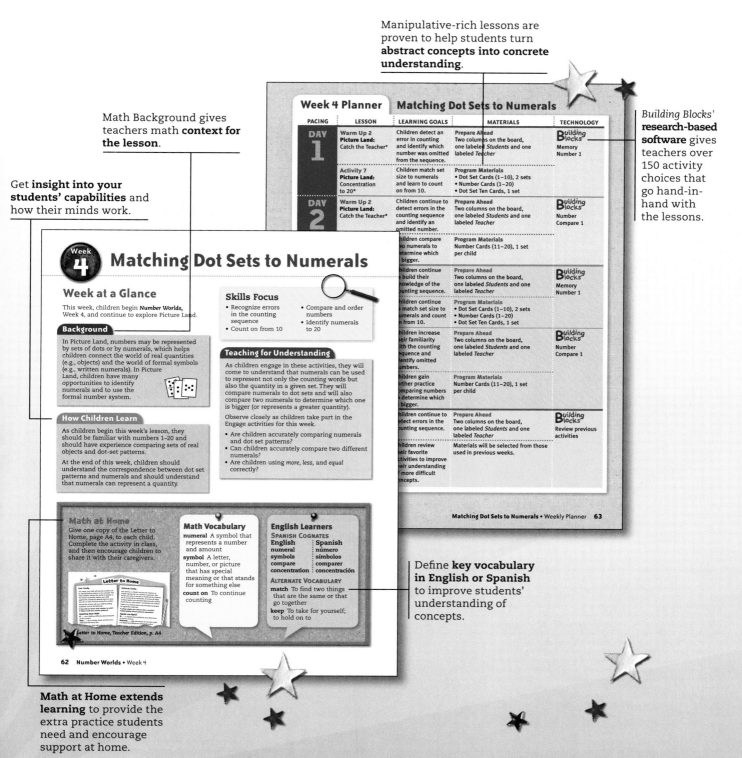

Thorough lesson plans guide you every step of the way

Every comprehensive *Number Worlds* lesson is divided into four distinct sections for simplified time management in the classroom. Whether it's time for concept building or skill building, in-depth discussion or assessment, *Number Worlds* always helps you keep learning objectives within reach.

FIN-tastic!

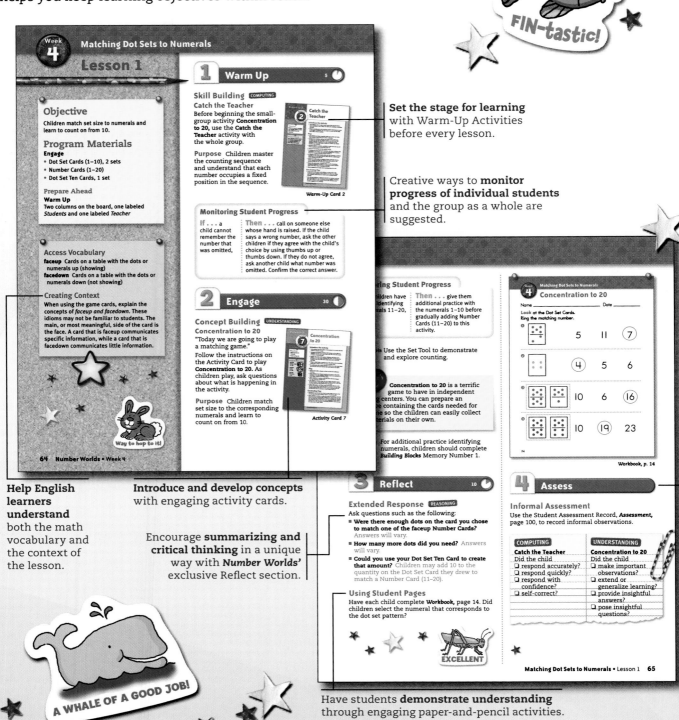

Week 4
Matching Dot Sets to Numerals

Lesson 1

Objective
Children match set size to numerals and learn to count on from 10.

Program Materials
Engage
- Dot Set Cards (1–10), 2 sets
- Number Cards (1–20)
- Dot Set Ten Cards, 1 set

Prepare Ahead
Warm Up
Two columns on the board, one labeled *Students* and one labeled *Teacher*

Access Vocabulary
faceup Cards on a table with the dots or numerals up (showing)
facedown Cards on a table with the dots or numerals down (not showing)

Creating Context
When using the game cards, explain the concepts of *faceup and facedown*. These idioms may not be familiar to students. The main, or most meaningful, side of the card is the face. A card that is faceup communicates specific information, while a card that is facedown communicates little information.

Way to hop to it!

64   Number Worlds • Week 4

1 Warm Up   5

Skill Building COMPUTING
Catch the Teacher
Before beginning the small-group activity **Concentration to 20**, use the **Catch the Teacher** activity with the whole group.

Purpose Children master the counting sequence and understand that each number occupies a fixed position in the sequence.

Warm-Up Card 2

Monitoring Student Progress

| If . . . a child cannot remember the number that was omitted, | Then . . . call on someone else whose hand is raised. If the child says a wrong number, ask the other children if they agree with the child's choice by using thumbs up or thumbs down. If they do not agree, ask another child what number was omitted. Confirm the correct answer. |

2 Engage   30

Concept Building UNDERSTANDING
Concentration to 20
"Today we are going to play a matching game."
Follow the instructions on the Activity Card to play **Concentration to 20**. As children play, ask questions about what is happening in the activity.

Purpose Children match set size to the corresponding numerals and learn to count on from 10.

Activity Card 7

Set the stage for learning with Warm-Up Activities before every lesson.

Creative ways to **monitor progress of individual students** and the group as a whole are suggested.

...ing Student Progress

| ...hildren have ...identifying ...als 11–20, | Then . . . give them additional practice with the numerals 1–10 before gradually adding Number Cards (11–20) to this activity. |

...s Use the Set Tool to demonstrate ... and explore counting.

... **Concentration to 20** is a terrific ... game to have in independent ... centers. You can prepare an ... containing the cards needed for ... so the children can easily collect ...erials on their own.

... For additional practice identifying ...numerals, children should complete ...Building Blocks Memory Number 1.

Week 4
Matching Dot Sets to Numerals
Concentration to 20
Name ____   Date ____
Look at the Dot Set Cards.
Ring the matching number.

	5	11	⑦
	④	5	6
	10	6	⑯
	10	⑲	23

Workbook, p. 14

Help English learners understand both the math vocabulary and the context of the lesson.

Introduce and develop concepts with engaging activity cards.

Encourage **summarizing and critical thinking** in a unique way with *Number Worlds'* exclusive Reflect section.

3 Reflect   10

Extended Response REASONING
Ask questions such as the following:
- Were there enough dots on the card you chose to match one of the faceup Number Cards? Answers will vary.
- How many more dots did you need? Answers will vary.
- Could you use your Dot Set Ten Card to create that amount? Children may add 10 to the quantity on the Dot Set Card they drew to match a Number Card (11–20).

Using Student Pages
Have each child complete **Workbook**, page 14. Did children select the numeral that corresponds to the dot set pattern?

EXCELLENT

4 Assess

Informal Assessment
Use the Student Assessment Record, **Assessment**, page 100, to record informal observations.

COMPUTING	UNDERSTANDING
Catch the Teacher	**Concentration to 20**
Did the child	Did the child
❏ respond accurately?	❏ make important observations?
❏ respond quickly?	❏ extend or generalize learning?
❏ respond with confidence?	❏ provide insightful answers?
❏ self-correct?	❏ pose insightful questions?

Matching Dot Sets to Numerals • Lesson 1   65

Have students **demonstrate understanding** through engaging paper-and-pencil activities.

A WHALE OF A GOOD JOB!

Assess student progress after each lesson.

Program Authors

Sharon Griffin, Ph.D., is a Professor of Education and Psychology at Clark University. She specializes in child development and mathematics education and has been studying how playing games that involve numbers helps children structure and understand the world. She conducted research on the development of math competence in the preschool and early school years and used this theoretical work as the basis to create the *Number Worlds* curriculum. Dr. Griffin has worked closely with teachers as they have introduced this curriculum into their classrooms. She has also worked collaboratively with schools around the country and in Canada as they have sought to systematically reform their mathematics programs.

Dr. Griffin's work has been widely published, and she is the author of the chapter "Fostering the Development of Whole-Number Sense: Teaching Mathematics in the Primary Grades" that appeared in a recent book published by the National Research Council. She is a member of the Mathematical Sciences Education Board in the National Academies of Science. She is also involved in an Organization of Economic Cooperation and Development project that brings together leading researchers in neuroscience and cognitive science from several countries to allow each to inform and advance the others' work.

Dr. Griffin holds a B.A. in Psychology from McGill University, a M.Ed. in Learning and Instruction from the University of New Hampshire, and a Ph.D. in Cognitive Science from the University of Toronto.

Douglas H. Clements, Professor of Early Childhood, Mathematics, and Computer Education at the University at Buffalo, State University of New York, has conducted research and published widely on the learning and teaching of geometry, computer applications in mathematics education, the early development of mathematical ideas, and the effects of social interactions on learning. Along with Julie Sarama, Dr. Clements has directed several research projects funded by the National Science Foundation and the U.S. Department of Education's Institute of Educational Sciences, one of which resulted in the mathematics software and activities included in *Building Blocks*.

Julie Sarama is an Associate Professor of Mathematics Education at the University at Buffalo, State University of New York. She conducts research on the implementation and effects of software and curricula in mathematics classrooms, young children's development of mathematical concepts and competencies, implementation and scale-up of educational reform, and professional development. Dr. Sarama has taught secondary mathematics and computer science, gifted mathematics at the middle school level, and preschool and Kindergarten mathematics methods and content courses for elementary to secondary teachers.

Sherry Booth is a mathematics curriculum specialist. Her past projects include the JASON web-based mathematics courses, the ATLAS Project, and the Math Partners project funded by the National Science Foundation. She has collaborated with researchers and designers to develop mathematics curricula that includes software, video, teacher guides, and student materials.

Contents

Contents

Contents

Contents

Appendix

Getting Started

Preparing to Use *Number Worlds*

This section provides an overview of classroom management issues and explanations of the Number Worlds program elements and how to use them.

Program Goal

Number Worlds was designed to foster the development of good intuitions about number and number environments. It also was designed to give children a desire to explore such environments and a sense of confidence in moving around within them. Children who develop a solid base of Number Sense have developed the core foundation on which all higher-order understandings are built.

Levels A–C (PreKindergarten, Kindergarten, and First Grade) are intended to serve as **Prevention** for later problems in mathematics. The thoroughly tested, engaging activities bring all students up to level so that they are ready for elementary mathematics. The activities are specifically targeted to address foundational understandings, but are engaging for all students.

Levels D–H (Grades 2–6) provide **Intervention** for students who are 1–2 years behind their peers in mathematics. Targeted units focus on computing, understanding, reasoning, applying, and engaging—the specific concepts and skills that build mathematics proficiency.

Program Objectives

- Teach concepts and skills that build a strong foundation for later learning
- Expose children to the major ways numbers are represented in the world—as objects, symbols, horizontal and vertical lines, and on dials
- Ensure that children acquire the interconnected knowledge that underlies number sense
- Include hands-on and computer activities that provide concrete representations of concepts
- Encourage communication using the language and vocabulary of formal mathematics
- Develop concept understanding with activities that are engaging and appropriate for children from all social and cultural backgrounds

Building Blocks

In addition to hands-on and workbook activities, *Number Worlds* includes **Building Blocks** software to provide additional exposure to and practice with foundational math concepts. **Building Blocks** software is an essential element of the *Number Worlds* curriculum.

Building Blocks software has these advantages

- It combines visual displays, animated graphics, and speech.
- It links "concrete" (graphical) and symbolic (e.g., numerals or spoken words) representations, which build understanding.
- It provides feedback.
- It provides opportunities to explore.
- It focuses children's attention and increases their motivation.
- It individualizes—gives children tasks at children's own ability levels.
- It provides undivided attention, proceeding at the child's pace.
- It keeps a variety of records.
- It provides more manageable manipulatives (e.g., manipulatives "snap" into position).
- It offers more flexible and extensible manipulatives (e.g., manipulatives can be cut apart).
- It provides more manipulatives (you never run out!).
- It stores and retrieves children's work so they can work on it again and again, which facilitates reflection and long-term projects.
- It records and replays children's actions.
- It presents clearer mathematics (e.g., using tools such as a "turn tool," helps children become aware of mathematical processes).

Building Blocks software stores records of how children are doing on **every** activity. It assigns them to just the right difficulty level. You can also view records of how the whole group or any individual is doing at any time.

The **Number Worlds** program is designed for flexible use. Each lesson is designed to take from 45–60 minutes. It is highly recommended that students spend at least one hour daily using the **Number Worlds** program.

Intervention Models

The program can be effectively used in each of the following environments:

• During Class Time

The Prevention program is designed for whole-class implementation at PreK–1. It can also be used in the same way as the Intervention program with a teacher or teacher's aide working with small groups apart from the rest of the class.

• Math Resource Room

Number Worlds is effective used in a math resource room with a teacher or teacher's aide working with a small group of students.

• After School

Because the activities are engaging, after school programs have used **Number Worlds** effectively for both intervention and regular education students. Both benefit from the intensive math activities.

• Intervention Classrooms

Number Worlds is very effective at every grade level for classes that need intervention in mathematics.

• Summer School

Number Worlds is ideal for summer school programs with an intensive focus on mathematics. Depending on the length of the summer school session, students can work through more than one lesson a day and make substantial progress.

• Tutoring

Number Worlds is also very effective when used in a one-on-one tutoring situation in which a teacher or aide participates in activities with the child.

Many teachers who have used the **Number Worlds** program have found that a teaching assistant—a student teacher, a teacher aide, or a parent volunteer—can help students become extremely familiar with the participatory structures that are required to teach the program so students are able to assume more control over their own learning and become less reliant on the help of a guide or coach. In fact, many **Number Worlds** teachers find it useful to institute "mini-teachers," students who have experience with an activity and understand its concepts, to help their peers.

Although **Number Worlds** activities can be effective with on-level students, they are truly beneficial for students who are not making adequate progress in their core program.

Supplemental Intervention

Number Worlds is geared for students identified with math deficiencies and who have not responded to reteaching efforts. It provides scientifically-based math instruction emphasizing the five critical elements of mathematics proficiency: understanding, computing, applying, reasoning, and engaging. This can be accomplished in a variety of environments for 45–90 minutes per day with a teacher or teacher's aide.

Intensive Intervention

Number Worlds is also effective for students with low skills and a sustained lack of adequate progress in mathematics. The program provides intensive focus on developing mathematical understanding and skills, and includes explicit instruction designed to meet the individual needs of struggling students. This can be accomplished when **Number Worlds** is used as replacement of core lessons for 60–90 minutes a day. The program must be implemented by a teacher or specialized math teacher or teacher's aide.

Understanding Students

For successful math intervention, it is imperative that teachers understand what students know and don't know about math. Teachers also need to know where students' current knowledge fits within the expected developmental sequence, what knowledge students have available to build upon, and what knowledge—the next steps in the sequence—students have yet to master. Finally, teachers need to become familiar with the problem-solving strategies students are using and the range of strategies they need to acquire to become efficient mathematics thinkers and problem-solvers. Using the Number Worlds program and the assessment tools it provides will help teachers acquire a rich understanding of their students along all of these dimensions.

Grades PreK–1 (Levels A–C)

Children with impoverished math backgrounds may come to preschool already 1–2 years behind their peers. Adults frequently overestimate children's understanding of number. These children may not have had experience playing board games, singing counting songs, or participating in number experiences at home. These children may not realize that the same number on a line and on a clock face represent the same quantity.

In the Levels A–C **Number Worlds** program, children explore five different ways number and quantity is represented. Each of the five "lands" of **Number Worlds** that children encounter in the lower grades exposes children to a different representation of number and helps children learn the language used to talk about number in that context.

In **Object Land** students explore the world of counting numbers by counting and comparing sets of objects or pictures of objects. In Object Land you might ask:

- **How many or few do you have?**
- **Which is bigger or smaller?**

In **Picture Land** numbers are represented as sets of stylized, semi-abstract dot-set patterns such as in a die and also as tally marks and numerals. In Picture Land you might ask:

- **What did you roll/pick?**
- **Which has more or less?**

In **Line Land** number is represented as a position on a path or a line. The language used for numbers in Line Land refers to a particular place on a line and also to the moves along a line. These types of questions are asked in Line Land:

- **Where are you now? How far did you go?**
- **Who is farther or less far along the line?**
- **Do you go forward or backward?**

In **Sky Land** number is represented as a position on a vertical scale such as on a thermometer or a bar graph. Sky Land inspires these questions:

- **How high or low are you now?**
- **What number or amount is higher or lower?**

In **Circle Land** number is represented as a point on a dial, such as a clock face or a sundial. In Circle Land you might ask:

- **How many times did you go around the dial?**
- **Which number is farther or less far around?**

Grades 2–6 (Levels D–H)

At grades 2–6 (Levels D–H) students may have difficulty with one, two, or many different math concepts. The goal of the upper grades is to develop foundational understandings in each concept so that students develop mathematical proficiency. Every lesson involves activities that actively engage students in understanding, computing, applying, reasoning, and engagement with the concept. At the end of each week, an assessment will help teachers determine whether a student has reached proficiency. The units are carefully sequenced to develop concepts across grade levels and can be used flexibly to meet student needs.

A variety of program materials are designed to help teachers provide a quality mathematics curriculum.

Teacher Edition

The **Teacher Edition** is the heart of the **Number Worlds** curriculum. It provides background for teachers and complete lesson plans with thorough instructions on how to develop math concepts. It explains when and how to use the program resources.

Activity Cards

These cards are packaged so that each can be carried around and used to direct or teach an activity. The activities, particularly at grades PreK–1 are employed again and again. The Activity Cards provide detailed descriptions of each activity, with suggested questions to ask during the activity.

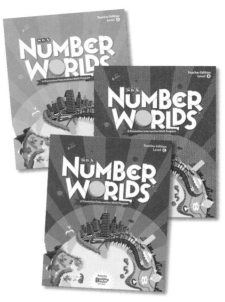

Activity Sheets

In the **Teacher Edition,** you will find numbered Activity Sheets in blackline master form that accompany activities developed for **Number Worlds.** These sheets can be copied in sufficient quantity for the number of children in your class who will be using that activity on a particular day.

Student Workbook

The **Student Workbook** includes developmental activities to help students develop higher-order thinking skills and practice basic skills.

Assessment

Number World's flexible assessment component allows for prescriptive placement and reliable curriculum and criterion assessment.

Each level of the program includes the following:

- **Placement Tests** to identify where students should begin the **Number Worlds** curriculum.
- **Weekly Tests** to measure student comprehension of the week's five daily lessons.
- **Unit Tests** to evaluate concept acquisition for the entire unit. These tests are available in both open-response and multiple-choice formats.
- **Rubrics** to informally evaluate student understanding of each lesson.
- The **Number Knowledge Test,** designed to measure students' intuitive knowledge of number—the knowledge that helps children make sense of quantitative problems.

Manipulatives and Technology

Number Worlds provides a wealth of manipulative resources. Establishing procedures for use of materials will simplify management issues and allow students to spend more time developing mathematical understanding.

Intervention Packages

A *Number Worlds Intervention Package* is available at every level and includes everything small groups need to use the program.

- Teacher Guide
- Assessment
- Student Workbooks
- Activity Manipulatives and Props
- Software

The package provides storage space for the program so it can be stored and moved as convenient. Activity game mats, manipulatives, demonstration props, playing cards, and pieces are all included. Students will benefit from a demonstration on how you want students to remove materials and return materials to the kit.

Manipulative Modules

Manipulatives are also available for specific topics: Counting, Base-Ten, Fractions, Geometry, Measurement, Money, and Time. Every unit of *Number Worlds* uses manipulative material. If these manipulatives are not already an integral part of your math curriculum, they are available from SRA.

Manipulative Topic Modules Grades K–6

Base-Ten: Cubes, Flats, Rods, Units
ISBN 0-07-605418-7

Geometry I: Pattern Blocks, Attribute Blocks, Solids, Mirror Cards
ISBN 0-07-605414-4

Geometry II: Geoboards, Protractors, Safety Compass, Gummed Tape
ISBN 0-07-605415-2

Measurement I: Rulers, Tape Measures, Measuring Cups, Liter Pitcher, Double-Pan Balance, Thermometer
ISBN 0-07-605416-0

Measurement II: Platform Scale, Metric Weight Set, Customary Weight Set
ISBN 0-07-605417-9

Money: Pennies, Nickles, Dimes, Quarters, Half Dollars, Money Packet
ISBN 0-07-605419-5

Counting: Craft Sticks, Rubber Bands, Panda Bear Counters, Math-Link Cubes
ISBN 0-07-605412-8

Probability: Spinners, Counters
ISBN 0-07-605420-9

Time: Clock Faces, Stopwatches
ISBN 0-07-605421-7

Fractions: Fraction Tiles, Fraction Circles
ISBN 0-07-605413-6

Technology Resources

For Students

Building Blocks activities are engaging, research-based software activities designed to reinforce levels of mathematical development in different strands of mathematics.

eMathTools are electronic tools to help students solve problems and explore and demonstrate concepts.

For Teachers

eAssess An assessment tool to grade, track, and report student progress

For more information on *Number Worlds* manipulatives and software see Appendix A.

Levels A–C
(Grades PreK–1)

Includes 30 weeks of daily lessons.

Levels D–H (Grades 2–6)

Each level includes 6 4-week units that address specific concepts and skills for a total of 24 weeks of instruction. Units are carefully sequenced to develop concepts for students who are one to two grade levels behind their peers. Placement Tests in the **Assessment** Book help teachers place students at the appropriate level and concept based on their demonstrated understanding.

Weekly Overview

- **Teaching for Understanding** provides the big ideas of the chapter.
- **Background** provides a refresher of the mathematics principles relevant to the chapter.
- **How Children Learn** offers insight into how children learn and gives research-based teaching strategies.
- **Weekly Planner** includes objectives that explain how the key concepts are developed lesson by lesson and which resources can be used with each lesson.

- **Lessons** provide overview, ideas for differentiating instruction, complete lesson plans, teaching strategies, and assessments that inform instruction.
- **Cumulative Tests** are provided incrementally to allow you to evaluate whether students are retaining previously developed concepts and skills.

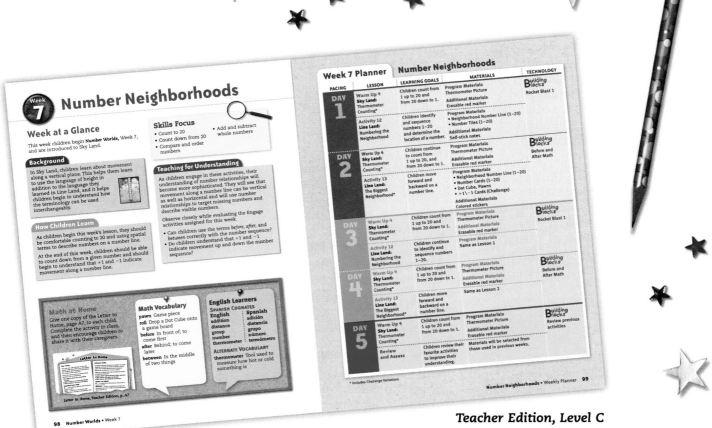

Teacher Edition, Level C

Routines

Lesson Plans

Every lesson throughout _Number Worlds_ is structured the same way.

1 Warm Up
2 Engage
3 Reflect
4 Assess

Routines for each part of the lesson are explained in the following discussions:

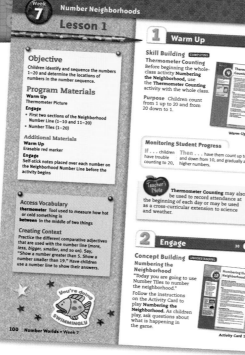
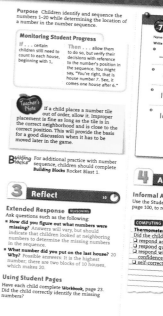

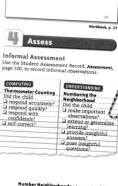

(Teacher Edition sample pages, Level C)

1 Warm Up

Warm Up exercises provide cumulative review and computation practice for students and give you opportunities to assess students' skills quickly. Warm Up is an essential component of **Number Worlds** because it helps students review concepts they will need in the Engage activities. It also gives students daily opportunities to sharpen their counting and mental math skills.

Begin every day with the lesson Warm Up with the entire group. These short activities are used in a whole-group format at the start of each day's lesson.

2 Engage

Engage is the heart of the lesson instruction. Here are suggestions for how to introduce lesson concepts, ideas for Guided Discussion, Skill Building, Game Demonstrations, and Strategy Building activities to develop student understanding. Before beginning the Engage activity,

familiarize yourself with the lesson objective, and make sure materials are available for student use. Introduce the activity, and use math talk to help students explore the lesson concepts.

Math Talk is expected in every **Number Worlds** lesson. When students speak the language of mathematics, they communicate mathematically, explain their thinking, and demonstrate understanding. Routines or rules for Math Talk established at the beginning of the year can make discussions more productive and promote listening and speaking skills.

1. **Pay attention to others.** Give your full attention to the person who is speaking. This includes looking at the speaker and nodding to show that you understand.
2. **Wait for speakers to answer and complete their thoughts.** Sometimes teachers and other students get impatient and move on and ask someone else or give the answer before someone has

Teacher Edition, Level C

a chance to think and speak. Giving students time to answer is a vital part of teaching for understanding.

3. **Listen.** Let yourself finish listening before you begin to speak. You can't listen if you are busy thinking about what you want to say next.
4. **Respect speakers.** Take turns and make sure that everyone gets a chance to speak and that no one dominates the conversation.
5. **Build on others' ideas.** Make connections, draw analogies, or expand on the idea.
6. **Ask questions.** Asking questions of another speaker shows that you were listening. Ask if you are not sure you understand what the speaker has said, or ask for clarification or explanation. It is a good idea to repeat in your own words what the speaker said so you can be sure your understanding is correct.

The **Activity Cards** include suggestions for some of the many questions that can be asked during an activity. The questions were selected because they are the kinds of questions children will be able to learn "by heart" and ask themselves. Eventually, the children can and should assume the role of asking the questions during a game. This will help them to become independent learners, responsible for their own learning.

The kinds of questions you ask will depend upon the way in which number is represented, the stage during the activity that you are asking the questions, and the level of difficulty the children are ready to explore. Whatever the combination of these three elements, you should ask the children as often as applicable, "How do you know?" and "How did you figure it out?"

Familiarizing children with these different ways of talking about number and quantity is a major goal of the **Number Worlds** program. It will enable children to realize, for example, that words such as *bigger, more, farther, higher,* and *further around* can all refer to an increase in quantity of the same magnitude, even though this change is expressed in different words, and very likely is represented in a different form. This understanding lies at the heart of number sense.

Using Student Pages

In every week of levels C–H, you will assign student pages to be completed during class. Students will finish at different times and should know what they can do to use their time productively until the Reflect part of the lesson. Students should not feel penalized for finishing early but should do something that is mathematically rewarding.

Student Workbook exercises in Number Worlds are primarily non-mechanical. Students cannot do all the problems on a page in a mechanical, non-thinking manner. Student workbook pages help the students learn to think about the problems.

Because student workbook exercises are non-mechanical, they sometimes require your active participation.

1. Make sure students know what pages to work on and any special requirements of those pages.
2. Tell students whether they should work independently or in small groups as they complete the pages.
3. Tell students how long they have to work on the student pages before you plan to begin the Reflect part of the lesson.

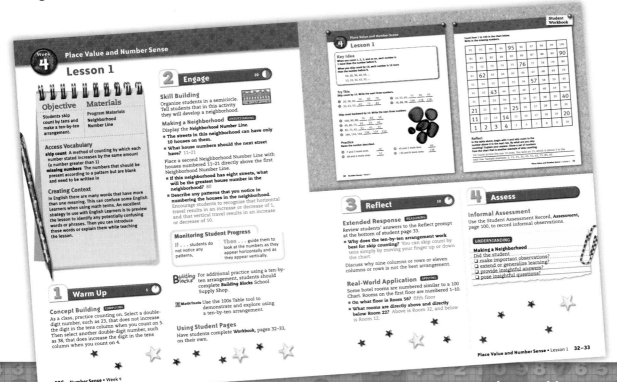

Teacher Edition, Level F

Routines

4. Tell students what their options are if they finish early. Technology resources are listed under the Assign Student Pages heading in each lesson. These activities further explore and develop lesson concepts. They include
   a. suggested **eMathTools** to use.
   b. **Building Blocks** activities to do.

5. As students work on the student pages, circulate around the room to monitor their progress. Use the Monitoring Student Progress suggestions for ideas on what to look for. Comment positively on student work, and stop to ask exploring, synthesizing, clarifying, or refocusing questions.

6. You may also use this time to work with English Learners or students who need intervention.

7. Because activities are an important and integral part of the program that provide necessary practice in traditional basic skills as well as higher-order thinking skills, when activities are included in a lesson, be sure to stop work on student pages early enough to leave enough time to play the game. Students may have to complete the student pages outside school in that case.

8. Complete the Informal Assessment Checklists on the last page of each lesson.

3 Reflect

Reflect is a vital part of the lesson that offers ways to help students summarize and reflect on their understanding of lesson concepts. Engaging children in reflection is as important as assessing—and a good way to assess. When children talk about their thinking, using their own words, they engage in mathematical generalizing and communicating. Allowing children to discuss what they did during an activity helps build mathematical reasoning but also develops social skills such as turn taking, listening, and speaking. At the designated time, have students stop working and direct their attention to reflecting on the lesson.

Use the suggested questions in Reflect or ask students to consider these ideas:

A. Summarize their ideas about the lesson concepts

B. Compare how the lesson concept or skill is like or different from other skills

C. Ask how students have seen or can apply the lesson in other curricular areas, other strands of mathematics, or in the world outside of school

D. Think about related matters that go beyond the scope of the lesson

Discuss student solutions to Extended Response questions.

Reflect Questions

A powerful reflection question is, "How do you know?" or "How did you figure that out?" Children may or may not answer you and often cannot provide reasons for their answers. They may shrug their shoulders and say, "I don't know," "Because," or, "Because I'm smart." As the year progresses, children become accustomed to explaining their ideas and their answers give more insight into their mathematical thinking. Young children who have such discussions with teachers and with each other begin to question and correct each other. Incorporate time for reflection into your classroom mathematics activities to develop deep understanding. The following are good questions and challenges.

- **How do you know?**
- **How did you figure that out?**
- **Why?**
- **Show me how you did that.**
- **Tell me about . . .**
- **How is that the same?**
- **How is that different?**

4 Assess

Assess helps you use informal and formal assessments to summarize and analyze evidence of student understanding and plan for differentiating instruction.

Goals of Assessment

1. **Improve instruction** by informing teachers about the effectiveness of their lessons
2. **Promote growth** of students by identifying where they need additional instruction and support
3. **Recognize accomplishments**

Phases of Assessment

Planning As you develop lesson plans, you can consider how you might assess the instruction, determining how you will tell if students have grasped the material.

Gather Evidence Throughout the instructional phase, you can informally and formally gather evidence of student understanding. The Informal Assessment Checklist and **Student Assessment Record** are provided to help you record data. The end of every lesson is designed to help in conducting meaningful assessments.

Summarize Findings Taking time to reflect on the assessments to summarize findings and make plans for follow-up is a critical part of any lesson.

Use Results Use the results of your findings to differentiate instruction or to adjust or confirm future lessons.

Number Worlds is rich in opportunities to monitor student progress to accomplish these goals.

Informal Daily Assessment

Informal Daily assessments evaluate students' math proficiencies in computational fluency, reasoning, understanding, applying, and engagement.

Warm-Up exercises, activities, and **Student Workbook** pages can be used for day-to-day observation and assessment of how well each student is learning skills and grasping concepts. Because of their special nature, these activities are an effective and convenient means of monitoring students.

Activities, for example, allow you to watch students practice particular skills under conditions more natural to them than most classroom activities. Warm-Up exercises allow you to see individual responses, give immediate feedback, and involve the entire class.

Simple rubrics enable teachers to record and track their observations. These can later be recorded by hand on the Student Assessment Record or in **eAssess** to help provide a more complete view of student proficiency.

Formal Assessment

The **Student Workbook** and the **Assessment Book** provide formal assessments for each chapter. Included are Entry Tests, Weekly Tests, and Cumulative Reviews to evaluate students' understanding of chapter concepts.

Routines

Classroom Management

The majority of the *Number Worlds* activities involve small groups and materials that will be new and intriguing to your students. Children as young as five years can learn to function well in semi-autonomous small learning groups. A clear-cut set of classroom expectations must be in place to support and encourage independent learning behaviors. Creating this set of expectations and preparing children to work effectively in small groups will be a task you may wish to address right from the start, as early as the first week of school.

Hints for Starting Small-Group Activities

The following suggestions will help you handle several small groups at once. You might decide to use alternative suggestions with different activities.

- Arrange the class into groups, and have the materials ready and divided ahead of time. Lead the activity with all the groups following along at the same time.
- Have the whole group sit in a semicircle around you. Have volunteers demonstrate the activity, giving many children an opportunity to participate. Once the children are familiar with the activity (perhaps on another day), organize them into small groups and have each group do the activity.
- Have a class helper lead a whole-group activity while you take one group at a time.
- Concentrate on one group while the other groups are involved with another, independent activity.

Because managing several small learning groups at once can be a challenge, several aids and devices have been included in the *Number Worlds* program to make this easier.

Activity Cards are available for each activity in the program. They can be carried around by the teacher or teacher's helper and used, on the spot, to direct the activity. The **Activity Cards** have a built-in categorization system that will tell you how difficult an activity is, how many children can play, what the children will be doing, what they can learn, and your role in the activity.

Activity Card 48, Level C

Activity Sheets are available in this *Teacher Edition.* These activity sheets accompany activities developed for each *Number Worlds* land. These sheets are intended to be copied in sufficient quantities for the number of children in your class who will be using that activity on a particular day.

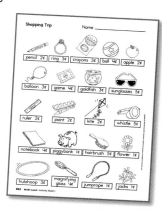

page B22, Level C

Manipulatives are included for all the activities in the five lands. Once the children are familiar with the materials, they can be responsible for gathering them for activities in small groups.

The materials themselves should be organized within the **Number Worlds Intervention Package** so the children can gather the supplies themselves. It is important that you establish a procedure for the management of the materials so that your students will be able to collect and return the materials efficiently.

Technology Resources for Teachers

Number Worlds includes several pieces of integrated technology for teachers designed to increase efficiency and effectiveness of instruction and assessment.

eAssess

- **Daily Records** Use *eAssess* to record daily formal and informal assessments.
- **Report Cards** Use *eAssess* to print student and class reports to determine grades.
- **Parent-Teacher Conferences** Use *eAssess* to print student reports to discuss with parents.

Technology Resources for Students

Number Worlds provides engaging technology resources to enrich, apply, and extend learning.

eMathTools are electronic tools that students can use to solve problems, test solutions, explore concepts, or demonstrate understanding.

Building Blocks activities are designed to reinforce key concepts and develop mathematics understanding.

Using Technology

A. Determine rules for computer use and communicate them to students. Rules should include
- sharing available computers. Some teachers have a computer sign-up chart for each computer. Some teachers have the students track these themselves.
- computer time. You might limit the amount of time students can be at the computer or allow students to work in pairs. Some teachers have students work until they complete an activity. Others allow students to continue on with additional activities.

B. Train students on your rules for proper use of computers, including how to turn computers on, load programs, and shut down the computer. Some teachers manage computers themselves; others have an aide or student in charge of computer management.

C. Using the suggestions for eMathTools and Building Blocks, make sure the computers are on and the programs are loaded and that students know how to access the software.

D. At the beginning of each week, demonstrate and discuss any new Building Blocks or eMath Tools computer activities.

E. Make sure all children work on the assigned computer activities individually at least twice per week for about 15 minutes each time.

F. When the children have finished the assigned activities, they should always get the chance to play and learn with the "free explore" activity. This might be individual but is also an excellent opportunity for children to explore cooperatively, posing problems for each other, solving problems together, or just learning through play.

G. Remember that preparation and followup are as necessary for computer activities as they are for any other activities. Do not omit critical whole group discussion sessions following computer work. Help children communicate their solution strategies and reflect on what they've learned.

H. Make sure children make sense of the mathematics.

For more information about Technology, see Appendix C.

Counting Objects

Week at a Glance

This week children begin **Number Worlds,** Week 1, and are introduced to Object Land.

Background

In Object Land, numbers are represented as groups of objects. This is the first way numbers were represented historically, and this is the first way children naturally learn about numbers. In Object Land, children work with real, tangible objects and with pictures of objects.

How Children Learn

As children begin this week's lesson, they may have already learned to count small groups of objects. They may also understand phrases such as *a lot* and *a little*.

At the end of this week, each child should begin to understand that when counting, objects are associated with numbers. Children should be able to identify the number of objects in a group as the last number when counting.

Skills Focus

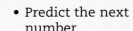

- Count to 5
- Count to 10
- Predict the next number

Teaching for Understanding

As children engage in these activities, they will learn to move back and forth among objects and numbers and will improve their ability to count up to 5. They will begin to understand that a group of objects can be labeled with a number to show the group's size.

Observe closely while evaluating the Engage activities assigned for this week.

- Are children counting to 5 fluently?
- Are children counting to 10?
- Are children predicting the correct number of objects in a set after counting the objects?

Math at Home

Give one copy of the Letter to Home, page A1, to each child. Complete the activity in class, and then encourage children to share it with their caregivers.

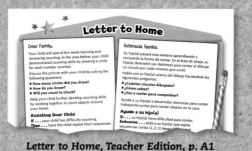

Letter to Home, Teacher Edition, p. A1

Math Vocabulary

Counter A small plastic circle used in activities

next The one after this one

count silently To think the numbers instead of saying them

English Learners

SPANISH COGNATES

English	Spanish
group	grupo
number	número
mathematics	matemáticas
objects	objetos

ALTERNATE VOCABULARY

clap To put your hands together to make a noise

count in your head To count without speaking

Week 1 Planner | Counting Objects

PACING	LESSON	LEARNING GOALS	MATERIALS	TECHNOLOGY
DAY 1	**Warm Up** **Object Land:** Count Up*	Children practice and improve counting skills.	No additional materials needed	**Building Blocks** Kitchen Counter
	Engage **Object Land:** Watch, Listen, and Count*	Children learn to associate objects and numbers.	**Program Materials** 10 Counters **Additional Materials** Wide-mouth container	**e MathTools** Set Tool
DAY 2	**Warm Up** **Object Land:** Count Up*	Children further develop counting skills.	No additional materials needed	**Building Blocks** Kitchen Counter
	Engage **Object Land:** Watch, Listen, and Count*	Children reinforce the concept that objects are associated with numbers.	**Program Materials** 10 Counters **Additional Materials** Wide-mouth container	
DAY 3	**Warm Up** **Object Land:** Count Up Variation 1*	Children gain additional counting experience.	No additional materials needed	**Building Blocks** Kitchen Counter
	Engage **Object Land:** Watch, Listen, and Count Variation*	Children further associate objects and numbers.	**Program Materials** 10 Counters **Additional Materials** Wide-mouth container	
DAY 4	**Warm Up** **Object Land:** Count Up Variation 1*	Children reinforce counting skills.	No additional materials needed	**Building Blocks** Kitchen Counter
	Engage **Object Land:** Watch, Listen, and Count Variation*	Children reinforce number concepts.	**Program Materials** 10 Counters **Additional Materials** Wide-mouth container	
DAY 5	**Warm Up** **Object Land:** Count Up*	Children expand their counting skills.	No additional materials needed	**Building Blocks** Review previous activities
	Review **Object Land:** Free-Choice Activity	Children become more comfortable associating objects and numbers.	**Program Materials** 10 Counters **Additional Materials** Wide-mouth container	

* Includes Challenge Variations

Lesson 1

Objective

Children practice counting and learn to associate objects and numbers.

Program Materials

Engage
10 Counters

Additional Materials

Engage
Wide-mouth container

Access Vocabulary

Bigger "More than something else." Open your arms wide to help explain.

Creating Context

It is important for children to develop the skills to make comparisons. Using something handy in the classroom, show children large amounts and small amounts of beans, paper clips, or blocks and ask which pile is bigger. Help English Learners to use *a lot/a little, many/few,* and other common comparing words.

Leaping Lizards!

---

1 Warm Up 5

Concept Building COMPUTING

Count Up

Before beginning the whole-class activity **Watch, Listen, and Count,** use the **Count Up** activity with the whole group.

Purpose **Count Up** helps children learn the order in which numbers appear and count easily and flawlessly from 1 to 5.

Warm-Up Card 1

Monitoring Student Progress

If . . . children are not participating,

Then . . . make sure you count slowly enough to give them time to think of the next number.

---

2 Engage 30

Skill Building UNDERSTANDING

Watch, Listen, and Count

"Today you will practice counting items you see and hear."

Follow the instructions on the Activity Card to play **Watch, Listen, and Count.** As children play, ask questions about what is happening in the activity.

Activity Card 1

Purpose **Watch, Listen, and Count** provides multiple cues to help children learn that when counting, each number is associated with an object. Counting the Counters that have been dropped reinforces the idea that the last number said identifies the size of the set.

 MathTools Use the Set Tool to demonstrate and explore counting.

Monitoring Student Progress

| **If . . .** children cannot tell which number will come next, | **Then . . .** remove the Counters from the container, recount them from 1, and continue adding one Counter at a time. |

Teacher's Note Rote counting is one of the earliest methods children use to make sense of numbers. This activity will help you assess the abilities of the children as they begin school. It also gives children experience associating numbers with objects as they count and exposes them to identifying set size and making predictions.

Building Blocks For additional practice with one-to-one correspondence, children should complete **Building Blocks** Kitchen Counter.

 Reflect 10

Extended Response REASONING

Ask questions such as the following:

- **What did we learn about numbers today?** Children should restate their specific learning.
- **How did you know the size of each set of Counters?** Possible answer: the size of each set is the last number counted.
- **How would you explain the** Watch, Listen, and Count **activity to someone who wasn't here?** Possible answer: You count one number for each Counter that drops into the container.

4 Assess

Informal Assessment

Use the Student Assessment Record, **Assessment,** page 100, to record informal observations.

COMPUTING	UNDERSTANDING
Count Up	**Watch, Listen, and Count**
Did the child	Did the child
❏ respond accurately?	❏ make important observations?
❏ respond quickly?	❏ extend or generalize learning?
❏ respond with confidence?	❏ provide insightful answers?
❏ self-correct?	❏ pose insightful questions?

Lesson 2

Objective

Children improve counting skills and reinforce the idea of associating objects with numbers.

Program Materials

Engage
10 Counters

Additional Materials

Engage
Wide-mouth container

Access Vocabulary

Counters Plastic circles used in games
set A group of objects that you can count

Creating Context

Use music to help English Learners rote count. There are many counting songs that can help provide rhythm and repetition so that children remember the counting sequence in English.

Doggone Good!

1 Warm Up   5

Concept Building   COMPUTING
Count Up

Before beginning the whole-class activity **Watch, Listen, and Count,** use the **Count Up** activity with the whole group.

Purpose **Count Up** helps children learn the order in which numbers appear and to count easily and flawlessly from 1 to 5.

Warm-Up Card 1

Monitoring Student Progress

If . . . children are not sure of the next number up, | **Then . . .** prompt them or ask a classmate to say the next number.

2 Engage   30

Skill Building   UNDERSTANDING
Watch, Listen and Count

"Today you will practice counting in your head. This means you count without talking."

Follow the instructions on the Activity Card to play **Watch, Listen, and Count.** As the children play, ask questions about what is happening in the activity.

Activity Card 1

Purpose **Watch, Listen, and Count** provides multiple cues to help children learn that, when counting, each number is associated with an object. Counting the Counters that have been dropped reinforces the idea that the last number said identifies the size of the set. Counting silently instead of aloud helps children continue to improve their counting skills.

Monitoring Student Progress

| **If . . .** children cannot tell which number will come next, | **Then . . .** remove the Counters from the container, recount them from 1, and continue adding one Counter at a time. |

Teacher's Note Encourage children to use this week's vocabulary words as they engage in the activities, discuss math concepts, and make predictions.

Building Blocks For additional practice with one-to-one correspondence, children should complete **Building Blocks** Kitchen Counter.

3 Reflect 10

Extended Response APPLYING

Ask questions such as the following:

- **What did we learn about numbers today?**
  Answers will vary.

- **Is counting silently different from counting aloud? Do you get the same answer each time?**
  Possible answer: You say the numbers in your head and don't talk.

4 Assess

Informal Assessment

Use the Student Assessment Record, **Assessment,** page 100, to record informal observations.

COMPUTING	UNDERSTANDING
Count Up Did the child	**Watch, Listen, and Count** Did the child
❏ respond accurately?	❏ make important observations?
❏ respond quickly?	❏ extend or generalize learning?
❏ respond with confidence?	❏ provide insightful answers?
❏ self-correct?	❏ pose insightful questions?

Lesson 3

Objective

Children gain additional counting experience.

Program Materials

Engage
10 Counters

Additional Materials

Engage
Wide-mouth container

Access Vocabulary

clap Putting your hands together to make a noise
container Something that you can put things in

Creating Context

Some young English Learners may already know how to count in their home language. This can reinforce the concept of counting in English. Invite children to share counting in their home language if they have learned how to do so.

Math-a-saurus

1 Warm Up 5

Concept Building COMPUTING

Count Up

Before beginning the whole-class activity **Watch, Listen, and Count,** use **Count Up** Variation 1: **Count and Clap** with the whole group.

Purpose Count and Clap helps children learn the order in which numbers appear and learn to associate actions with numbers.

Warm-Up Card 1

Monitoring Student Progress

If . . . children have trouble counting and clapping,

Then . . . have them try a different action such as tapping their fingers or blinking their eyes.

2 Engage 30

Skill Building UNDERSTANDING

Watch, Listen, and Count

"Today you will practice counting the sounds you hear."

Follow the instructions on the Activity Card to introduce and play **Watch, Listen, and Count** Variation: **Listen and Count.** As children play, ask questions about what is happening in the activity.

Activity Card 1

Purpose **Listen and Count** provides multiple cues to help children learn that each number can be associated with a sound when counting. Counting the Counters that have been dropped reinforces the idea that the last number said identifies the size of the set.

Monitoring Student Progress

| If . . . children cannot predict which number will come next, | Then . . . remove the Counters from the container, have the class recount them from 1, and continue adding one Counter at a time. |

 For additional practice with one-to-one correspondence, children should complete **Building Blocks** Kitchen Counter.

3 Reflect 10 ▶

Extended Response APPLYING

Ask questions such as the following:

- **What was the biggest number you counted to?** 5
- **Was it easier or harder to count without looking? Why?** Accept all reasonable answers.

4 Assess

Informal Assessment

Use the Student Assessment Record, **Assessment**, page 100, to record informal observations.

COMPUTING	UNDERSTANDING
Count Up	**Watch, Listen, and Count**
Did the child	Did the child
❑ respond accurately?	❑ make important observations?
❑ respond quickly?	❑ extend or generalize learning?
❑ respond with confidence?	❑ provide insightful answers?
❑ self-correct?	❑ pose insightful questions?

Lesson 4

Objective
Children reinforce and expand counting skills and number concepts.

Program Materials
Engage
10 Counters

Additional Materials
Engage
Wide-mouth container

Access Vocabulary
Counters Plastic circles used in the game
clap Putting your hands together to make a noise
tap Touching your fingers or toes to the ground to make a quiet noise

Creating Context
In Week 1, certain actions are used to reinforce counting skills. English Learners may need to learn the meaning of words such as clap and tap.

1 Warm Up 5

Concept Building COMPUTING
Count Up
Before beginning the whole-class activity **Watch, Listen, and Count,** use **Count Up** Variation 1: **Count and Clap** with the whole group.

Purpose **Count and Clap** helps children learn to count easily and flawlessly from 1 to 10. Adding clapping to the activity helps children learn that each number can be associated with an action as well as an object.

Warm-Up Card 1

Monitoring Student Progress
If . . . children are not participating,

Then . . . make sure you count slowly enough to give them time to think of the next number.

2 Engage 30

Skill Building UNDERSTANDING
Watch, Listen, and Count
"Today you will keep practicing counting items you hear but don't see."

Follow the instructions on the Activity Card to play **Watch, Listen, and Count** Variation: **Listen and Count** and focus on numbers 6–10. As children play, ask questions about what is happening in the activity.

Activity Card 1

Purpose **Listen and Count** provides multiple cues to help children learn that each number is associated with an object when counting. Counting without watching helps children understand that the same counting principles apply in different situations.

Monitoring Student Progress

If . . . children are counting fluently,

Then . . . have them count silently as they listen.

Building Blocks For additional practice with one-to-one correspondence, children should complete **Building Blocks** Kitchen Counter.

3 Reflect 10 ▶

Extended Response REASONING

Ask questions such as the following:

- **What was the biggest number we counted to when I dropped the Counters into the container? What was the smallest number?** Possible answers: 10; 6

- **Were there more Counters in the container when we counted to 5 or when we counted to 10?** when we counted to 10 **How do you know? How can we see if you are right?** Possible answer: Put 10 Counters in the container while counting to 10, and then dump them out. Put 5 Counters in the container while counting and dump them out. Compare the piles.

- **Can we say that the pile of 10 Counters has more than the pile of 5 Counters? Can we say that 10 is a bigger number than 5?** yes; yes

4 Assess

Informal Assessment

Use the Student Assessment Record, **Assessment,** page 100, to record informal observations.

COMPUTING	UNDERSTANDING
Count Up	**Watch, Listen, and Count**
Did the child	Did the child
❏ respond accurately?	❏ make important observations?
❏ respond quickly?	❏ extend or generalize learning?
❏ respond with confidence?	❏ provide insightful answers?
❏ self-correct?	❏ pose insightful questions?

Lesson 5

Review

Objective

Children review the counting sequence and become more comfortable with associating numbers and objects.

Program Materials

Engage
10 Counters

Additional Materials

Engage
Wide-mouth container

Creating Context

It is important that your assessment include questions that check the understanding of English Learners at varying proficiency levels. Beginning English Learners often do well with questions that let the children show their understanding through actions or demonstrations instead of lengthy verbal responses.

High Flyer!

1 Warm Up 5

Concept Building COMPUTING

Count Up

Before beginning the whole-class activity **Watch, Listen, and Count,** use the **Count Up** activity with the whole group and practice counting to 10.

Purpose Count Up helps children learn the order of the counting sequence and to count easily and flawlessly from 1 to 10.

Warm-Up Card 1

Monitoring Student Progress

If . . . children are not counting correctly,	Then . . . have other children model counting to 5 and to 10 in different situations throughout the day.

2 Engage 20

Skill Building UNDERSTANDING

Free-Choice Activity

For the last day of the week, allow the children to choose an activity from the previous four days. Some activities they may choose include the following:

- Object Land: **Watch, Listen, and Count**
- Object Land: **Watch, Listen, and Count** Variation 1: **Listen and Count**

Make a note of the activities children select. Do they prefer easy or challenging activities? If you believe your children would benefit from extra practice on specific skills, choose an activity for them.

3 Reflect 10

Extended Response REASONING

Ask questions such as the following:

- **What did you like about playing** Watch, Listen, and Count?
- **Was there anything about playing this game you didn't like?**
- **Did this game help you do something you couldn't do before? What did it help you do?**
- **What was easy when you were playing** Watch, Listen, and Count?
- **What was hard when you were playing** Watch, Listen, and Count?
- **What do you remember most about this game?**
- **What was your favorite part?**

4 Assess 10

A Gather Evidence

Formal Assessment

Have students complete the weekly test on *Assessment,* page 25. Record formal assessment scores on the Student Assessment Record, *Assessment,* page 100.

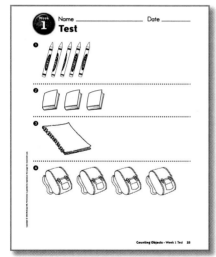

Assessment, p. 25

B Summarize Findings

Review the Student Assessment Records. Determine whether children have Minimal, Basic, or Secure understanding of the concepts presented in Week 1.

C Differentiate Instruction

Based on your observations, use these teaching strategies next week to follow up.

Minimal Understanding

- Repeat the Warm-Up and Engage activities to develop number awareness and counting skills.
- Use **Building Blocks** computer activities beginning with Kitchen Counter to reinforce counting and sequencing concepts.

Basic Understanding

- Repeat Engage activities in subsequent weeks to give children further practice with counting concepts and one-to-one correspondence.
- Use **Building Blocks** computer activities beginning with Kitchen Counter to reinforce this week's concepts.

Secure Understanding

- Use Challenge variations of activities.
- Use computer activities to extend children's understanding of the Week 1 concepts.

Counting and Sorting Objects

Week at a Glance

This week children begin **Number Worlds,** Week 2, and continue to explore Object Land.

Background

In Object Land, numbers are represented as groups of objects. This is the first way numbers were represented historically, and this is the first way children naturally learn about numbers. In Object Land, children work with real, tangible objects and with pictures of objects.

How Children Learn

As children begin this week's lesson, they should be more comfortable with counting objects, but may not know how many objects are in a set or how to compare sets.

At the end of the week, each child should begin to understand sorting and counting skills to compare quantities.

Skills Focus

- Count to 5
- Count to 10
- Predict the next number up, when counting
- Sort objects by attribute

Teaching for Understanding

As the children engage in these activities, they will develop the skills needed to discuss and compare quantities.

Observe closely while evaluating the Engage activities assigned for this week.

- Are children counting to 5?
- Are children counting to 10?
- Are children counting in the correct order?

Math at Home

Give one copy of the Letter to Home, page A2, to each child. Complete the activity in class, and then encourage children to share it with their caregivers.

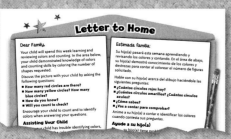

Letter to Home, Teacher Edition, p. A2

Math Vocabulary

count out To take a number of items out of a group of items

primary colors Red, yellow, and blue

recount To count again

English Learners

SPANISH COGNATES

English	Spanish
color	color
explore	explorer
concepts	conceptos

ALTERNATE VOCABULARY

. . . count in their heads Count silently, thinking the names of the numbers without saying them

sort Separate into groups of objects that are alike

Week 2 Planner — Counting and Sorting Objects

PACING	LESSON	LEARNING GOALS	MATERIALS	TECHNOLOGY
DAY 1	**Warm Up** Object Land: Count Up*	Children review counting skills.	No additional materials needed	**Building Blocks** Party Time Free Explore
	Engage Object Land: Identifying Colors	Children practice sorting and counting skills.	**Program Materials** 40 Counters in four colors, 1 set per group **Additional Materials** Bowls or containers, 1 per group	
DAY 2	**Warm Up** Object Land: Count Up*	Children further develop counting skills.	No additional materials needed	**Building Blocks** Party Time Free Explore
	Engage Object Land: Identifying Colors	Children practice counting and social skills.	**Program Materials** 40 Counters in four colors, 1 set per group **Additional Materials** Bowls or containers, 1 per group	
DAY 3	**Warm Up** Object Land: Count Up Variation 1*	Children gain additional counting experience.	No additional materials needed	**Building Blocks** Party Time Free Explore
	Engage Object Land: Identifying Colors	Children continue to practice counting and social skills.	**Program Materials** 40 Counters in four colors, 1 set per group **Additional Materials** Bowls or containers, 1 per group	**MathTools** Set Tool
DAY 4	**Warm Up** Object Land: Count Up Variation 1*	Children increase their understanding of one-to-one correspondence.	No additional materials needed	**Building Blocks** Party Time Free Explore
	Engage Object Land: Identifying Colors	Children reinforce sorting and counting skills.	**Program Materials** 40 Counters in four colors, 1 set per group **Additional Materials** Bowls or containers, 1 per group	
DAY 5	**Warm Up** Object Land: Count Up*	Children expand their counting skills.	No additional materials needed	**Building Blocks** Review previous activities
	Review Object Land: Free-Choice Activity	Children review their favorite activities or are assigned activities to improve their understanding of more difficult concepts.	Materials will be selected from those used in Week 1 and Week 2.	

* Includes Challenge Variations

Lesson 1

Objective

Children practice sorting and counting skills and begin to compare sets.

Program Materials

Engage
40 Counters with 10 Counters each in red, yellow, blue, and green, 1 set per group of four children

Additional Materials

Engage
Bowls or containers, 1 per group of four children

Access Vocabulary

primary colors Blue, red and yellow; the colors used to make other colors
recount To count again

Creating Context

English Learners may benefit when checking for understanding includes questions or tasks that allow them to show what they know. For example, when we ask "Are there some other ways you could sort the Counters?" we can ask ELL students to "show another way to sort the Counters."

Making Tracks

1 Warm Up 5

Concept Building COMPUTING

Count Up
Before beginning the **Identifying Colors** activity, use the **Count Up** activity with the whole group.

Purpose **Count Up** helps children learn that numbers appear in the counting sequence in a fixed order and that this order makes it possible to predict the next number up.

Warm-Up Card 1

Monitoring Student Progress

| **If . . .** a child does not know the next number, | **Then . . .** have another child help or prompt him or her and have the class recount from 1. |

2 Engage 30

Skill Building ENGAGING

Identifying Colors

"Today you will work with your classmates to separate Counters and practice counting."

Follow the instructions on the Activity Card to play **Identifying Colors.** As children play, ask questions about what is happening in the activity.

Activity Card 2

Purpose **Identifying Colors** helps children to recognize colors and practice counting.

Monitoring Student Progress

If . . . children cannot tell which number will come next,

Then . . . have them remove the Counters from the bowl, recount them from 1, and continue counting one Counter at a time.

 Teacher's Note Encourage children to use this week's vocabulary words as they engage in the activities, discuss math concepts, and make predictions.

Building Blocks For additional counting practice, children should complete **Building Blocks** Party Time Free Explore.

3 Reflect · 10

Extended Response · APPLYING

Ask the following questions:

- **How can you describe the groups of Counters? How are they different? How are they alike or the same?** Possible answers: They are different colors; they are the same size; there are 5 Counters in each group.

- **Are there other ways you could sort the Counters?** Answers will vary.

4 Assess

Informal Assessment

Use the Student Assessment Record, **Assessment**, page 100, to record informal observations.

COMPUTING	ENGAGING
Count Up	**Identifying Colors**
Did the child	Did the child
❑ respond accurately?	❑ pay attention to the contributions of others?
❑ respond quickly?	
❑ respond with confidence?	❑ contribute information and ideas?
❑ self-correct?	❑ improve on a strategy?
	❑ reflect on and check accuracy of work?

Lesson 2

Objective

Children continue to practice sorting and counting skills and continue to develop social skills for group work.

Program Materials

Engage
40 Counters with 10 Counters each in red, yellow, blue, and green, 1 set per group of four children

Additional Materials

Engage
Bowls or containers, 1 per group of four children

Access Vocabulary

What is happening in the game? "Describe what we are doing."
figure out Solve the problem; come up with a way to find the answer

Creating Context

Children learning English will benefit from counting concrete objects to help them learn the counting numbers and the concept of one-to-one correspondence. Using different colors keeps this exercise interesting and helps develop children's English vocabulary.

---

1 Warm Up 5

Concept Building COMPUTING

Count Up

Before beginning the **Identifying Colors** activity, use the **Count Up** activity with the whole group.

Purpose Count Up helps children learn that numbers appear in the counting sequence in a fixed order and that this order makes it possible to predict the next number up.

Warm-Up Card 1

Monitoring Student Progress

| **If . . .** children are having trouble remembering the sequence of numbers, | **Then . . .** model counting for them throughout the day. |

---

2 Engage 30

Skill Building ENGAGING

Identifying Colors

"Today you and your classmates will sort and count Counters."

Follow the instructions on the Activity Card to play **Identifying Colors.** As children play, ask questions about what is happening in the activity.

Purpose Identifying Colors gives children opportunities to explore number concepts such as number relationships and helps children develop social skills by working in groups.

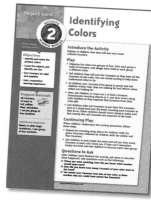

Activity Card 2

Monitoring Student Progress

| **If . . .** children cannot remember how many Counters of a certain color are in the bowl, | **Then . . .** have them remove the Counters from the bowl and recount them from 1. |

 Building Blocks For additional practice counting, children should complete **Building Blocks** Party Time Free Explore.

3 Reflect 10

Extended Response REASONING

Ask the following questions:

- **How did you figure out how many red Counters were in the bowl?** Possible answer: The last number counted was 5 so there are 5 Counters in the bowl.

- **How can you check your answer?** Possible answer: I can recount the Counters.

4 Assess

Informal Assessment

Use the Student Assessment Record, **Assessment**, page 100, to record informal observations.

COMPUTING	ENGAGING
Count Up	**Identifying Colors**
Did the child	Did the child
❏ respond accurately?	❏ pay attention to the contributions of others?
❏ respond quickly?	❏ contribute information and ideas?
❏ respond with confidence?	
❏ self-correct?	❏ improve on a strategy?
	❏ reflect on and check accuracy of work?

stand-up JOB

Whooo did a great job? YOU!

High Flyer!

Lesson 3

Objective

Children continue to practice sorting and counting skills and continue to develop social skills for group work.

Program Materials

Engage
40 Counters with 10 Counters each in red, yellow, blue, and green, 1 set per group of four children

Additional Materials

Engage
Bowls or containers, 1 per group of four children

Access Vocabulary

color piles Counters sorted into groups by color
patterns Repeated colors, shapes, or objects that form a design

Creating Context

Including movement and activity helps English Learners practice and remember the counting numbers in sequence. Make sure children clap only when a number is counted.

Good for Ewe!

1 Warm Up 5

Concept Building COMPUTING
Count Up

Before beginning the **Identifying Colors** activity, use **Count Up** Variation 1: **Count and Clap** with the whole group.

Purpose **Count and Clap** helps children learn that numbers appear in the counting sequence in a fixed order; adding clapping helps children learn that each number can be associated with an action.

Warm-Up Card 1

Monitoring Student Progress

If . . . children are having trouble counting and clapping,

Then . . . begin by counting and clapping to 3 and slowly work up to counting and clapping to 5.

2 Engage 30

Skill Building ENGAGING
Identifying Colors

"Today you will cooperate to count and sort Counters."

Follow the instructions on the Activity Card to play **Identifying Colors**. As children play, ask questions about what is happening in the activity.

Purpose **Identifying Colors** gives children opportunities to explore number concepts such as number relationships and helps children develop social skills by working in groups.

Activity Card 2

MathTools Use the Set Tool to demonstrate and explore counting.

Monitoring Student Progress

| **If . . .** children need more practice sorting and counting, | **Then . . .** have them count out 5 more Counters from their color piles and drop them into the bowl. |

Building Blocks For additional counting practice, children should complete **Building Blocks** Party Time Free Explore.

3 Reflect   10

Extended Response   APPLYING

Ask the following questions:

- **What patterns do you see?** Answers will vary.
- **What does this remind you of?** Answers will vary.
- **Did all of you put the same number of Counters into the bowl? How do you know? How can you figure this out?** Possible answer: Yes; we all counted to 5; we can count together.

4 Assess

Informal Assessment

Use the Student Assessment Record, **Assessment,** page 100, to record informal observations.

COMPUTING	ENGAGING
Count Up	**Identifying Colors**
Did the child	Did the child
❏ respond accurately?	❏ pay attention to the contributions of others?
❏ respond quickly?	
❏ respond with confidence?	❏ contribute information and ideas?
❏ self-correct?	
	❏ improve on a strategy?
	❏ reflect on and check accuracy of work?

Lesson 4

Objective

Children deepen their understanding of sorting and counting and will identify groups by set size.

Program Materials

Engage
40 Counters with 10 Counters each in red, yellow, blue, and green, 1 set per group of four children

Additional Materials

Engage
Bowls or containers, 1 per group of four children

Access Vocabulary

sort by color To separate objects of the same color into smaller groups

Creating Context

Working with a partner is helpful to English Learners because partner work often provides opportunities to interact more easily. A partner who is bilingual or proficient in English can help an ELL student respond more quickly and comfortably.

Leaping Lizards!

1 Warm Up 5

Concept Building COMPUTING
Count Up

Before beginning the **Identifying Colors** activity, use **Count Up** Variation 1: **Count and Clap** with the whole group.

Purpose **Count and Clap** helps children learn that numbers appear in the counting sequence in a fixed order; adding clapping helps children learn that each number can be associated with an action.

Warm-Up Card 1

Monitoring Student Progress

| If . . . children are not participating, | Then . . . invite them to stamp their feet instead of clapping. |

2 Engage 30

Skill Building ENGAGING
Identifying Colors

"Today you will count and sort Counters with your classmates."

Follow the instructions on the Activity Card to introduce and play **Identifying Colors.** As children play, ask questions about what is happening in the activity.

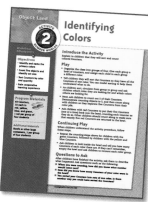
Activity Card 2

Purpose **Identifying Colors** gives children opportunities to explore number concepts such as one-to-one correspondence and helps children learn cooperatively.

Monitoring Student Progress

| **If . . .** children are not participating in their groups, | **Then . . .** work on inclusive strategies, and remind the class that in a group, everyone works together to help each other learn. |

 Building Blocks For additional counting practice, children should complete *Building Blocks* Party Time Free Explore.

3 Reflect 10

Extended Response APPLYING

Ask the following questions:

- **How would you explain this activity to someone who had never played it before?** Possible answer: You sort the counters by color, then make a pile of 5 and drop them into the bowl.

- **Can you think of other times at home or at school when you count things?** Answers will vary.

4 Assess

Informal Assessment

Use the Student Assessment Record, **Assessment,** page 100, to record informal observations.

COMPUTING	ENGAGING
Count Up	**Identifying Colors**
Did the child	Did the child
❏ respond accurately?	❏ pay attention to the contributions of others?
❏ respond quickly?	❏ contribute information and ideas?
❏ respond with confidence?	❏ improve on a strategy?
❏ self-correct?	❏ reflect on and check accuracy of work?

Lesson 5

Review

Objective

Children review the material presented in Weeks 1 and 2 of *Number Worlds*.

Program Materials

See Activity Cards for materials.

Additional Materials

See Activity Cards for materials.

Creating Context

Working in small groups is an excellent way for English Learners to increase English fluency and better understand concepts. Small groups allow English Learners to practice both with proficient speakers of English and with children who are fluent in the same primary language.

---

1 Warm Up 5

Concept Building COMPUTING

Count Up

Before beginning the **Identifying Colors** activity, use the **Count Up** activity with the whole group and count from 1 to 10.

Purpose **Count Up** helps children develop automaticity in counting so that they can use the counting sequence to solve math problems.

Warm-Up Card 1

Monitoring Student Progress

| **If . . .** children are having trouble remembering which number comes next, | **Then . . .** model counting for them throughout the day. As children become more confident, have them lead this activity. |

---

2 Engage 20

Skill Building ENGAGING

Free-Choice Activity

For the last day of the week, allow children to choose an activity from this week or the previous week. These include the following:

- Object Land: **Watch, Listen, and Count**
- Object Land: **Watch, Listen, and Count** Variation 1: **Listen and Count**
- Object Land: **Identifying Colors**

Make a note of the activities children select. Do they prefer easy or challenging activities? If you believe your children would benefit from extra practice on specific skills, choose an activity for them.

Reflect 10

Extended Response APPLYING

Ask the following questions:

- **What did you like about playing** Identifying Colors?
- **Was there anything about playing this game you didn't like?**
- **Did this game help you do something you couldn't do before? What did it help you do?**
- **What was easy when you were playing** Identifying Colors?
- **What was hard when you were playing** Identifying Colors?
- **What do you remember most about this game?**
- **What was your favorite part?**

 Assess 10

A Gather Evidence

Formal Assessment

Have students complete the weekly test, **Assessment,** page 27. Record formal assessment scores on the Student Assessment Record, **Assessment,** page 100.

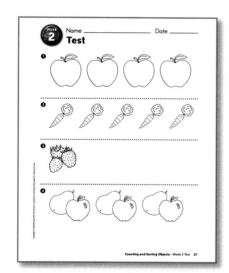

Assessment, p. 27

B Summarize Findings

Review the Student Assessment Records. Determine whether children have Minimal, Basic, or Secure understanding of the concepts presented in Week 2.

C Differentiate Instruction

Based on your observations, use these teaching strategies next week to follow up.

Minimal Understanding

- Repeat the Warm-Up and Engage activities to develop number awareness and counting skills.
- Use **Building Blocks** computer activities beginning with Kitchen Counter to reinforce counting and sequencing concepts.

Basic Understanding

- Repeat Engage activities in subsequent weeks to give children further practice with sorting and counting concepts.
- Use **Building Blocks** computer activities beginning with Party Time Free Explore to reinforce this week's concepts.

Secure Understanding

- Use Challenge variations of activities.
- Use computer activities to extend children's understanding of the Week 2 concepts.

More Sorting and Counting Objects

Week at a Glance

This week children begin **Number Worlds,** Week 3, and continue to explore Object Land.

Background

In Object Land, numbers are represented as groups of objects. This is the first way numbers were represented historically, and this is the first way children naturally learn about numbers. In Object Land, children work with real, tangible objects and with pictures of objects.

How Children Learn

As children begin this week's lesson, they should be more familiar with sorting and the number sequence.

By the end of the week, each child should have a firmer grasp on one-to-one correspondence and set size.

Skills Focus

- Count from 1 to 5
- Count from 1 to 10
- Predict the next number up, when counting up
- Identify primary colors

Teaching for Understanding

As children engage in these activities, they reinforce counting skills and expand their number knowledge.

Observe closely while evaluating the Engage activities assigned for this week.

- Are children counting in the correct order?
- Are children predicting the next number in sequence when asked?
- Are children sorting by color?

Math at Home

Give one copy of the Letter to Home, page A3, to each child. Complete the activity in class, and then encourage children to share it with their caregivers.

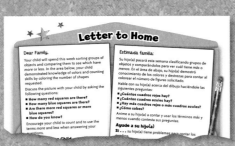

Letter to Home, Teacher Edition, p. A3

Math Vocabulary

amount A number of objects

sort To group things that are alike together

rectangle A shape with four sides

English Learners

SPANISH COGNATES

English	Spanish
count	contar
number	número
object	objetos
group	grupo
describe	describer

ALTERNATE VOCABULARY

blastoff Blastoff is when a rocket goes up in the air

More Sorting and Counting Objects

PACING	LESSON	LEARNING GOALS	MATERIALS	TECHNOLOGY
DAY 1	**Warm Up** **Object Land:** Count Up*	Children review counting skills.	No additional materials needed	**Building Blocks** Count and Race
	Engage **Object Land:** Using the Sorting Mats*	Children practice sorting and counting skills.	**Program Materials** • Color Sorting Mats, 1 per group • 40 Counters in four colors, 1 set per group **Additional Materials** Bowls or containers, 1 per group of four children	
DAY 2	**Warm Up** **Object Land:** Count Up Variation 1*	Children gain additional practice with the number sequence.	No additional materials needed	**Building Blocks** Count and Race
	Engage **Object Land:** Using the Sorting Mats*	Children gain additional practice identifying objects by attribute.	**Program Materials** • Color Sorting Mats, 1 per group • 40 Counters in four colors, 1 set per group **Additional Materials** Bowls or containers, 1 per group	**e MathTools** Set Tool
DAY 3	**Warm Up** **Object Land:** Blastoff!*	Children practice counting backward.	No additional materials needed	**Building Blocks** Count and Race
	Engage **Object Land:** Using the Sorting Mats*	Children continue to practice counting and sorting skills.	**Program Materials** • Color Sorting Mats, 1 per group • 40 Counters in four colors, 1 set per group **Additional Materials** Bowls or containers, 1 per group of four children	
DAY 4	**Warm Up** **Object Land:** Count Up Variation 2*	Children increase their understanding of one-to-one correspondence.	No additional materials needed	**Building Blocks** Count and Race
	Engage **Object Land:** Using the Sorting Mats*	Children identify set size.	**Program Materials** • Color Sorting Mats, 1 per group • 40 Counters in four colors, 1 set per group of four children **Additional Materials** Bowls or containers, 1 per group	
DAY 5	**Warm Up** **Object Land:** Blastoff!*	Children gain further practice counting backward.	No additional materials needed	**Building Blocks** Review previous activities
	Review **Object Land:** Free-Choice Activity	Children review and reinforce skills and concepts from this and previous weeks.	See materials listed on Activity Cards.	

* Includes Challenge Variations

Lesson 1

Objective

Children learn that each number can be associated with an object and that a number can show the size of a set of objects.

Program Materials

Engage

- Color Sorting Mats, 1 per group of four children
- 40 Counters in four colors, 1 set per group of four children

Additional Materials

Engage

Bowls or containers, 1 per group of four children

Access Vocabulary

What is this activity about? What are we practicing and learning? Why do we do this activity?

Creating Context

To play the game in this lesson, the children need to know the words we use to make comparisons. Introduce the words *more, less,* and *same.* Give children many examples to help them understand these terms. Demonstrate the meanings of the words using classroom objects.

Bonus!

1 Warm Up 5

Concept Building COMPUTING

Count Up

Before beginning **Using the Sorting Mats,** use the **Count Up** activity with the whole group.

Purpose **Count Up** helps children learn that numbers appear in the counting sequence in a fixed order and that this order makes it possible to predict the next number up.

Warm-Up Card 1

Monitoring Student Progress

If . . . children have trouble remembering the sequence of numbers,

Then . . . model counting for them throughout the day.

2 Engage 30

Skill Building UNDERSTANDING

Using the Sorting Mats

"Today you will share Sorting Mats that will help you sort and count your Counters."

Follow the instructions on the Activity Card to play **Using the Sorting Mats.** As children play, ask questions about what is happening in the activity.

Purpose **Using the Sorting Mats** helps children practice comparing sets of objects to identify which sets have more, less, or the same amount.

Activity Card 3

Monitoring Student Progress

If . . . children have trouble correctly identifying the color squares,	**Then . . .** review the color names with them throughout the day.

Teacher's Note Classifying and sorting objects by attributes such as color is one of children's earliest experiences with data analysis. Counting objects helps children expand their number knowledge from saying numbers in sequence to associating each number with an object or action.

As children become proficient at counting, gradually reduce the extent of your modeling until you are saying only the first number aloud and mouthing the remaining numbers, or remaining silent as children continue the counting sequence by themselves.

Building Blocks For additional practice counting, children should complete **Building Blocks** Count and Race.

3 Reflect 10

Extended Response APPLYING

Ask the following questions:

- **What is this game about?** Possible answer: You have to put the right number of Counters on the right color.

- **What does this remind you of?** Answers will vary.

4 Assess

Informal Assessment

Use the Student Assessment Record, **Assessment,** page 100, to record informal observations.

COMPUTING	**UNDERSTANDING**
Count Up	**Using the Sorting Mats**
Did the child	Did the child
❏ respond accurately?	❏ make important observations?
❏ respond quickly?	❏ extend or generalize learning?
❏ respond with confidence?	❏ provide insightful answers?
❏ self-correct?	❏ pose insightful questions?

Doggone Good!

Pretty Good!

Week 3

Lesson 2

Objective

Children will gain additional practice counting and identifying objects by attribute.

Program Materials

Engage

- Color Sorting Mats, 1 per group of four children
- 40 Counters in four colors, 1 set per group of four children

Additional Materials

Engage

Bowls or containers, 1 per group of four children

Access Vocabulary

How would you describe this activity? What words would you use to tell someone about this activity?

Creating Context

Using music and rhythm is an excellent way to help English Learners practice, remember, and acquire the vocabulary of counting. Find ways to incorporate songs about numbers into different parts of the day.

1 Warm Up 5

Concept Building COMPUTING
Count Up

Before beginning **Using the Sorting Mats,** use the **Count Up** activity Variation 1: **Count and Clap** with the whole group.

Purpose **Count and Clap** paves the way for children to learn that each number can be associated with an object and that a group of objects can be labeled with a number to show the group's size.

Warm-Up Card 1

Monitoring Student Progress

| If . . . children are counting fluently, | Then . . . have them count silently as they clap. Verify that children are counting correctly by asking them to tell you how many times they clapped. |

2 Engage 30

Skill Building UNDERSTANDING
Using the Sorting Mats

"Today we will work with Counters and Sorting Mats to practice what we learned yesterday."

Follow the instructions on the Activity Card to play **Using the Sorting Mats.** As the children play, ask questions about what is happening in the activity.

Activity Card 3

Purpose **Using the Sorting Mats** helps children gain additional counting practice and practice comparing sets of objects to identify which sets have more, less, or the same amount.

MathTools Use the Set Tool to demonstrate and explore counting.

Monitoring Student Progress

If . . . children cannot tell how many Counters are on the mat,

Then . . . have them recount from one as a small group.

Teacher's Note Encourage children to use this week's vocabulary words as they engage in the activities, discuss math concepts, and make predictions.

Building Blocks For additional practice counting, children should complete **Building Blocks** Count and Race.

3 Reflect 10

Extended Response REASONING

Ask the following questions:

■ **How would you tell someone how to play this game?** Possible answer: We each get Counters. We take turns putting them on the square that is the same color. Then we count the Counters.

■ **What patterns do you see?** Possible answer: You put the same number of Counters in each square each time.

4 Assess

Informal Assessment

Use the Student Assessment Record, **Assessment,** page 100, to record informal observations.

COMPUTING	UNDERSTANDING
Count Up	**Using the Sorting Mats**
Did the child	Did the child
❏ respond accurately?	❏ make important observations?
❏ respond quickly?	❏ extend or generalize learning?
❏ respond with confidence?	❏ provide insightful answers?
❏ self-correct?	❏ pose insightful questions?

Lesson 3

Objective

Children continue to practice identifying set size by sorting and counting.

Program Materials

Engage
- Color Sorting Mats, 1 per group of four children
- 40 Counters in four colors, 1 set per group of four children

Additional Materials

Engage
Bowls or containers, 1 per group of four children

Access Vocabulary

blastoff The time when a rocket leaves the launch site is called the blastoff
countdown In a rocket launch, the countdown is counting back from 10 to 0 before the blastoff
check your answer Make sure that your answer is correct

Creating Context

Explain that when a rocket takes off for outer space it creates a huge, noisy, and dangerous explosion. Scientists count down to zero to indicate when the rocket is ready to launch or blast off.

You've got it DOWN

---

1 Warm Up 5

Concept Building COMPUTING
Blastoff!
Before beginning **Using the Sorting Mats,** use the **Blastoff!** activity with the whole group.

Purpose **Blastoff!** gives children opportunities to practice counting backward and lays the foundation for later work with subtraction.

Warm-Up Card 2

Monitoring Student Progress

| **If . . .** children have difficulty saying the numbers backward, | **Then . . .** have them repeat short sequences after you. |

2 Engage 30

Skill Building UNDERSTANDING
Using the Sorting Mats
"Today we will practice counting and sorting."
Follow the instructions on the Activity Card to play **Using the Sorting Mats.** As children play, ask questions about what is happening in the activity.

Activity Card 3

Purpose **Using the Sorting Mats** helps children practice counting and comparing sets of objects to identify which sets have more, less, or the same number of Counters.

Monitoring Student Progress

If . . . children have trouble counting the Counters on the mat,

Then . . . have them place the Counters in lines on the mat instead of groups.

Building Blocks For additional practice counting, children should complete **Building Blocks** Count and Race.

3 Reflect 10

Extended Response REASONING

Ask the following questions:

- **How did you know how many Counters are on each square?** The last number said tells how many Counters.
- **Does each square have the same number of Counters?** Yes.
- **How do you know? How can you figure this out?** Possible answer: We counted together.

4 Assess

Informal Assessment

Use the Student Assessment Record, **Assessment,** page 100, to record informal observations.

COMPUTING	UNDERSTANDING
Blastoff!	**Using the Sorting Mats**
Did the child	Did the child
❏ respond accurately?	❏ make important observations?
❏ respond quickly?	❏ extend or generalize learning?
❏ respond with confidence?	❏ provide insightful answers?
❏ self-correct?	❏ pose insightful questions?

Lesson 4

Objective

Children deepen their understanding of sorting and counting skills.

Program Materials

Engage

- Color Sorting Mats, 1 per group of four children
- 40 Counters in four colors, 1 set per group of four children

Additional Materials

Engage

Bowls or containers, 1 per group of four children

Access Vocabulary

Counters Plastic circles used when counting

count silently Count inside your head and not out loud

Creating Context

Collaborating in small groups is an excellent way for English Learners to practice English and better understand math concepts. Letting English Learners interact with other children who speak the same primary language is one way to check understanding. Working with children who are proficient in English helps English Learners practice speaking the new language as they acquire English vocabulary.

Warm Up   5

Concept Building COMPUTING
Count Up

Before beginning **Using the Sorting Mats,** use **Count Up** Variation 2: **Count Silently and Clap** with the whole group.

Purpose Count Silently and Clap with clapping helps children learn that each number can be associated with an action as well as an object.

Warm-Up Card 1

Monitoring Student Progress

If . . . children have trouble counting and clapping,

Then . . . begin by counting and clapping to 3, and slowly work up to counting and clapping to 5.

2   Engage   30

Skill Building ENGAGING
Using the Sorting Mats

"Today we will work together to compare groups of Counters."

Follow the instructions on the Activity Card to play **Using the Sorting Mats.** As children play, ask questions about what is happening in the activity.

Purpose Using the Sorting Mats gives students an opportunity to improve social skills as they practice counting and sorting.

Activity Card 3

Monitoring Student Progress

| **If . . .** children have trouble placing the correct number of Counters onto the mat, | **Then . . .** have them recount from 1. |

 Building Blocks For additional practice counting, children should complete **Building Blocks** Count and Race.

3 Reflect   10

Extended Response   REASONING

Ask the following questions:

- **How did you know how many Counters are on the mat altogether?** Possible answer: We counted them together.
- **What did you like about this game?** Answers will vary.
- **What was the biggest number of Counters on a color on the mat?** 4
- **What was the smallest number of Counters on a color on the mat?** 1

4 Assess

Informal Assessment

Use the Student Assessment Record, **Assessment,** page 100, to record informal observations.

COMPUTING	**ENGAGING**
Count Up	**Using the Sorting Mats**
Did the child	Did the child
❏ respond accurately?	❏ pay attention to the contributions of others?
❏ respond quickly?	
❏ respond with confidence?	❏ contribute information and ideas?
❏ self-correct?	
	❏ improve on a strategy?
	❏ reflect on and check accuracy of work?

Lesson 5

Review

Objective

Children will review and practice skills from the past three weeks.

Program Materials

See Activity Cards for Materials.

Additional Materials

See Activity Cards for Materials.

Creating Context

Using puppets is an excellent way for children to practice speaking English. Puppets put the focus on the puppet and not on the student. Have children work in pairs to "teach" a puppet to count up.

1 Warm Up 5

Concept Building COMPUTING

Blastoff!

Before beginning the **Free-Choice** activity, use the **Blastoff!** activity with the whole class.

Purpose **Blastoff!** helps children build skills that they will use later to solve subtraction problems.

Warm-Up Card 2

Monitoring Student Progress

If . . . children cannot remember which number comes next,

Then . . . allow them to begin the sequence again. Some children may count forward quietly from 1 to identify the correct number.

2 Engage 20

Skill Building ENGAGING

Free-Choice Activity

For the last day of the week, allow children to choose an activity from this week or the previous weeks. These include the following:

- Object Land: **Watch, Listen, and Count**
- Object Land: **Identifying Colors**
- Object Land: **Using the Sorting Mats**

Make a note of the activities children select. Do they prefer easy or challenging activities? If you believe your children would benefit from extra practice on specific skills, choose an activity for them.

Reflect 10

Extended Response REASONING

Ask the following questions:

- **What did you like best about playing** Using the Sorting Mats?
- **Was there anything about playing this game you didn't like?**
- **Did this game help you do something you couldn't do before? What did it help you do?**
- **What was easy when you were playing** Using the Sorting Mats?
- **What was hard when you were playing** Using the Sorting Mats?
- **What do you remember most about this game?**
- **What was your favorite part?**

 Assess 10

A Gather Evidence

Formal Assessment

Have students complete the weekly test on *Assessment,* page 29. Record formal assessment scores on the Student Assessment Record, *Assessment,* page 100.

Assessment, p. 29

B Summarize Findings

Review the Student Assessment Records. Determine whether children have Minimal, Basic, or Secure understanding of the concepts presented in Week 3.

C Differentiate Instruction

Based on your observations, use these teaching strategies next week to follow up.

Minimal Understanding

- Repeat the Warm-Up and Engage activities to provide additional practice with number awareness and counting skills.
- Use **Building Blocks** computer activities beginning with Count and Race to develop and reinforce counting and sorting concepts.

Basic Understanding

- Repeat the Engage activities in subsequent weeks to give children further practice with sorting and counting concepts.
- Use **Building Blocks** counting computer activities beginning with Count and Race to reinforce this week's concepts.

Secure Understanding

- Use Challenge variations of activities.
- Use computer activities to extend children's understanding of the Week 3 concepts.

Sequencing Sets

Week at a Glance

This week children begin **Number Worlds,** Week 4, and continue to explore Object Land.

Background

In Object Land, numbers are represented as groups of objects. This is the first way numbers were represented historically, and this is the first way children naturally learn about numbers. In Object Land children work with real, tangible objects and with pictures of objects.

How Children Learn

As children begin this week's lesson, they should have a firmer grasp of the order numbers follow and the one-to-one relationship between objects and numbers.

At the end of the week, children should begin to understand that pictures can represent numbers as well.

Skills Focus

- Count from 1 to 5
- Count from 1 to 10
- Identify primary colors
- Identify quantity in dot-set patterns
- Sequence sets

Teaching for Understanding

As children engage in these activities, they begin to count and compare groups of dots, as well as three-dimensional Counters. This helps prepare them for the representations of numbers they will encounter in a few weeks when they visit Picture Land.

Observe closely while evaluating the Engage activities assigned for this week.

- Are children counting to 5 and to 10?
- Are children recognizing the number of dots in the dot-set patterns?
- Are children ordering groups in the correct number sequence?

Math at Home

Give one copy of the Letter to Home, page A4, to each child. Complete the activity in class, and then encourage children to share it with their caregivers.

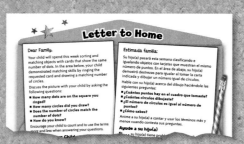

Letter to Home, Teacher Edition, p. A4

Math Vocabulary

pattern Something that repeats the same way each time

dot-set pattern Dots that you can count to find a number

English Learners

SPANISH COGNATES

English	Spanish
sequence	sequencia
compare	comparer
contrast	contrastar
identify	identificar
basic	básico

ALTERNATE VOCABULARY

stack A pile of objects with one on top of the other

. . . rest of the group. All the other people in your group

Week 4 Planner | Sequencing Sets

PACING	LESSON	LEARNING GOALS	MATERIALS	TECHNOLOGY
DAY 1	**Warm Up** Object Land: Count Up*	Children expand their counting skills.	No additional materials needed	**Building Blocks** Party Time 2
	Engage Object Land: Matching Colors and Quantities	Children extend their knowledge of number concepts.	**Program Materials** • Dot Set Cards (1–5), 1 set per student • Counters in four colors, 80 per group of four **Additional Materials** Bowls, 1 per group of four	
DAY 2	**Warm Up** Object Land: Blastoff!*	Children continue to practice counting down.	No additional materials needed	**Building Blocks** Party Time 2
	Engage Object Land: Matching Colors and Quantities	Children continue to work with objects and patterns.	**Program Materials** • Dot Set Cards (1–5), 1 set per student • Counters in four colors, 80 per group of four **Additional Materials** Bowls, 1 per group of four	
DAY 3	**Warm Up** Object Land: Blastoff!*	Children lay the groundwork for subtraction.	No additional materials needed	**Building Blocks** Party Time 2
	Engage Object Land: Matching Colors and Quantities	Children begin sequencing sets.	**Program Materials** • Dot Set Cards (1–5), 1 set per student • Counters in four colors, 80 per group of four **Additional Materials** Bowls, 1 per group of four	
DAY 4	**Warm Up** Object Land: Count Up Variation 1*	Children further expand their counting skills.	No additional materials needed	**Building Blocks** Party Time 2
	Engage Object Land: Matching Colors and Quantities	Children continue to explore relationships between objects, patterns, and numbers.	**Program Materials** • Dot Set Cards (1–5), 1 set per student • Counters in four colors, 80 per group of four **Additional Materials** Bowls, 1 per group of four	**e MathTools** Set Tool
DAY 5	**Warm Up** Object Land: Blastoff!*	Children gain further practice counting backward.	No additional materials needed	**Building Blocks** Review previous activities
	Review Object Land: Free-Choice Activity	Children review and reinforce skills and concepts from this week and from previous weeks.	See materials listed on Activity Cards.	

* Includes Challenge Variations

Lesson 1

Objective

Children will begin to identify quantity and sequence dot-set patterns.

Program Materials

Engage
- Dot Set Cards (1–5), 1 set per child
- 80 Counters, 20 in each of four colors, 1 set per group of four children

Additional Materials

Engage
Bowls or containers, 1 per group of four children

Access Vocabulary

pretend Imagine and act like (*Pretend that you are on a rocket ship* means to act like you are inside a rocket ship, even though we all know that you are really in class.)

most/fewest Most means the biggest amount, and fewest means the smallest amount

Creating Context

Review the color words with English Learners. Name a color and have them point to something of that color in the classroom, or have children sit in a circle and stand up if they are wearing something with that color.

you're doing SWIMMINGLY

---

1 Warm Up 5

Concept Building COMPUTING

Count Up

Before beginning **Matching Colors and Quantities,** use the **Count Up** activity with the whole group. Have the children count to 10.

Purpose Count Up

helps children expand their counting skills by counting to 10.

Warm-Up Card 1

Monitoring Student Progress

If . . . children have trouble counting to 10,

Then . . . give them more practice counting to 5 and gradually increase the numbers.

---

2 Engage 30

Skill Building UNDERSTANDING

Matching Colors and Quantities

"Today we will practice putting numbers in order using pictures of dots."

Follow the instructions on the Activity Card to play **Matching Colors and Quantities.** As children play, ask questions about what is happening in the activity.

Activity Card 4

Purpose Matching Colors and Quantities helps children

extend their knowledge of number concepts from concrete objects to include two-dimensional images such as dot sets.

Monitoring Student Progress

If ... children cannot tell how many dots are on a card,

Then ... have them count the dots together as a small group.

Teacher's Note Encourage children to use this week's vocabulary words as they engage in the activities, discuss math concepts, and make predictions.

Building Blocks For additional practice counting, children should complete **Building Blocks** Party Time 2.

3 Reflect 10

Extended Response REASONING

Ask the following questions:

- **Describe this activity.** We match the right color Counters to the cards.

- **Tell me something you know about these cards.** Possible answers: They have different numbers of dots on them; the dots are shaped like the Counters.

- **Have you ever seen the patterns on the Dot Set Cards anywhere else? Where?** Possible answer: on dot cubes and dominoes

4 Assess

Informal Assessment

Use the Student Assessment Record, **Assessment,** page 100, to record informal observations.

COMPUTING	UNDERSTANDING
Count Up	**Matching Colors and Quantities**
Did the child	Did the child
❑ respond accurately?	❑ make important observations?
❑ respond quickly?	❑ extend or generalize learning?
❑ respond with confidence?	❑ provide insightful answers?
❑ self-correct?	❑ pose insightful questions?

Lesson 2

Objective

Children will extend their knowledge of number concepts to include patterns.

Program Materials

Engage
- Dot Set Cards (1–5), 1 set per student
- 80 Counters, 20 in each of four colors, 1 set per group of four children

Additional Materials

Engage
Bowls or containers, 1 per group of four children

Access Vocabulary

dot-set patterns Dots that form a pattern that makes it easy to recognize the number of dots

matching colors Putting objects that are the same color into groups

Creating Context

Using models and concrete objects is an excellent way for English Learners to make connections between new concepts and the associated vocabulary. English Learners may find it helpful to practice counting back using objects.

---

1 Warm Up

Concept Building COMPUTING

Blastoff!
Before beginning the **Matching Colors and Quantities** activity, use the **Blastoff!** activity with the whole group.

Purpose **Blastoff!** is played in each of the five **Number Worlds** lands in a format appropriate to each land to give children ample practice with counting backward.

Warm-Up Card 2

Monitoring Student Progress

If . . . children have trouble "blasting off" at the correct time,

Then . . . practice counting down from 3 instead of 5.

---

2 Engage

Skill Building UNDERSTANDING

Matching Colors and Quantities

"Today we will work on matching sets of Counters with dots on a card."

Follow the instructions on the Activity Card to play **Matching Colors and Quantities.** As children play, ask questions about what is happening in the activity.

Activity Card 4

Purpose **Matching Colors and Quantities** allows children to extend their knowledge of number concepts from concrete objects to include dot sets.

Monitoring Student Progress

| **If . . .** children have trouble counting the dots on the Dot Set Cards, | **Then . . .** have them touch each dot as they count. |

 Building Blocks For additional practice counting, children should complete **Building Blocks** Party Time 2.

3 Reflect 10

Extended Response REASONING

Ask the following questions:

- **What patterns do you see?** Accept all reasonable answers.
- **How do you know how many Counters go with each card?** Possible answer: by counting the dots on the card.

4 Assess

Informal Assessment

Use the Student Assessment Record, **Assessment,** page 100, to record informal observations.

COMPUTING	UNDERSTANDING
Blastoff!	**Matching Colors and Quantities**
Did the child	Did the child
❏ respond accurately?	❏ make important observations?
❏ respond quickly?	❏ extend or generalize learning?
❏ respond with confidence?	❏ provide insightful answers?
❏ self-correct?	❏ pose insightful questions?

Lesson 3

Objective

Children will count and sequence objects and patterns.

Program Materials

Engage
- Dot Set Cards (1–5), 1 set per student
- 80 Counters, 20 in each of four colors, 1 set per group of four children

Additional Materials

Engage
Bowls or containers, 1 per group of four children

Access Vocabulary

. . . say the numbers backward Count from high numbers to low numbers

Creating Context

Role playing is a great way for English Learners to practice new vocabulary. Have children work together to demonstrate different ways we count down.

Concept Building COMPUTING

Blastoff!
Before beginning the **Matching Colors and Quantities** activity, use the **Blastoff!** activity with the whole class.

Purpose **Blastoff!** teaches children to count backward and paves the way for later work with subtraction.

Warm-Up Card 2

Monitoring Student Progress

If . . . children have difficulty saying the number words backward,

Then . . . have them repeat short sequences after you say them.

Skill Building UNDERSTANDING

Matching Colors and Quantities
"Today we will work on matching colors and numbers."

Follow the instructions on the Activity Card to play **Matching Colors and Quantities.** As children play, ask questions about what is happening in the activity.

Activity Card 4

Purpose **Matching Colors and Quantities** extends children's knowledge of number concepts from concrete objects to two-dimensional images.

Monitoring Student Progress

If . . . children have trouble counting the dots on the Dot Set Cards,

Then . . . have them touch each dot as they count.

 Building Blocks For additional practice counting, children should complete **Building Blocks** Party Time 2.

3 Reflect   10

Extended Response   REASONING

Ask the following questions:

- **How do you know if your cards are in the right order?** Possible answer: They start at one dot and get bigger.

- **Is there another way you could sort the cards?** Possible answer: from 5 down to 1.

4 Assess

Informal Assessment

Use the Student Assessment Record, **Assessment,** page 100, to record informal observations.

COMPUTING	UNDERSTANDING
Blastoff!	**Matching Colors and Quantities**
Did the child	Did the child
❏ respond accurately?	❏ make important observations?
❏ respond quickly?	❏ extend or generalize learning?
❏ respond with confidence?	❏ provide insightful answers?
❏ self-correct?	❏ pose insightful questions?

Lesson 4

Objective

Children will continue to explore the relationship between quantities and patterns.

Program Materials

Engage

- Dot Set Cards (1–5), 1 set per child
- 80 Counters, 20 in each of four colors, 1 set per group of four children

Additional Materials

Engage

Bowls or containers, 1 per group of four children

Access Vocabulary

larger numbers Bigger or larger amounts, not the size of the numerals

Creating Context

English Learners can show what they know by drawing a picture or illustration in response to a question. Give children four Counters and have them draw the Counters on a piece of paper and then count. This way you can check to see if they understand that the picture is a representation of the objects.

Pretty Good!

1 Warm Up 5

Concept Building COMPUTING

Count Up

Before beginning the **Matching Colors and Quantities** activity, use **Count Up** Variation 1: **Count and Clap** with the whole group.

Purpose Count and Clap expands children's counting skill and helps them relate numbers to a set of sounds.

Warm-Up Card 1

Monitoring Student Progress

If . . . children have trouble counting and clapping,

Then . . . have them stamp their feet or slap their knees.

2 Engage 30

Skill Building UNDERSTANDING

Matching Colors and Quantities

"Today we will continue to play **Matching Colors and Quantities.**"

Follow the instructions on the Activity Card to play **Matching Colors and Quantities.** As children play, ask questions about what is happening in the activity.

Activity Card 4

Purpose Matching Colors and Quantities extends children's knowledge of number concepts from concrete objects to two-dimensional images.

⊘MathTools Use the Set Tool to demonstrate and explore counting.

Monitoring Student Progress

| **If . . .** children have trouble sorting their Dot Set Cards, | **Then . . .** have them sort sets of Counters and match the cards to the Counters. |

 Building Blocks For additional practice counting, children should complete **Building Blocks** Party Time 2.

3 Reflect 10

Extended Response **APPLYING**

Ask the following questions:

- **What would happen if you used different-colored Counters?** Possible answer: The number of Counters needed for each card would still be the same.

- **What did you think would happen during the activity? Did it happen?** Accept all reasonable answers.

4 Assess

Informal Assessment

Use the Student Assessment Record, **Assessment,** page 100, to record informal observations.

COMPUTING	**UNDERSTANDING**
Count Up	**Matching Colors and Quantities**
Did the child	Did the child
❑ respond accurately?	❑ make important observations?
❑ respond quickly?	❑ extend or generalize learning?
❑ respond with confidence?	❑ provide insightful answers?
❑ self-correct?	❑ pose insightful questions?

Whooo did a great job? **YOU!**

Bonus!

Lesson 5
Review

Objective

Children will review and practice skills from previous weeks.

Program Materials

See Activity Cards for materials.

Additional Materials

See Activity Cards for materials.

Creating Context

Pictorial representations can help English Learners remember the words needed to describe the concepts they are learning in class. Have children draw a large picture of a rocket ship taking off to help them picture the count down.

A WHALE OF A GOOD JOB!

Grrrr8!

1 Warm Up 5

Concept Building COMPUTING

Blastoff!

Before beginning the activity, use the **Blastoff!** activity with the whole class.

Purpose **Blastoff!** helps children learn to count backward.

Warm-Up Card 2

Monitoring Student Progress

If . . . children have trouble counting down,

Then . . . model counting down from 5 for them throughout the day. You can count down as you distribute books in sets of five, as children line up for lunch in sets of five, and so on.

2 Engage 20

Skill Building ENGAGING

Free-Choice Activity

For the last day of the week, allow children to choose an activity from the previous weeks. Some activities they may choose include the following:

- Object Land: **Watch, Listen, and Count**
- Object Land: **Identifying Colors**
- Object Land: **Matching Colors and Quantities**

Note which activity they select. Do they prefer easy or challenging activities? Do they prefer activities that involve movement? If you believe your children would benefit from extra practice on specific skills, choose an activity for them.

Reflect 10

Extended Response REASONING

Ask the following questions:

- **What did you like best about playing** Matching Colors and Quantities?
- **Was there anything about playing this game you didn't like?**
- **Did this game help you do something you couldn't do before? What did it help you do?**
- **What was easy when you were playing** Matching Colors and Quantities?
- **What was hard when you were playing** Matching Colors and Quantities?
- **What do you remember most about this game?**
- **What was your favorite part?**

 Assess 10

A Gather Evidence

Formal Assessment

Have students complete the weekly test, **Assessment,** page 31. Record formal assessment scores on the Student Assessment Record, **Assessment,** page 100.

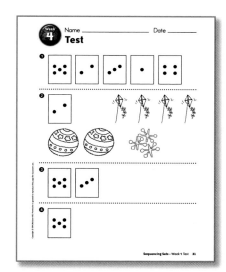

Assessment, p. 31

B Summarize Findings

Review the Student Assessment Records. Determine whether children have Minimal, Basic, or Secure understanding of the concepts presented in Week 4.

C Differentiate Instruction

Based on your observations, use these teaching strategies next week to follow up.

Minimal Understanding

- Repeat the Warm-Up and Engage activities to provide additional practice with number awareness and counting skills.
- Use **Building Blocks** computer activities beginning with Party Time 1 to reinforce counting and sequencing concepts.

Basic Understanding

- Repeat the Engage activities in subsequent weeks to give children additional practice with sequencing and matching sets.
- Use **Building Blocks** computer activities beginning with Party Time 2 to reinforce this week's concepts.

Secure Understanding

- Use Challenge variations of activities.
- Use computer activities to extend children's understanding of the Week 4 concepts.

Week at a Glance

This week, children begin **Number Worlds,** Week 5, and continue to explore Object Land.

Background

In Object Land, numbers are represented as groups of objects. This is the first way numbers were represented historically, and this is the first way children naturally learn about numbers. In Object Land, children work with real, tangible objects and with pictures of objects.

How Children Learn

As children begin this week's lesson, they may know the names and appearance of basic shapes.

By the end of the week, each child should be able to describe, compare, and contrast different shapes.

Skills Focus

- Identify basic shapes
- Classify shapes
- Compare and contrast shapes

Teaching for Understanding

As children engage in these activities, they develop spatial reasoning and geometry skills. This influences children's learning in mathematics and other subjects.

Observe closely while evaluating the Engage activities assigned for this week.

- Are children recognizing basic shapes?
- Are children sorting shapes into appropriate categories?
- Are children seeing similarities and differences when they look at shapes?

Bonus!

Math at Home

Give one copy of the Letter to Home, page A5, to each child. Complete the activity in class, and then encourage children to share it with their caregivers.

Letter to Home, Teacher Edition, p. A5

Math Vocabulary

circle A round shape ◯

square A shape with four sides that are all the same size ▢

triangle A shape with three sides △

English Learners

SPANISH COGNATES

English	Spanish
class	clase
practice	practicar
geometry	geometría

ALTERNATE VOCABULARY

count up When we begin with a number and count up or count forward

Week 5 Planner | Shapes

PACING	LESSON	LEARNING GOALS	MATERIALS	TECHNOLOGY
DAY 1	**Warm Up** **Object Land:** Count Up*	Children gain further practice counting.	No additional materials needed	**B**uilding **B**locks Mystery Pictures e **MathTools** Shape Tool
	Engage **Object Land:** Identifying Shapes	Children begin to identify basic shapes.	**Additional Materials** Attribute blocks, 2 sets	
DAY 2	**Warm Up** **Object Land:** Blastoff!*	Children build on their number knowledge.	No additional materials needed	**B**uilding **B**locks Mystery Pictures 1
	Engage **Object Land:** Identifying Shapes	Children use geometric terms to identify and describe shapes.	**Additional Materials** Attribute blocks, 2 sets	
DAY 3	**Warm Up** **Object Land:** Blastoff!*	Children gain confidence counting backward.	No additional materials needed	**B**uilding **B**locks Mystery Pictures 1
	Engage **Object Land:** Identifying Shapes	Children continue to explore and discuss shapes.	**Additional Materials** Attribute blocks, 2 sets	
DAY 4	**Warm Up** **Object Land:** Count Up*	Children reinforce counting skills.	No additional materials needed	**B**uilding **B**locks Mystery Pictures 1
	Engage **Object Land:** Identifying Shapes Variation	Children compare and contrast shapes.	**Additional Materials** Attribute blocks, 2 sets	
DAY 5	**Warm Up** **Object Land:** Blastoff!*	Children develop subtraction readiness.	No additional materials needed	**B**uilding **B**locks Review previous activities
	Review **Object Land:** Free-Choice Activity	Children may review their favorite activities or be assigned activities to improve their understanding of more difficult concepts.	See materials listed on Activity Cards.	

* Includes Challenge Variations

Lesson 1

Objective

Children begin to identify basic shapes.

Program Materials

No program materials needed

Additional Materials

Engage
Attribute blocks, 2 sets

Access Vocabulary

basic shapes Common shapes such as circle, triangle, and rectangle

Creating Context

For some English Learners the pronunciation of the names of the shapes may be challenging. For example, the word *square* begins with a sound combination that may not be used in some languages, such as Spanish. Children may naturally add a vowel syllable before the word *square,* for example *es-SQUARE.* Help children hear and practice the correct pronunciation.

Stand-up JOB

---

1 Warm Up 5

Concept Building COMPUTING

Count Up

Before beginning **Identifying Shapes,** use the **Count Up** activity with the whole group.

Purpose **Count Up** helps children learn that numbers appear in the counting sequence in a fixed order and that this order makes it possible to predict the next number up.

Warm-Up Card 1

Monitoring Student Progress

If . . . children are not ready to count to 10, | **Then . . .** give them further practice counting to 5 and slowly increase the numbers.

---

2 Engage 30

Skill Building UNDERSTANDING

Identifying Shapes

"Today we will look at shapes and practice their names."

Follow the instructions on the Activity Card to play **Identifying Shapes.** As the children play, ask questions about what is happening in the activity.

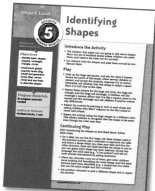

Activity Card 5

Purpose **Identifying Shapes** introduces children to basic geometric shapes.

MathTools Use the Shape Tool to demonstrate and explore shapes.

Monitoring Student Progress

If . . . children have trouble identifying the shapes or discussing their attributes,

Then . . . allow them to touch the attribute blocks as well as look at them.

 Teacher's Note Exposing children to a variety of shapes in different sizes, colors, and configurations helps them build a solid foundation for future geometric learning. Geometry and spatial sense will not only help children in math class, but will also help them as they learn about direction, distance, location, and maps.

When children can fluently identify and discuss the shapes, invite them to construct shapes using a variety of materials.

Building Blocks For additional practice with shapes, children should complete **Building Blocks** Mystery Pictures 1.

3 Reflect 10

Extended Response REASONING

Ask the following questions:

- **What is this activity about?** Possible answer: naming and describing shapes.
- **What did you notice about these shapes?** Possible answers: They were the same color; some have straight sides, and one had curved sides.

4 Assess

Informal Assessment

Use the Student Assessment Record, **Assessment,** page 100, to record informal observations.

COMPUTING	UNDERSTANDING
Count Up	**Identifying Shapes**
Did the child	Did the child
❑ respond accurately?	❑ make important observations?
❑ respond quickly?	❑ extend or generalize learning?
❑ respond with confidence?	❑ provide insightful answers?
❑ self-correct?	❑ pose insightful questions?

Lesson 2

Objective

Children begin to use geometric terms to identify and describe basic shapes.

Program Materials

No program materials needed

Additional Materials

Engage
Attribute blocks, 2 sets

Access Vocabulary

contrast Describe how two or more objects are different from one another
pretend Role play; to act as though you are doing some other task, even though we all know that you are really in class

Creating Context

English Learners may need to develop some basic English vocabulary to describe shapes. With the whole group, brainstorm some descriptive words and then use them in a sentence as a model.

---

1 Warm Up   5

Concept Building   COMPUTING

Blastoff!
Before beginning **Identifying Shapes,** use the **Blastoff!** activity with the whole group.

Purpose Blastoff! builds on children's number knowledge by counting down from 5.

Warm-Up Card 2

Monitoring Student Progress

If . . . children have trouble counting down from 5,

Then . . . practice counting down from 3 and gradually increase the numbers.

---

2 Engage   30

Skill Building   UNDERSTANDING

Identifying Shapes
"Today we will continue talking about shapes."

Follow the instructions on the Activity Card to play **Identifying Shapes.** As children play, ask questions about what is happening in the activity.

Activity Card 5

Purpose Identifying Shapes builds children's knowledge of the language of geometry while teaching them to identify and describe basic shapes.

Teacher's Note It may be helpful to make a chart listing the words and phrases your children use to describe the shapes. Even if the children do not read, this provides a print-rich environment and gives you a reference when children have trouble identifying shapes.

Building Blocks For additional practice with shapes, children should complete **Building Blocks** Mystery Pictures 1.

 3 Reflect 10

Extended Response   APPLYING

Ask the following questions:

- **Pretend the shapes are small enough to fit in your hand. Would they still be a square/circle/triangle/rectangle?** Yes.
- **What shapes do you see in your neighborhood?** Accept all reasonable answers.
- **What shapes do you see at school?** Accept all reasonable answers.

4 Assess

Informal Assessment

Use the Student Assessment Record, **Assessment,** page 100, to record informal observations.

COMPUTING	UNDERSTANDING
Blastoff!	**Identifying Shapes**
Did the child	Did the child
❑ respond accurately?	❑ make important observations?
❑ respond quickly?	
❑ respond with confidence?	❑ extend or generalize learning?
❑ self-correct?	❑ provide insightful answers?
	❑ pose insightful questions?

Lesson 3

Objective

Children continue to explore and discuss the characteristics of basic shapes.

Program Materials

No program materials needed

Additional Materials

Engage
Attribute blocks, 2 sets

Access Vocabulary

figure out Think about a problem and solve it; discover the answer
"guesser" The person in the game who tries to figure out, or "guess," which shape is being described

Creating Context

To check children's understanding, let them show what they know. For example, describe one of the shapes using some of the attributes you have discussed, and let them draw the shape, size, and color you describe.

BEARY NICE

---

1 Warm Up   5

Concept Building COMPUTING
Blastoff!
Before beginning **Identifying Shapes,** use the **Blastoff!** activity with the whole group.

Purpose **Blastoff!** gives children opportunities to practice counting backward and lays the foundation for later work with subtraction.

Warm-Up Card 2

Monitoring Student Progress

If . . . children have trouble remembering the numbers,   **Then . . .** have them repeat short sequences after you say them.

2 Engage   30

Skill Building UNDERSTANDING
Identifying Shapes
"Today we will continue to look at and describe shapes."

Follow the instructions on the Activity Card to play **Identifying Shapes.** As children play, ask questions about what is happening in the activity.

Purpose Identifying Shapes allows children to explore and discuss the attributes of shapes in different sizes, colors, and configurations, helping them build a foundation for future learning.

Activity Card 5

Monitoring Student Progress

| **If . . .** children have trouble describing the shapes, | **Then . . .** work with them throughout the day to identify and describe shapes in the classroom. |

 Building Blocks For additional practice with shapes, children should complete **Building Blocks** Mystery Pictures 1.

3 Reflect 10

Extended Response REASONING

Ask the following questions:

■ **When you were a "guesser," how did you figure out which shape was being described?** Accept all reasonable answers.

■ **What did you find out about shapes today? Did you learn something new?** Answers will vary.

4 Assess

Informal Assessment

Use the Student Assessment Record, **Assessment,** page 100, to record informal observations.

COMPUTING	UNDERSTANDING
Blastoff!	**Identifying Shapes**
Did the child	Did the child
❏ respond accurately?	❏ make important observations?
❏ respond quickly?	❏ extend or generalize learning?
❏ respond with confidence?	❏ provide insightful answers?
❏ self-correct?	❏ pose insightful questions?

Lesson 4

Objective

Children develop their ability to analyze objects by comparing and contrasting basic shapes.

Program Materials

No program materials needed

Additional Materials

Engage
Attribute blocks, 2 sets

Access Vocabulary

"describer" The person who gives clues or describes the shape in the game **Identifying Shapes**

Creating Context

Graphic organizers are an excellent way to reinforce the vocabulary English Learners need to describe the new concepts. Make a two-column chart on chart paper and have children help compare and contrast the attributes of basic shapes.

KEEP IT UP!

---

1 Warm Up 5

Concept Building `COMPUTING`
Count Up

Before beginning **Identifying Shapes,** use the Count Up activity with the whole group.

Purpose **Count Up**
reinforces counting skills from 1 to 10.

Warm-Up Card 1

Monitoring Student Progress

If . . . children are counting fluently,

Then . . . invite them to start counting from a number other than 1 and continue counting up to 10.

2 Engage 30

Skill Building `UNDERSTANDING`
Identifying Shapes

"Today we will talk about how some shapes are like each other and how some shapes are different. This is called *comparing and contrasting.*"

Follow the instructions on the Activity Card to play **Identifying Shapes** Variation: **Comparing and Contrasting Shapes.** As children play, ask questions about what is happening in the activity.

Activity Card 5

Purpose **Comparing and Contrasting Shapes** develops children's data-analysis and communication skills.

Monitoring Student Progress

| **If . . .** children have trouble describing or comparing the shapes, | **Then . . .** create shapes on the floor with masking tape. Have children walk around the different shapes and describe the experience. |

Teacher's Note Encourage children to use this week's vocabulary words as they engage in the activities, discuss math concepts, and make predictions.

Building Blocks For additional practice with shapes, children should complete **Building Blocks** Mystery Pictures 1.

3 Reflect 10

Extended Response REASONING

Ask the following questions:

- **What patterns do you see?** Accept all reasonable answers.
- **How would you describe these shapes to someone who had never seen them before?** Accept all reasonable answers.
- **What are some other shapes? Describe them to me.** Answers will vary.

Informal Assessment

Use the Student Assessment Record, **Assessment,** page 100, to record informal observations.

COMPUTING	**UNDERSTANDING**
Count Up	**Identifying Shapes**
Did the child	Did the child
❏ respond accurately?	❏ make important observations?
❏ respond quickly?	❏ extend or generalize learning?
❏ respond with confidence?	❏ provide insightful answers?
❏ self-correct?	❏ pose insightful questions?

Lesson 5
Review

Objective

Children will review and practice skills from this week and previous weeks.

Program Materials

Engage
See Activity Cards for Materials.

Additional Materials

Engage
See Activity Cards for Materials.

Creating Context

Real-life application of new vocabulary can help English Learners remember new concepts. Work with the class to identify familiar objects in the classroom or on the playground that have the shape of a circle, square, rectangle, and triangle.

1 Warm Up   5

Concept Building   COMPUTING

Blastoff!
Before beginning the Free-Choice Activity, use the **Blastoff!** activity with the whole class.

Purpose **Blastoff!** helps children build skills that they will use later to solve subtraction problems.

Warm-Up Card 2

Monitoring Student Progress

If . . . a child is having trouble,

Then . . . ask another child to help by modeling counting down from 5.

2 Engage   20

Skill Building   ENGAGING

Free-Choice Activity
For the last day of the week, allow children to choose an activity from this week or the previous weeks. These include:

- Object Land: **Watch, Listen, and Count**
- Object Land: **Identifying Colors**
- Object Land: **Using the Sorting Mats**
- Object Land: **Matching Colors and Quantities**

Note which activity they select. Do they prefer easy or challenging activities? Do they prefer activities that involve movement? If you believe your children would benefit from extra practice on specific skills, choose an activity for them.

3 Reflect 10

Extended Response REASONING

Ask the following questions:

- **What do you like best about playing** Identifying Shapes?
- **Was there anything about playing this game you didn't like?**
- **Did this game help you do something you couldn't do before? What did it help you do?**
- **What was easy when you were playing** Identifying Shapes?
- **What was hard when you were playing** Identifying Shapes?
- **What do you remember most about this game?**
- **What was your favorite part?**

4 Assess 10

A Gather Evidence

Formal Assessment

Have students complete the weekly test, *Assessment,* page 33. Record formal assessment scores on the Student Assessment Record, *Assessment,* page 100.

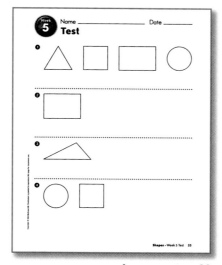

Assessment, p. 33

B Summarize Findings

Review the Student Assessment Records. Determine whether children have Minimal, Basic, or Secure understanding of the concepts presented in Week 5.

C Differentiate Instruction

Based on your observations, use these teaching strategies next week to follow up.

Minimal Understanding

- Throughout the day, point out and discuss shapes and their attributes with children. You can focus on shapes in and around the school, such as the board, classroom windows, playground equipment, bicycle tires, and so on.
- Use **Building Blocks** computer activities beginning with Mystery Shapes 1 to reinforce concepts about basic shapes.

Basic Understanding

- Repeat the **Identifying Shapes** activity in subsequent weeks to reinforce concepts about basic shapes.
- Use **Building Blocks** computer activities beginning with Mystery Shapes 2 to reinforce this week's concepts.

Secure Understanding

- Draw shapes on the board and include less familiar shapes. Ask students to identify the shapes they know and the ones they do not know.
- Use computer activities to extend children's understanding of the Week 5 concepts.

More Shapes

Week at a Glance

This week children begin **Number Worlds,** Week 6, and continue to explore Object Land.

Background

In Object Land, numbers are represented as groups of objects. This is the first way numbers were represented historically, and this is the first way children naturally learn about numbers. In Object Land, children work with real, tangible objects and with pictures of objects.

How Children Learn

As children begin this week's lesson, they should be familiar with basic shapes—circles, squares, triangles, and rectangles.

By the end of the week, each child should have increased his or her ability to describe, compare, and contrast different shapes. They will also continue to build on their social skills as they work with other children to complete this week's activities.

Skills Focus

- Identify basic shapes
- Identify and name primary colors

Teaching for Understanding

As children engage in these activities, they use multiple senses to develop their understanding of shapes and how to identify, compare, and contrast the characteristics of shapes.

Observe closely while evaluating the Engage activities assigned for this week.

- Are children recognizing basic shapes?
- Are children identifying colors by name?
- Are children seeing similarities and differences when they look at shapes?

Math at Home

Give one copy of the Letter to Home, page A6, to each child. Complete the activity in class, and then encourage children to share it with their caregivers.

Letter to Home, Teacher Edition, p. A6

Math Vocabulary

match To put things that are alike together

pretend Make believe

English Learners

SPANISH COGNATES

English	Spanish
circle	círculo
triangle	triángulo
rectangle	rectángulo
examples	ejemplos

ALTERNATE VOCABULARY

primary colors Red, blue, and yellow

basic shapes Common shapes such as circle, triangle, and rectangle

Week 6 Planner — More Shapes

PACING	LESSON	LEARNING GOALS	MATERIALS	TECHNOLOGY
DAY 1	**Warm Up** **Object Land:** Count Up*	Children continue to build number skills.	No additional materials needed	**Building Blocks** Memory Geometry 1
	Engage **Object Land:** Matching Colors and Shapes*	Children gain further practice identifying shapes and colors.	**Program Materials** 2 Color Sorting Mats **Additional Materials** Attribute blocks, 3 sets	
DAY 2	**Warm Up** **Object Land:** Blastoff!*	Children gain further practice counting down from 5.	No additional materials needed	**Building Blocks** Memory Geometry 1
	Engage **Object Land:** Matching Colors and Shapes*	Children reinforce their knowledge of basic shapes.	**Program Materials** 2 Color Sorting Mats **Additional Materials** Attribute blocks, 3 sets	**e MathTools** Shape Tool
DAY 3	**Warm Up** **Object Land:** Count Up Variation 1*	Children reinforce counting skills.	No additional materials needed	**Building Blocks** Memory Geometry 1
	Engage **Object Land:** Matching Colors and Shapes*	Children continue to explore and discuss shapes.	**Program Materials** 2 Color Sorting Mats **Additional Materials** Attribute blocks, 3 sets	
DAY 4	**Warm Up** **Object Land:** Blastoff!*	Children expand their counting skills as they count down from 10.	No additional materials needed	**Building Blocks** Memory Geometry 1
	Engage **Object Land:** Matching Colors and Shapes*	Children work cooperatively while exploring shapes.	**Program Materials** 2 Color Sorting Mats **Additional Materials** Attribute blocks, 3 sets	
DAY 5	**Warm Up** **Object Land:** Blastoff!*	Children continue to practice counting down from 10.	No additional materials needed	**Building Blocks** Review previous activities
	Review **Object Land:** Free-Choice Activity	Children reinforce number and geometry skills.	See materials listed on Activity Cards.	

* Includes Challenge Variations

Lesson 1

Objective
Children gain further experience identifying basic shapes.

Program Materials
Engage
2 Color Sorting Mats

Additional Materials
Engage
Attribute blocks, 3 sets

Access Vocabulary
matching colors Pairing up colors that are the same
matching shapes Pairing up shapes that are the same
teammate A partner, someone in the same small group

Creating Context
A strategy that helps English Learners rapidly expand their English vocabulary is to use cognates. These are words in English that are almost identical to the children's home language because they have the same shared root from Latin, Greek, or Arabic. For example, the English shape words *circle*, *rectangle*, and *triangle* are *círculo*, *rectángulo*, and *triángulo* in Spanish.

Purr-fect

---

1 Warm Up
5

Concept Building COMPUTING
Count Up
Before beginning **Matching Colors and Shapes**, use the **Count Up** activity to count up to 10 with the whole group.

Purpose **Count Up** helps children reinforce their number skills as they gain further practice counting.

Warm-Up Card 1

Monitoring Student Progress
If . . . children are not ready to count to 10, **Then . . .** slowly increase the numbers.

---

2 Engage
30

Skill Building UNDERSTANDING
Matching Colors and Shapes
"Today we will match shapes and colors."
Follow the instructions on the Activity Card to play **Matching Colors and Shapes.** As children play, ask questions about what is happening in the activity.

Purpose **Matching Colors and Shapes** gives children further practice identifying and matching geometric shapes.

Activity Card 6

Monitoring Student Progress

| If . . . children have trouble identifying colors, | Then . . . work with them through the day to identify colors in the classroom. |

 Teacher's Note You can also use food such as crackers to explore shapes. Children love to eat and are sometimes motivated by the opportunity to do so.

Building Blocks For additional practice matching shapes, children should complete **Building Blocks** Memory Geometry 1.

3 Reflect 10

Extended Response REASONING

Ask the following questions:

- **What is this activity about?** Possible answer: matching shapes and colors

- **What facts do you know about the shapes?** Possible answers: There are different colors and sizes; the triangles have three sides. Accept all reasonable answers.

4 Assess

Informal Assessment

Use the Student Assessment Record, **Assessment,** page 100, to record informal observations.

COMPUTING	UNDERSTANDING
Count Up Did the child ❏ respond accurately? ❏ respond quickly? ❏ respond with confidence? ❏ self-correct?	**Matching Colors and Shapes** Did the child ❏ make important observations? ❏ extend or generalize learning? ❏ provide insightful answers? ❏ pose insightful questions?

Lesson 2

Objective

Children reinforce their knowledge of basic shapes.

Program Materials

Engage
2 Color Sorting Mats

Additional Materials

Engage
Attribute blocks, 3 sets

Access Vocabulary

block A piece of wood or plastic that looks like a shape

plan The thinking you do before solving a problem

Creating Context

Repetition through rhyme and song is an excellent way for English Learners to get the kind of extra practice that can help them learn new English vocabulary. There are many counting songs and rhymes about shapes and colors. Listen carefully to be sure that English Learners are clearly learning the words, not just approximating the pronunciation.

Way to hop to it!

---

1 Warm Up 5

Concept Building COMPUTING

Blastoff!

Before beginning **Matching Colors and Shapes,** use the **Blastoff!** activity with the whole group.

Purpose **Blastoff!** builds on children's number knowledge by counting down from 5.

Warm-Up Card 2

Monitoring Student Progress

If . . . children have trouble counting down from 5,

Then . . . practice counting down from 3 and gradually increase the numbers.

---

2 Engage 30

Skill Building ENGAGING

Matching Colors and Shapes

"Today we will continue matching colors and shapes."

Follow the instructions on the Activity Card to introduce and play **Matching Colors and Shapes.** As children play, ask questions about what is happening in the activity.

Activity Card 6

Purpose **Matching Colors and Shapes** reinforces children's knowledge of basic shapes.

eMathTools Use the Shape Tool to demonstrate and explore shapes.

Monitoring Student Progress

If . . . children have difficulty identifying colors or shapes,

Then . . . invite a group-mate to help by discussing the characteristics that define the shape.

 Building Blocks For additional practice matching shapes, children should complete *Building Blocks* Memory Geometry 1.

Teacher's Note Encourage children to use this week's vocabulary words as they engage in the activities, discuss math concepts, and make predictions.

3 Reflect   10

Extended Response REASONING

Ask the following questions:

- **What plan did you use to match the shape of the teacher's block? Why did you choose that plan?** Accept all reasonable answers.

- **Did you think of any other ways to solve this problem? Which ones? Why didn't you use those plans?** Accept all reasonable answers.

4 Assess

Informal Assessment

Use the Student Assessment Record, **Assessment,** page 100, to record informal observations.

COMPUTING	ENGAGING
Blastoff!	**Matching Colors and Shapes**
Did the child	Did the child
❑ respond accurately?	❑ pay attention to the contributions of others?
❑ respond quickly?	
❑ respond with confidence?	❑ contribute information and ideas?
❑ self-correct?	
	❑ improve on a strategy?
	❑ reflect on and check accuracy of work?

Good for Ewe!

Your Work

Hogs the Spotlight

Lesson 3

Objective

Children continue to explore and discuss the characteristics of basic shapes.

Program Materials

Engage
2 Color Sorting Mats

Additional Materials

Engage
Attribute blocks, 3 sets

Access Vocabulary

trade shapes with the teacher Give your shape to the teacher and the teacher gives you his or hers

volunteer A person who chooses to do something without being made to do it

Creating Context

Be sure to include questions that allow English Learners to demonstrate their understanding of what they have learned. For example, when asking children to explain the activity, allow English Learners to demonstrate their knowledge by having children show you a color or shape when asked.

Stand-up JOB

---

1 Warm Up

5

Concept Building COMPUTING
Count Up

Before beginning **Matching Colors and Shapes,** use **Count Up** Variation: **Count and Clap** with the whole group.

Purpose **Count and Clap** reinforces number skills and reminds children that actions can be associated with numbers.

Warm-Up Card 1

Monitoring Student Progress

If . . . children have trouble counting and clapping to 10,

Then . . . have them count and clap to 5 and gradually increase the numbers.

2 Engage

30

Skill Building ENGAGING
Matching Colors and Shapes

"Today we will continue to work in teams to recognize shapes."

Follow the instructions on the Activity Card to introduce and play **Matching Colors and Shapes.** As children play, ask questions about what is happening in the activity.

Activity Card 6

Purpose **Matching Colors and Shapes** allows children to explore and discuss the attributes of shapes using the language of numbers and geometry.

Monitoring Student Progress

If . . . children have trouble identifying the shapes,

Then . . . work with them throughout the day to identify and describe shapes in the classroom. For example, when it is time for cleanup, have children look for circles to put away. Discuss how they knew the objects were circles.

 Building Blocks For additional practice matching shapes, children should complete **Building Blocks** Memory Geometry 1.

3 Reflect 10

Extended Response APPLYING

Ask the following questions:

- **How do you know which shapes match?** Accept all reasonable answers.

- **How would you describe this game to someone who has never played it before?** Possible answer: The teacher chooses a shape; we have to say the shape's name and color, then find the same shape and color in our pile; we trade shapes with the teacher and put the shape on the right color on the mat.

4 Assess

Informal Assessment

Use the Student Assessment Record, **Assessment,** page 100, to record informal observations.

COMPUTING	ENGAGING
Count Up	**Matching Colors and Shapes**
Did the child	Did the child
❏ respond accurately?	❏ pay attention to the contributions of others?
❏ respond quickly?	
❏ respond with confidence?	❏ contribute information and ideas?
❏ self-correct?	
	❏ improve on a strategy?
	❏ reflect on and check accuracy of work?

Lesson 4

Objective

Children develop their ability to work cooperatively to achieve goals.

Program Materials

Engage
2 Color Sorting Mats

Additional Materials

Engage
Attribute blocks, 3 sets

Access Vocabulary

features the parts of a shape that help us know what shape it is

work cooperatively Work together in a group, like a team, to do a job or solve a problem

Creating Context

When forming small work groups, matching English Learners in groups with proficient English speakers gives ELL children an opportunity to practice their English vocabulary and build fluency. Working with other children who speak the same primary language allows them to interact and discuss concepts more easily. If possible, give children opportunities to work with English speakers and other children who share the same language.

No "lion"– Math is fun!

---

1 Warm Up   5

Concept Building   COMPUTING

Blastoff!

Before beginning **Matching Colors and Shapes,** use the **Blastoff!** activity to count down from 10 with the whole group.

Purpose   **Blastoff!** helps children expand their number skills as they count down from 10.

Warm-Up Card 2

Monitoring Student Progress

If . . . children have trouble counting down from 10,

Then . . . give them further practice counting down from 5 and gradually increase the numbers.

2 Engage   30

Skill Building   UNDERSTANDING

Matching Colors and Shapes

"Today we will practice matching shapes and colors."

Follow the instructions on the Activity Card to play **Matching Colors and Shapes.** As children play, ask questions about what is happening in the activity.

Activity Card 6

Purpose   **Matching Colors and Shapes** teaches children to work cooperatively as they explore the attributes of basic shapes.

Monitoring Student Progress

| **If . . .** children have trouble identifying the shapes, | **Then . . .** review the attributes of each shape with the children. |

Teacher's Note When children can easily identify shapes and colors, start introducing new shapes with the attribute blocks. Describe and discuss the attributes of the new shapes with the class, and invite the children to make their own observations. Then you can add the new shapes to the game.

Building Blocks For additional practice matching shapes, children should complete **Building Blocks** Memory Geometry 1.

3 Reflect 10

Extended Response APPLYING

Ask the following questions:

- **What did you think would happen during the activity? Did that happen?** Accept all reasonable answers.

- **Can you think of other times at school or at home when you use shapes?** Accept all reasonable answers.

4 Assess

Informal Assessment

Use the Student Assessment Record, **Assessment**, page 100, to record informal observations.

COMPUTING	UNDERSTANDING
Blastoff! Did the child ❑ respond accurately? ❑ respond quickly? ❑ respond with confidence? ❑ self-correct?	**Matching Colors and Shapes** Did the child ❑ pay attention to the contributions of others? ❑ contribute information and ideas? ❑ improve on a strategy? ❑ reflect on and check accuracy of work?

Lesson 5

Review

Objective

Children reinforce number and geometry skills.

Program Materials

Engage
See Activity Cards for Materials.

Additional Materials

Engage
See Activity Cards for Materials.

Creating Context

Throughout this program, many games emphasize language practice in addition to concept learning. Remind children to say their answers out loud so they can practice and receive feedback.

✓ Cumulative Assessment

After Lesson 5 is complete, children should complete the Cumulative Assessment, **Assessment,** pages 85–86. Using the key on **Assessment,** page 84, identify incorrect responses. Reteach and review the activities to reinforce concept understanding.

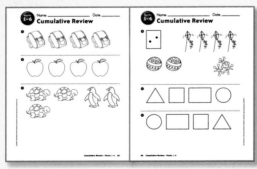

Assessment, pp. 85–86

---

1 Warm Up 5

Concept Building COMPUTING

Blastoff!
Before beginning the **Free-Choice** activity, use the **Blastoff!** activity to count down from 10 with the whole class.

Purpose **Blastoff!** helps children gain further practice counting down from 10.

Warm-Up Card 2

Monitoring Student Progress

| If . . . children are having trouble counting down from 10, | Then . . . have them repeat shorter number sequences after you say them. |

---

2 Engage 20

Skill Building ENGAGING

Free-Choice Activity
For the last day of the week, allow children to choose an activity from this week or the previous weeks. These include the following:

- Object Land: **Watch, Listen, and Count**
- Object Land: **Identifying Colors**
- Object Land: **Using the Sorting Mats**
- Object Land: **Matching Colors and Quantities**
- Object Land: **Identifying Shapes**

Make a note of the activities children select. Do they prefer easy or challenging activities? If you believe your children would benefit from extra practice on specific skills, choose an activity for them.

Reflect
10

Extended Response REASONING

Ask the following questions:

- **What do you like best about playing** Matching Colors and Shapes?
- **Was there anything about playing this game you didn't like?**
- **Did this game help you do something you couldn't do before? What did it help you do?**
- **What was easy when you were playing** Matching Colors and Shapes?
- **What was hard when you were playing** Matching Colors and Shapes?
- **What do you remember most about this game?**
- **What was your favorite part?**

Assess
10

A Gather Evidence
Formal Assessment

Have students complete the weekly test, **Assessment,** page 35. Record formal assessment scores on the Student Assessment Record, **Assessment,** page 100.

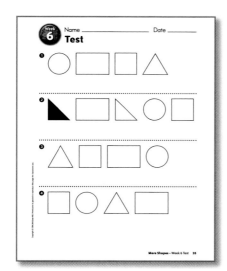

Assessment, p. 35

B Summarize Findings

Review the Student Assessment Records. Determine whether children have Minimal, Basic, or Secure understanding of the concepts presented in Week 6.

C Differentiate Instruction

Based on your observations, use these teaching strategies next week to follow up.

Minimal Understanding

- Repeat the Warm-Up and Engage activities with children; the number knowledge they gain from these activities will help them as they discuss the number of sides and angles that each shape has.
- Use **Building Blocks** computer activities beginning with Memory Geometry 1 to develop and reinforce concepts about basic shapes.

Basic Understanding

- Repeat Engage activities in subsequent weeks to reinforce concepts about basic shapes.
- Use **Building Blocks** computer activities starting with Memory Geometry 2 to reinforce this week's concepts.

Secure Understanding

- Begin to introduce new shapes to the children. Start with rhombus and trapezoid and move on to hexagon when ready.
- Use computer activities to extend children's understanding of the Week 6 concepts.

Week 7

Learning about Set Size

Week at a Glance

This week children begin **Number Worlds,** Week 7, and begin to explore Picture Land.

Background

In Picture Land, numbers are represented as sets of dots. When a child is asked, "What number do you have?" he or she might count each dot to determine the set size or identify the amount by recognizing the pattern which marks that amount. Thinking about numbers as a product of counting and as a set that has a particular value will become intertwined in the child's sense of number.

How Children Learn

As children begin this week's lesson, they should be familiar with the idea that counting can help predict the next number up and should know how to identify set size.

By the end of the week, each child should understand that numbers can be represented as images and other symbols.

Skills Focus

- Count from 1 to 5
- Count from 1 to 10
- Represent numbers using fingers
- Predict the next number
- Compute set size

Teaching for Understanding

As children engage in these activities, they use dot-set patterns to begin to transition between using movable objects and abstract symbols when counting. They are exposed to numerals along with the dot-set patterns to further build these linkages.

Observe closely while evaluating the Engage activities assigned for this week.

- Are children showing numbers with their fingers when asked?
- Are children using their knowledge of counting to predict the next number in sequence?
- Are children able to find the correct set size?

Math at Home

Give one copy of the Letter to Home, page A7, to each child. Complete the activity in class, and then encourage children to share it with their caregivers.

Letter to Home, Teacher Edition, p. A7

Math Vocabulary

Dot Set Card A card that shows a number of dots

Action Card A card that shows something you will do

wink To open and shut one eyelid

English Learners

SPANISH COGNATES

English	Spanish
pattern	patrón
comparison	comparación

ALTERNATE VOCABULARY

symbol A letter, number, or picture that has special meaning or stands for something else

Picture Land Where numbers are represented by dots

Week 7 Planner — Learning about Set Size

PACING	LESSON	LEARNING GOALS	MATERIALS	TECHNOLOGY
DAY 1	**Warm Up** Picture Land: Count Up*	Children review counting skills.	No additional materials needed	**Building Blocks** Count and Race
	Engage Picture Land: Tapping and Clapping*	Children learn about the magnitude of sets.	**Program Materials** • Large Dot Set Cards (1–5) • Action Cards • Large Dot Set Cards (1–10) for the Challenge	
DAY 2	**Warm Up** Picture Land: Count Up*	Children build a foundation for mental addition.	No additional materials needed	**Building Blocks** Count and Race
	Engage Picture Land: Tapping and Clapping*	Children explore dot-set patterns and set size.	**Program Materials** • Large Dot Set Cards (1–5) • Action Cards • Large Dot Set Cards (1–10) for the Challenge	
DAY 3	**Warm Up** Picture Land: Count Up*	Children learn that adding one and counting up give the same result.	No additional materials needed	**Building Blocks** Count and Race
	Engage Picture Land: Tapping and Clapping*	Children deepen their understanding of set size.	**Program Materials** • Large Dot Set Cards (1–5) • Action Cards • Large Dot Set Cards (1–10) for the Challenge	
DAY 4	**Warm Up** Picture Land: Count Up*	Children further refine concepts of number sequence and set size.	No additional materials needed	**Building Blocks** Count and Race
	Engage Picture Land: Tapping and Clapping Variation 1*	Children practice silent counting.	**Program Materials** • Large Dot Set Cards (1–5) • Action Cards • Large Dot Set Cards (1–10) for the Challenge	**e MathTools** Coins and Money
DAY 5	**Warm Up** Picture Land: Count Up*	Children continue to practice counting with their fingers.	No additional materials needed	**Building Blocks** Review previous activities
	Review Picture Land: Free-Choice Activity	Children reinforce number skills.	See materials listed on Activity Cards.	

* Includes Challenge Variations

Lesson 1

Objective

Children learn about the magnitude of sets as they perform an action sequence based on set size.

Program Materials

Engage
- Large Dot Set Cards (1–5)
- Action Cards
- Large Dot Set Cards (1–10) for Challenge

Additional Materials
No additional materials needed

Access Vocabulary

recount To count again
dots Circles seen in a pattern

Creating Context

Working with concrete objects helps English Learners connect new vocabulary with concept learning. Have children add additional objects one by one to a set of three objects. Have them count the total number in the set after each addition. Guide children to the observation that the set gets bigger each time we add an object.

STEADY JOB!

1 Warm Up   5

Concept Building   COMPUTING
Count Up

Before beginning **Tapping and Clapping,** use the **Count Up** activity to count up to 5 with the whole group.

Purpose   Count Up helps children review counting skills and learn to associate counting up with increased set size as the children count from 1 to 5 with their fingers.

Warm-Up Card 3

Monitoring Student Progress

If . . . children are not participating,   **Then . . .** make sure you count slowly enough to give them time to think of the next number and show it with their fingers.

2 Engage   30

Skill Building   UNDERSTANDING
Tapping and Clapping

"We will play a counting activity and move different ways to show the numbers we count."

Follow the instructions on the Activity Card to play **Tapping and Clapping.** As children play, ask questions about what is happening in the activity.

Activity Card 14

Purpose   Tapping and Clapping gives the children a concrete example of the size of a set by having them perform an action sequence that corresponds in length to the size of the set.

Monitoring Student Progress

If . . . children have trouble counting the dot sets on the cards,

Then . . . distribute Dot Set Cards to them and allow children to touch each dot as they count.

Teacher's Note Encourage children to use this week's vocabulary words as they engage in the activities, discuss math concepts, and make predictions.

Building Blocks For additional practice counting, children should complete **Building Blocks** Count and Race.

3 Reflect 10

Extended Response REASONING

Ask questions such as the following:

- **What is this activity about?** Possible answer: You count the dots on a Dot Card and do the action on the Action Card that many times.

- **What do you know about the Dot Cards?** Possible answers: They have different numbers of dots on them; the dots are in different patterns. Accept all reasonable answers.

4 Assess

Informal Assessment

Use the Student Assessment Record, **Assessment**, page 100, to record informal observations.

COMPUTING	UNDERSTANDING
Count Up	**Tapping and Clapping**
Did the child	Did the child
❏ respond accurately?	❏ make important observations?
❏ respond quickly?	❏ extend or generalize learning?
❏ respond with confidence?	❏ provide insightful answers?
❏ self-correct?	❏ pose insightful questions?

Lesson 2

1 Warm Up
5

Concept Building COMPUTING

Count Up

Before beginning **Tapping and Clapping,** use the **Count Up** activity with the whole group.

Purpose **Count Up** allows children to further explore number concepts as they build a foundation for mental addition, and to see that counting up by 1 and adding 1 to a set, then recounting, yield the same result.

Warm-Up Card 3

Monitoring Student Progress

If . . . children have trouble counting while holding up their fingers,

Then . . . model the finger displays that are used for each number, and remind children that their thumb can be used to hold their fingers down.

Objective

Children continue to explore dot-set patterns and set size.

Program Materials

Engage
- Large Dot Set Cards (1–5)
- Action Cards
- Large Dot Set Cards (1–10) for Challenge

Additional Materials

No additional materials needed

Access Vocabulary

dots Circles seen in a pattern
tapping A movement of fingers or toes against the floor or a desk

Creating Context

Help English Learners build vocabulary through the use of synonyms. In this lesson, we see that larger sets use higher numbers. Some children may only know the word *high* as it is used to describe a location. Help these children understand that higher numbers are greater quantities or larger sets.

2 Engage
30

Skill Building ENGAGING

Tapping and Clapping

"Today we will continue playing **Tapping and Clapping.**"

Follow the instructions on the Activity Card to play **Tapping and Clapping.** As children play, ask questions about what is happening in the activity.

Activity Card 14

Purpose **Tapping and Clapping** helps children continue to explore dot-set patterns and set size.

Monitoring Student Progress

| If . . . children have trouble completing the activity, | Then . . . have them perform a consistent action, such as clapping. As children become more comfortable, gradually introduce different actions to the activity. |

 Building Blocks For additional practice counting, children should complete **Building Blocks** Count and Race.

3 Reflect 10

Extended Response REASONING

Ask questions such as the following:

- **What plan did you use to find out how many dots are on the Dot Card? Why did you choose that plan?** Accept all reasonable answers.

- **Did you think of other ways of doing this? Which ones? Why didn't you use those plans?** Accept all reasonable answers.

4 Assess

Informal Assessment

Use the Student Assessment Record, **Assessment,** page 100, to record informal observations.

COMPUTING	ENGAGING
Count Up	**Tapping and Clapping**
Did the child	Did the child
❑ respond accurately?	❑ pay attention to the contributions of others?
❑ respond quickly?	
❑ respond with confidence?	❑ contribute information and ideas?
❑ self-correct?	❑ improve on a strategy?
	❑ reflect on and check accuracy of work?

Lesson 3

Objective

Children identify set size and learn that higher numbers represent larger or longer sets.

Program Materials

Engage

- Large Dot Set Cards (1–5)
- Action Cards
- Large Dot Set Cards (1–10) for Challenge

Additional Materials

No additional materials needed

Access Vocabulary

count up Beginning with a number and counting up or forward; adding one number each time

count silently Counting without making a sound; saying the numbers inside your head and not out loud

Creating Context

Reading from a picture book is an excellent way for English Learners to acquire new concept vocabulary, especially when the pictures and words are very closely associated. There are many wonderful counting books available that employ pictures of items in a particular category and practice counting 0–10 or higher. Books on tape are even more helpful because children can listen to the language again and again.

1 Warm Up 5

Concept Building COMPUTING

Count Up

Before beginning **Tapping and Clapping,** use the **Count Up** activity with the whole group.

Purpose Count Up reinforces number skills and teaches children that counting up by one unit and adding one object to a set give the same result.

Warm-Up Card 3

Monitoring Student Progress

If . . . children have trouble doing the activity,

Then . . . model counting and showing set size with your fingers throughout the day. As you do this, have children practice with you.

2 Engage 30

Skill Building ENGAGING

Tapping and Clapping

"Today we will continue to work on counting while we use Action Cards and Dot Cards."

Follow the instructions on the Activity Card to play **Tapping and Clapping.** As children play, ask questions about what is happening in the activity.

Purpose Tapping and Clapping deepens children's understanding of set size.

Activity Card 14

Monitoring Student Progress

If . . . children have trouble counting the dot sets,

Then . . . let them touch each dot on the card as they count.

 Building Blocks For additional practice counting, children should complete **Building Blocks** Count and Race.

3 Reflect 10

Extended Response REASONING

Ask questions such as the following:

- **What patterns did you see?** Accept all reasonable answers.

- **Which Dot Card has the biggest number?** 5 **How do you know?** Possible answer: It has the most dots.

- **Which Dot Card has the smallest number?** 1 **How do you know?** Possible answer: It comes first when I count up.

4 Assess

Informal Assessment

Use the Student Assessment Record, **Assessment,** page 100, to record informal observations.

COMPUTING	ENGAGING
Count Up	**Tapping and Clapping**
Did the child	Did the child
❏ respond accurately?	❏ pay attention to the contributions of others?
❏ respond quickly?	
❏ respond with confidence?	❏ contribute information and ideas?
❏ self-correct?	
	❏ improve on a strategy?
	❏ reflect on and check accuracy of work?

Lesson 4

Objective
Children practice silent counting as they continue to explore set size.

Program Materials

Engage
- Large Dot Set Cards (1–5)
- Action Cards
- Large Dot Set Cards (1–10) for Challenge

Additional Materials
No additional materials needed

Access Vocabulary
Counters Plastic circles used in activities
How do you know when you are done?
How do you know that you have finished?

Creating Context
Prepositions help English Learners build concepts and express their ideas. Use the words *next, before,* and *after* by having children line up at least three items in a horizontal line. Then ask children to point to various items using the words *before, after,* and *next.*

You've got it DOWN

---

1 Warm Up   5

Concept Building [COMPUTING]
Count Up
Before beginning **Tapping and Clapping,** use the **Count Up** activity with the whole group.

Purpose **Count Up** helps children further refine concepts of number sequence and set size.

Warm-Up Card 3

Monitoring Student Progress

| **If . . .** children cannot use the appropriate finger display to predict what number will come next, | **Then . . .** have them recount from 1 with their fingers. |

---

2 Engage   30

Skill Building [ENGAGING]
Tapping and Clapping
"Today we will count silently as we play **Tapping and Clapping.**"

Follow the instructions on the Activity Card to play **Tapping and Clapping** Variation 1: **Silent Counting, Tapping, and Clapping.** As children play, ask questions about what is happening in the activity.

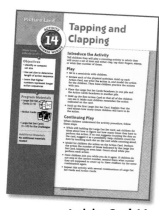

Activity Card 14

Purpose **Silent Counting, Tapping, and Clapping** teaches children to count silently as they explore set size.

MathTools Use Coins and Money to demonstrate and explore counting using pennies.

Monitoring Student Progress

| **If . . .** children have trouble counting silently, | **Then . . .** begin with numbers from 1 to 3 and gradually increase the numbers to 5. |

Building Blocks For additional practice counting, children should complete **Building Blocks** Count and Race.

3 Reflect   10

Extended Response   APPLYING

Ask questions such as the following:

■ **How do you know when you are done counting?** Possible answer: When you've counted the same number as the dots on the card.

■ **What does this remind you of?** Accept all reasonable answers.

4 Assess

Informal Assessment

Use the Student Assessment Record, **Assessment,** page 100, to record informal observations.

COMPUTING	ENGAGING
Count Up	**Tapping and Clapping**
Did the child	Did the child
❏ respond accurately?	❏ pay attention to the contributions of others?
❏ respond quickly?	❏ contribute information and ideas?
❏ respond with confidence?	
❏ self-correct?	❏ improve on a strategy?
	❏ reflect on and check accuracy of work?

Lesson 5

Review

Objective

Children reinforce number skills as they engage in the activity of their choice.

Program Materials

See Activity Cards for Materials.

Additional Materials

See Activity Cards for Materials.

Creating Context

Using physical movement to punctuate counting is an excellent way for English Learners to show that they understand the counting sequence. Demonstrate how to wink, stamp, and clap, and use these words frequently so that children learn the vocabulary and respond with the correct motion.

No "lion"– Math is fun!

You earned Your stripes!

1 Warm Up   5

Concept Building COMPUTING
Count Up
Before beginning the **Free-Choice** activity, use the **Count Up** activity with the whole class.

Purpose **Count Up** helps children build a foundation for mental addition as they count with their fingers.

Warm-Up Card 3

Monitoring Student Progress

If . . . children are having trouble counting with their fingers,

Then . . . have them repeat short sequences after you.

2 Engage   20

Skill Building ENGAGING
Free-Choice Activity

For the last day of the week, allow children to choose an activity from this week or the previous weeks. These include the following activities:
- Object Land: **Watch, Listen, and Count**
- Object Land: **Using the Sorting Mats**
- Object Land: **Matching Colors and Quantities**
- Object Land: **Identifying Shapes**
- Object Land: **Matching Colors and Shapes**

Make a note of the activities children select. Do they prefer easy or challenging activities? If you believe your children would benefit from extra practice on specific skills, choose an activity for them.

3 Reflect 10

Extended Response REASONING

Ask questions such as the following:

- **What did you like about playing** Tapping and Clapping?
- **Was there anything about playing this game you didn't like?**
- **Did this game help you do something you couldn't do before? What did it help you do?**
- **What was easy when you were playing?** Tapping and Clapping?
- **What was hard when you were playing?** Tapping and Clapping?
- **What do you remember most about this game?**
- **What was your favorite part?**

4 Assess 10

A Gather Evidence

Formal Assessment

Have students complete the weekly test on **Assessment,** page 37. Record formal assessment scores on the Student Assessment Record, **Assessment,** page 100.

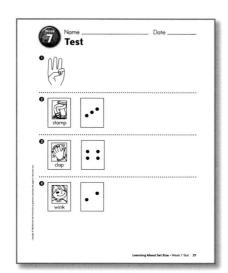

Assessment, p. 37

B Summarize Findings

Review the Student Assessment Records. Determine whether children have Minimal, Basic, or Secure understanding of the concepts presented in Week 7.

C Differentiate Instruction

Based on your observations, use these teaching strategies next week to follow up.

Minimal Understanding

- Repeat the Warm-Up and Engage activities to provide additional practice with number awareness and counting skills.
- Use **Building Blocks** computer activities beginning with Count and Race to reinforce counting and sequencing concepts.

Basic Understanding

- Repeat the Engage activities in subsequent weeks to give children further practice with number and set size concepts.
- Use **Building Blocks** computer activities beginning with Count and Race to reinforce this week's concepts.

Secure Understanding

- Use Challenge variations of **Tapping and Clapping.**
- Use computer activities to extend children's understanding of the Week 7 concepts.

Week 8 More Sets

Week at a Glance

This week children begin **Number Worlds,** Week 8, and work in both Picture Land and Object Land.

Background

In Object Land, numbers are represented as groups of objects. This is the first way numbers were represented, and this is the first way children naturally learn about numbers. In Object Land, children work with tangible objects and with pictures of objects. In Picture Land, children work with patterns.

How Children Learn

As children begin this week's lesson, they should be developing the skills to represent the numbers 1–5 with their fingers while they reinforce their counting skills.

By the end of the week, each child should be able to make decisions about how to match amounts so that when told that three people need pencils, they will know that three pencils are needed and that four pencils would be too many.

Skills Focus

- Count from 1 to 5
- Count from 1 to 10
- Represent numbers using fingers
- Predict the next number
- Compute and compare set size

Teaching for Understanding

As children engage in these activities, they use dot-set patterns and their knowledge of counting to begin basic problem solving. They will expand their sense of "a lot" and "a little" by checking to see if there is too much, too little, or enough of an object.

Observe closely while evaluating the Engage activities assigned for this week.

- Are children showing numbers with their fingers when asked?
- Are children able to find the set size?
- Are children able to solve problems that involve comparing set size?

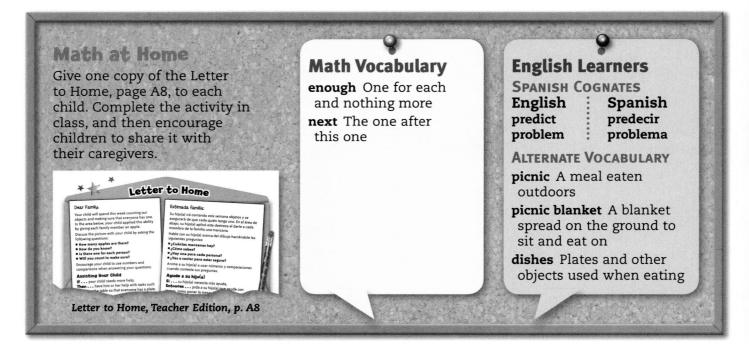

Math at Home

Give one copy of the Letter to Home, page A8, to each child. Complete the activity in class, and then encourage children to share it with their caregivers.

Letter to Home, Teacher Edition, p. A8

Math Vocabulary

enough One for each and nothing more

next The one after this one

English Learners

SPANISH COGNATES

English	Spanish
predict	predecir
problem	problema

ALTERNATE VOCABULARY

picnic A meal eaten outdoors

picnic blanket A blanket spread on the ground to sit and eat on

dishes Plates and other objects used when eating

PACING	LESSON	LEARNING GOALS	MATERIALS	TECHNOLOGY
DAY 1	**Warm Up** **Picture Land:** Count Up*	Children continue to review counting skills.	No additional materials needed	**Building Blocks** Party Time 3
	Engage **Object Land:** Picnic Party*	Children gain additional practice counting and comparing sets.	**Additional Materials** • Sets of paper, plastic, or children's dishes • Small tables or picnic blankets	
DAY 2	**Warm Up** **Picture Land:** Count Up*	Children further explore counting up.	No additional materials needed	**Building Blocks** Party Time 3
	Engage **Object Land:** Picnic Party*	Children explore counting in a real-world, problem-solving context.	**Additional Materials** • Sets of paper, plastic, or children's dishes • Small tables or picnic blankets	**e MathTools** Set Tool
DAY 3	**Warm Up** **Picture Land:** Count Up*	Children further refine concepts of number sequence and set size.	No additional materials needed	**Building Blocks** Party Time 3
	Engage **Object Land:** Picnic Party*	Children make use of counting to determine if a set has enough, not enough, or too much of an object.	**Additional Materials** • Sets of paper, plastic, or children's dishes • Small tables or picnic blankets	
DAY 4	**Warm Up** **Picture Land:** Count Up*	Children continue to build a foundation for mental addition.	No additional materials needed	**Building Blocks** Party Time 3
	Engage **Object Land:** Picnic Party Variation 1	Children prepare for later work with addition as they increase set size by 1.	**Additional Materials** • Sets of paper, plastic, or children's dishes • Small tables or picnic blankets	
DAY 5	**Warm Up** **Picture Land:** Count Up*	Children learn that counting up by one and adding one to a set are equivalent.	No additional materials needed	**Building Blocks** Review previous activities
	Review **Picture Land:** Free-Choice Activity	Children continue to explore number and set relationships.	See materials listed on Activity Cards.	

* Includes Challenge Variations

Lesson 1

Objective

Children count and compare sets of objects to solve a problem.

Program Materials

No program materials needed

Additional Materials

- Sets of paper, plastic, or children's dishes
- Small tables or picnic blankets

Access Vocabulary

paper plates Plates made of paper that are thrown away when you finish eating

enough At least one for each person

Creating Context

Not all children may have experience with picnics. Read aloud a book or use a video that shows a picnic scene so that all children can describe a picnic and what is needed to have a picnic party. Ask children to describe experiences they have had with picnics. Make sure to review the vocabulary of picnic items and which items each guest should have.

---

1 Warm Up 5

Concept Building COMPUTING

Count Up

Before beginning **Picnic Party,** use the **Count Up** activity to count up to 5 with the whole group.

Purpose Count Up has children count from 1 to 5 with their fingers to review counting skills and to help children associate counting up with increased set size.

Warm-Up Card 3

Monitoring Student Progress

If . . . children have trouble remembering the number sequence while counting with their fingers,

Then . . . give them additional practice repeating the numbers after you.

2 Engage 30

Skill Building UNDERSTANDING

Picnic Party

"Today we will get ready to have a picnic."

Follow the instructions on the Activity Card to play **Picnic Party.** As children play, ask questions about what is happening in the activity.

Purpose Picnic Party helps children gain additional practice counting and comparing sets of objects.

Activity Card 7

Monitoring Student Progress

| **If . . .** children are not participating in their groups, | **Then . . .** work on inclusive strategies, and remind them that everyone works together in a group to help one another learn. |

Teacher's Note Encourage children to use this week's vocabulary words as they engage in the activities, discuss math concepts, and make predictions.

Building Blocks For additional practice counting, children should complete **Building Blocks** Party Time 3.

3 Reflect 10

Extended Response REASONING

Ask questions such as the following:

- **What is this activity about?** Possible answer: counting to make sure everyone at a picnic has enough dishes
- **What facts do you need to know to play** Picnic Party? Possible answers: how many people are coming; how many dishes we have

4 Assess

Informal Assessment

Use the Student Assessment Record, **Assessment,** page 100, to record informal observations.

COMPUTING	UNDERSTANDING
Count Up Did the child	**Picnic Party** Did the child
❏ respond accurately?	❏ make important observations?
❏ respond quickly?	❏ extend or generalize learning?
❏ respond with confidence?	❏ provide insightful answers?
❏ self-correct?	❏ pose insightful questions?

Lesson 2

Objective

Children build on their counting skills to solve problems.

Program Materials

No program materials needed

Additional Materials

Engage

- Sets of paper, plastic, or children's dishes
- Small tables or picnic blankets

Access Vocabulary

silverware Knives, forks, and spoons
picnic blanket A blanket spread on the ground to sit on and place food on during a picnic

Creating Context

Graphic organizers are excellent tools for expanding English vocabulary. Draw a word web with *picnic* in the center and have the class generate a list of words that describe a picnic—foods, activities, things to bring, and places to go.

Off to a Running Start!

---

1 Warm Up 5

Concept Building COMPUTING

Count Up

Before beginning **Picnic Party,** use the **Count Up** activity with the whole group.

Purpose Count Up helps children further explore counting up as they build a foundation for mental addition.

Warm-Up Card 3

Monitoring Student Progress

If . . . children are counting fluently, | **Then . . .** gradually increase the numbers to 10 and use the appropriate finger displays.

2 Engage 30

Skill Building ENGAGING

Picnic Party

"Today we will continue playing Picnic Party."

Follow the instructions on the Activity Card to play **Picnic Party.** As children play, ask questions about what is happening in the activity.

Purpose Picnic Party invites children to explore counting in a real-world, problem-solving context as they determine whether a set of objects is large enough to meet a given need.

Activity Card 7

ⓔMathTools Use the Set Tool to demonstrate and explore counting.

Monitoring Student Progress

If . . . children have trouble counting the objects,

Then . . . have them place the objects in a line or move the objects from one pile to another as they count.

 For additional practice counting, children should complete **Building Blocks** Party Time 3.

3 Reflect 10

Extended Response **APPLYING**

Ask questions such as the following:

- **How did you figure out if you had enough plates for everyone at the picnic?** Accept all reasonable answers.

- **Can you think of other times at school or at home when this strategy would be helpful?** Accept all reasonable answers.

4 Assess

Informal Assessment

Use the Student Assessment Record, **Assessment,** page 100, to record informal observations.

COMPUTING	ENGAGING
Count Up	**Picnic Party**
Did the child	Did the child
❏ respond accurately?	❏ pay attention to the contributions of others?
❏ respond quickly?	
❏ respond with confidence?	❏ contribute information and ideas?
❏ self-correct?	
	❏ improve on a strategy?
	❏ reflect on and check accuracy of work?

Lesson 3

Objective

Children build problem-solving tools and improve number skills.

Program Materials

No program materials needed

Additional Materials
Engage
- Sets of paper, plastic, or children's dishes
- Small tables or picnic blankets

Access Vocabulary

picnic blanket A blanket spread on the ground to sit on and place food on during a picnic

paper plates Plates made of paper that are thrown away when you finish eating

predict To make a guess using what you know; to say what you think will happen

Creating Context

English Learners may have favorite foods that are typical in their family. Have the class discuss and draw pictures of their favorite foods. Sort the drawings into desserts and main dishes, and then have children count them. Use the information to make a graph.

You earned your stripes!

1 Warm Up 5

Concept Building COMPUTING
Count Up
Before beginning **Picnic Party,** use the **Count Up** activity with the whole group.

Purpose Count Up further refines children's concepts of number sequence and set size as they count from 1 to 5 with their fingers.

Warm-Up Card 3

Monitoring Student Progress

If . . . children cannot predict what number comes next,

Then . . . have them recount from 1 with their fingers.

2 Engage 30

Skill Building ENGAGING
Picnic Party
"Today we will make sure everyone has the dishes they need to have a picnic."

Follow the instructions on the Activity Card to play **Picnic Party.** As children play, ask questions about what is happening in the activity.

Activity Card 7

Purpose Picnic Party provides an opportunity for children to make increasingly sophisticated use of counting to determine if a set has enough, not enough, or too much of something.

Monitoring Student Progress

If . . . children need more practice sorting and counting,

Then . . . encourage them to put all of the dinnerware into a pile, sort out the correct number of one object, and place the objects on the table before sorting the next object.

 Building Blocks For additional practice counting, children should complete **Building Blocks** Party Time 3.

3 Reflect 10

Extended Response REASONING

Ask questions such as the following:

- **What steps did you use to solve the problem? What did you do first? What came next?** Accept all reasonable answers.

- **Did you have more people or more plates at the picnic? How do you know?** Accept all reasonable answers.

4 Assess

Informal Assessment

Use the Student Assessment Record, **Assessment,** page 100, to record informal observations.

COMPUTING	ENGAGING
Count Up	**Picnic Party**
Did the child	Did the child
❏ respond accurately?	❏ pay attention to the contributions of others?
❏ respond quickly?	
❏ respond with confidence?	❏ contribute information and ideas?
❏ self-correct?	
	❏ improve on a strategy?
	❏ reflect on and check accuracy of work?

High Flyer!

Whooo did a great job? YOU!

Stand-up JOB

Lesson 4

Objective

Children prepare for later work in addition as they increase set size by 1.

Program Materials

No program materials needed

Additional Materials

Engage

• Sets of paper, plastic, or children's dishes
• Small tables or picnic blankets

Access Vocabulary

picnic basket A large basket used to carry food, plates, and silverware to a picnic
hold up the correct number of fingers To show the correct number of fingers
place settings Plate, napkin, cup, and silverware for each person at the table

Creating Context

One way to help English Learners acquire new vocabulary is to have them participate in a real situation or experience. A field trip or an art project with other students can provide a rich language experience. To help children fully participate in the various counting and sorting activities this week, have a picnic with the class. This will help reinforce the activities.

Way to hop to it!

---

1 Warm Up
5

Concept Building   COMPUTING

Count Up

Before beginning **Picnic Party,** use the **Count Up** activity with the whole group.

Purpose Count Up helps children continue to build a foundation for mental addition as they count from 1 to 5 with their fingers. This gives children a concrete example of the association between counting up and increased set size.

Warm-Up Card 3

Monitoring Student Progress

| If . . . children have trouble counting with their fingers, | Then . . . give them opportunities to practice this form of counting themselves. |

2 Engage
30

Skill Building   ENGAGING

Picnic Party

"Today we will see what happens when more people come to our picnic."

Follow the instructions on the Activity Card to introduce **Picnic Party** Variation 1: **Who's Coming to Our Picnic?** As children play, ask questions about what is happening in the activity.

Activity Card 7

Purpose Who's Coming to Our Picnic? prepares children for later work with addition as they increase set size by 1.

Monitoring Student Progress

If . . . children have trouble adding 1 to the existing set,

Then . . . encourage them to recount the entire set from 1 to determine the total number.

 Building Blocks For additional practice counting, children should complete **Building Blocks** Party Time 3.

3 Reflect 10

Extended Response REASONING

Ask questions such as the following:

- **How did you figure out how many dishes you would need when another person came to the picnic?** Possible answer: I counted up 1.

- **How can you check your answers?** Possible answers: Count the whole set starting from 1; set the table to see if anyone is missing dishes.

4 Assess

Informal Assessment

Use the Student Assessment Record, **Assessment,** page 100, to record informal observations.

COMPUTING	ENGAGING
Count Up	**Picnic Party**
Did the child	Did the child
❏ respond accurately?	❏ pay attention to the contributions of others?
❏ respond quickly?	
❏ respond with confidence?	❏ contribute information and ideas?
❏ self-correct?	❏ improve on a strategy?
	❏ reflect on and check accuracy of work?

Lesson 5

Review

Objective

Children continue to explore number and set relationships as they do the activity of their choice.

Program Materials

Engage
See Activity Cards for Materials.

Additional Materials

Engage
See Activity Cards for Materials.

Creating Context

In order for English Learners to discuss comparisons, they will need to know the terms to use. Use small objects to demonstrate the difference between *more, less,* and *the same.*

FIN-tastic!

---

1 Warm Up 5

Concept Building COMPUTING

Count Up

Before beginning the **Free-Choice** activity, use the **Count Up** activity with the whole class.

Purpose **Count Up** teaches children that both counting up by one unit and adding one object to a set and recounting give equivalent results. This understanding is necessary before children can substitute mental counting for physical addition in an insightful way.

Warm-Up Card 3

Monitoring Student Progress

If . . . children have trouble counting with their fingers,

Then . . . have them practice short sequences after you.

---

2 Engage 20

Skill Building ENGAGING

Free-Choice Activity

For the last day of the week, allow children to choose an activity from this week or the previous weeks. These include the following:

- Object Land: **Watch, Listen, and Count**
- Object Land: **Identifying Colors**
- Object Land: **Using the Sorting Mats**
- Object Land: **Matching Colors and Quantities**
- Picture Land: **Tapping and Clapping**

Make a note of the activities children select. Do they prefer easy or challenging activities? If you believe your children would benefit from extra practice on specific skills, choose an activity for them.

3 Reflect   10

Extended Response REASONING

Ask questions such as the following:

- **What did you like about playing** Picnic Party?
- **Was there anything about playing this game you didn't like?**
- **Did this game help you do something you couldn't do before? What did it help you do?**
- **What was easy when you were playing** Picnic Party?
- **What was hard when you were playing** Picnic Party?
- **What do you remember most about this game?**
- **What was your favorite part?**

4 Assess   10

A Gather Evidence

Formal Assessment

Have children complete the weekly test on **Assessment,** page 39. Record formal assessment scores on the Student Assessment Record, **Assessment,** page 100.

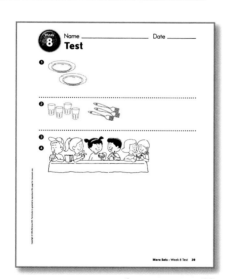

Assessment, p. 39

B Summarize Findings

Review the Student Assessment Records. Determine whether children have Minimal, Basic, or Secure understanding of the concepts presented in Week 8.

C Differentiate Instruction

Based on your observations, use these teaching strategies next week to follow up.

Minimal Understanding

- Repeat the Warm-Up and Engage activities to provide additional practice with number awareness and counting skills.
- Use **Building Blocks** computer activities beginning with Party Time 1 to reinforce counting and sequencing concepts.

Basic Understanding

- Repeat the Engage activities in subsequent weeks to give children further practice with number and set size concepts.
- Use **Building Blocks** computer activities beginning with Party Time 3 to reinforce this week's concepts.

Secure Understanding

- Use Challenge variations of **Picnic Party.**
- Use computer activities to extend children's understanding of the Week 8 concepts.

Easy as... 1-2-3

Week 9

Number Sequence

Week at a Glance

This week children begin **Number Worlds,** Week 9, and continue to explore Picture Land.

Background

In Picture Land, numbers are represented as sets of dots. When a child is asked, "What number do you have?" he or she might count each dot to determine the set size or recognize the pattern that marks that amount. Thinking about numbers as a product of counting and as a set that has a particular value will become intertwined in the child's sense of number.

How Children Learn

As children begin this week's lesson, they should be familiar with the sequence of numbers both when counting up and when counting backward.

By the end of the week, each child should be developing problem-solving skills and increasing the number skills needed to recognize missing numbers.

Skills Focus

- Count from 1 to 5 with appropriate finger displays
- Predict the next number
- Identify a missing number
- Compare quantities
- Recognize numerals

Teaching for Understanding

As children engage in these activities, they use dot-set patterns to begin to transition between using movable objects and abstract symbols when counting. They are exposed to numerals along with dot-set patterns and tally marks to further build these linkages.

Observe closely while evaluating the Engage activities assigned for this week.

- Are children showing numbers with their fingers when asked?
- Are children using their knowledge of counting to predict the next number in the sequence?
- Are children able to find the set size?

Math at Home

Give one copy of the Letter to Home, page A9, to each child. Complete the activity in class, and then encourage children to share it with their caregivers.

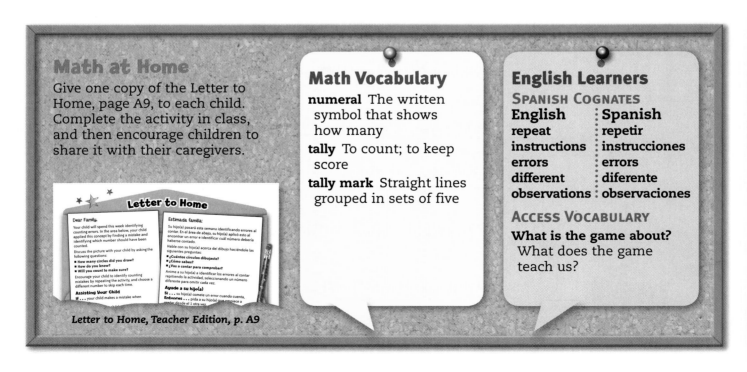

Letter to Home, Teacher Edition, p. A9

Math Vocabulary

numeral The written symbol that shows how many

tally To count; to keep score

tally mark Straight lines grouped in sets of five

English Learners

SPANISH COGNATES

English	Spanish
repeat	repetir
instructions	instrucciones
errors	errors
different	diferente
observations	observaciones

ACCESS VOCABULARY

What is the game about? What does the game teach us?

Number Sequence

PACING	LESSON	LEARNING GOALS	MATERIALS	TECHNOLOGY
DAY 1	**Warm Up** **Picture Land:** Count Up*	Children continue to refine their counting skills and associate counting with set size.	No additional materials needed	**Building Blocks** Build Stairs 3
	Engage **Picture Land:** Catch the Teacher*	Children strengthen their understanding of the counting sequence.	**Prepare Ahead** Two columns on the board, one labeled *Students*, one labeled *Teacher*	
DAY 2	**Warm Up** **Picture Land:** Name That Numeral*	Children build numeral recognition.	**Program Materials** Large Number Cards	**Building Blocks** Build Stairs 3
	Engage **Picture Land:** Catch the Teacher*	Children work with tally marks and gain practice with the counting sequence.	**Prepare Ahead** Two columns on the board, one labeled *Students*, one labeled *Teacher*	
DAY 3	**Warm Up** **Picture Land:** Name That Numeral*	Children learn to recognize numerals.	**Program Materials** Large Number Cards	**Building Blocks** Build Stairs 3 **e MathTools** Set Tool
	Engage **Picture Land:** Catch the Teacher*	Children identify counting errors.	**Prepare Ahead** Two columns on the board, one labeled *Students*, one labeled *Teacher*	
DAY 4	**Warm Up** **Picture Land:** Count Up*	Children gain additional practice with set size.	No additional materials needed	**Building Blocks** Build Stairs 3
	Engage **Picture Land:** Catch the Teacher Variation*	Children gain practice with reverse number sequence.	**Prepare Ahead** Two columns on the board, one labeled *Students*, one labeled *Teacher*	
DAY 5	**Warm Up** **Picture Land:** Name That Numeral Variation*	Children compete to name numerals.	**Program Materials** Large Number Cards	**Building Blocks** Review previous activities
	Review **Picture Land:** Free-Choice activity	Children reinforce number skills.	See materials listed on Activity Cards.	

* Includes Challenge Variations

Lesson 1

Objective

Children will associate counting with set size and will strengthen their understanding of the number sequence.

Program Materials

No additional materials needed

Prepare Ahead

Engage
2 columns on the board, one labeled *Students*, one labeled *Teacher*

Access Vocabulary

catch the teacher Find a mistake that the teacher makes

tally marks Lines that keep track of a score or action

Creating Context

Adjusting your rate of speech when working with English Learners can give them needed time to process what is being said. Articulate clearly and count deliberately so that children can match the numbers being counted with the correct finger displays.

Easy as . . . 1 - 2 - 3

---

1 Warm Up | 5

Concept Building COMPUTING

Count Up

Before beginning **Catch the Teacher,** use the **Count Up** activity with the whole group.

Purpose **Count Up** helps children refine counting skills and helps them learn to associate counting with set size as the children count with their fingers.

Warm-Up Card 3

Monitoring Student Progress

If . . . children are not participating,

Then . . . make sure you count slowly enough to give them time to think of the next number and to coordinate their fingers.

2 Engage | 30

Skill Building COMPUTING

Catch the Teacher

"Today you will try to catch me when I make mistakes counting."

Follow the instructions on the Activity Card to play **Catch the Teacher.** As children play, ask questions about what is happening in the activity.

Purpose **Catch the Teacher** strengthens children's understanding of the counting sequence as they identify omitted numbers and count tally marks.

Activity Card 15

Monitoring Student Progress

| **If** . . . children cannot identify a missing number, | **Then** . . . invite them to count aloud from 1 to 5 and to listen as you repeat the sequence and omit the same number. |

 Teacher's Note Encourage children to use this week's vocabulary words as they engage in the activities, discuss math concepts, and make predictions.

Building Blocks For additional practice identifying missing numbers, children should complete **Building Blocks** Build Stairs 3.

3 Reflect 10

Extended Response REASONING

Ask questions such as the following:

- **What is this activity about?** Possible answer: Finding the missing number when the teacher counts, then counting the points to see who wins.
- **How can you tell if a number is missing? Why did you do it that way?** Accept all reasonable answers.

4 Assess

Informal Assessment

Use the Student Assessment Record, **Assessment,** page 100, to record informal observations.

COMPUTING	COMPUTING
Count Up	**Catch the Teacher**
Did the child	Did the child
❏ respond accurately?	❏ respond accurately?
❏ respond quickly?	❏ respond quickly?
❏ respond with confidence?	❏ respond with confidence?
❏ self-correct?	❏ self-correct?

Lesson 2

Objective

Children will continue work with tally marks and will gain further practice with the counting sequence.

Program Materials

Warm Up
Large Number Cards

Prepare Ahead
Engage
2 columns on the board, one labeled *Students,* one labeled *Teacher*

Access Vocabulary

alike Almost the same or the same
see who won To determine who finished in first place or which person had the most points

Creating Context

In English it is common to use homophones, words that sound the same but mean something different and are spelled differently. Because young children are unlikely to rely on the spelling of new words to distinguish meaning, it may be helpful to explain the context and meaning of the words *one* and *won*.

---

1 Warm Up 5

Concept Building COMPUTING

Name That Numeral

Before beginning **Catch the Teacher,** use the **Name That Numeral** activity with the whole group.

Purpose **Name That Numeral** builds children's numeral recognition skills.

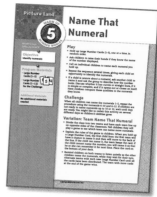

Warm-Up Card 5

Monitoring Student Progress

If . . . children have trouble recognizing numerals, | **Then . . .** point out how they are used around the classroom and the school.

---

2 Engage 30

Skill Building ENGAGING

Catch the Teacher

"Today you get another chance to beat me by figuring out what number I skip when counting."

Follow the instructions on the Activity Card to play **Catch the Teacher.** As children play, ask questions about what is happening in the activity.

Purpose **Catch the Teacher** expands children's knowledge of ways to write numbers and gives them further practice with the counting sequence as they look for errors.

Activity Card 15

Monitoring Student Progress

| If . . . children are unfamiliar with tally marks, | Then . . . explain that you make one mark for each point; and for the fifth point, you make a slash mark across the other marks to form a group of five. Demonstrate making marks up to 10. For further practice, have children tally other items throughout the day. |

 Building Blocks For additional practice identifying missing numbers, children should complete **Building Blocks** Build Stairs 3.

3 Reflect 10

Extended Response REASONING

Ask questions such as the following:

- **How can you tell who won?** You count the tally marks.

- **How are the tally marks like the dots on the Dot Set Cards? How are they different?** Possible answers: They are both sets; they tell you how many you have; they are different shapes. Accept all reasonable answers.

4 Assess

Informal Assessment

Use the Student Assessment Record, **Assessment**, page 100, to record informal observations.

COMPUTING	ENGAGING
Name That Numeral	**Catch the Teacher**
Did the child	Did the child
❏ respond accurately?	❏ pay attention to the contributions of others?
❏ respond quickly?	
❏ respond with confidence?	❏ contribute information and ideas?
❏ self-correct?	
	❏ improve on a strategy?
	❏ reflect on and check accuracy of work?

Week 9

Number Sequence

Lesson 3

Objective

Children gain further practice identifying counting errors.

Program Materials

Warm Up
Large Number Cards

Prepare Ahead
Engage
2 columns on the board, one labeled *Students,* one labeled *Teacher*

Access Vocabulary

tally marks Lines used to keep track of a score or action

What is the activity about? What do we learn when we do the activity?

Creating Context

Idiomatic expressions can make language interesting and add meaning if the listener is familiar with the expression. English Learners may be curious about the title **Catch the Teacher.** Explain that this is not a game of tag in which children might actually catch or tag the teacher. Instead, children listen carefully and see if the teacher makes a mistake; when they hear and explain the mistake, they "catch" the teacher.

---

1 Warm Up 5

Concept Building COMPUTING

Name That Numeral

Before beginning **Catch the Teacher,** use the **Name That Numeral** activity with the whole group.

Purpose Name That Numeral helps children learn to recognize numerals.

Warm-Up Card 5

Monitoring Student Progress

If . . . children have trouble naming the numerals,

Then . . . introduce only a few at a time.

---

2 Engage 30

Skill Building ENGAGING

Catch the Teacher

"Today we will continue to play **Catch the Teacher.**"

Follow the instructions on the Activity Card to play **Catch the Teacher.** As children play, ask questions about what is happening in the activity.

Purpose Catch the Teacher allows children to continue to identify counting errors and to work with sets of symbols.

Activity Card 15

ⓔMathTools Use the Set Tool to demonstrate and explore counting.

Monitoring Student Progress

If . . . children have trouble remembering the correct number sequence,

Then . . . spend some time counting from 1 to 5 as a group before beginning the activity.

 Building Blocks For additional practice identifying missing numbers, children should complete **Building Blocks** Build Stairs 3.

3 Reflect 10

Extended Response APPLYING

Ask questions such as the following:

- **Who has the most points? How do you know?** Possible answers: We have more marks; we have 4 and the teacher only has 1.
- **What does this activity remind you of?** Accept all reasonable answers.
- **Can you think of other ways we could keep track of the score?** Accept all reasonable answers.

4 Assess

Informal Assessment

Use the Student Assessment Record, **Assessment**, page 100, to record informal observations.

COMPUTING	ENGAGING
Name That Numeral	**Catch the Teacher**
Did the child	Did the child
❑ respond accurately?	❑ pay attention to the contributions of others?
❑ respond quickly?	
❑ respond with confidence?	❑ contribute information and ideas?
❑ self-correct?	
	❑ improve on a strategy?
	❑ reflect on and check accuracy of work?

FIN-tastic!

A WHALE OF A GOOD JOB!

Grrrr8!

Lesson 4

Objective

Children will gain further practice with the reverse counting sequence.

Program Materials

No additional materials needed

Prepare Ahead

Engage

2 columns on the board, one labeled *Students,* one labeled *Teacher*

Access Vocabulary

recount To count again

Creating Context

English Learners may find it helpful to work with partners who are either bilingual or proficient English speakers. This will help the English Learners interact and respond more easily. This way, response time does not become a barrier for ELL children.

1 Warm Up 5

Concept Building COMPUTING

Count Up

Before beginning **Catch the Teacher,** use the **Count Up** activity with the whole group.

Purpose **Count Up** gives children additional practice with set size as they count from 1 to 5 with their fingers.

Warm-Up Card 3

Monitoring Student Progress

If . . . children cannot predict what number comes next,

Then . . . have them recount from 1 with their fingers.

2 Engage 30

Skill Building ENGAGING

Catch the Teacher

"Today we will play **Catch the Teacher,** but I will count backward."

Follow the instructions on the Activity Card to play **Catch the Teacher** Variation: **Catch the Teacher Backward.** As children play, ask questions about what is happening in the activity.

Activity Card 15

Purpose **Catch the Teacher Backward** gives children further practice with both reverse-number sequence and the use of tally marks to keep track of quantity.

Monitoring Student Progress

If . . . children have trouble remembering the number sequence for counting down,

Then . . . spend some time counting down as a group before beginning the activity.

Teacher's Note To let all the children focus on the missing number and the countdown sequence, tell the children to remain quiet and to keep their hands in their laps until you have finished counting and have asked for volunteers to identify the missing number.

Building Blocks For additional practice identifying missing numbers, children should complete **Building Blocks** Build Stairs 3.

3 Reflect   10

Extended Response APPLYING

Ask questions such as the following:

- **Can you think of other times at school or at home when you could use tally marks?** Accept all reasonable answers.

- **Did anything about this activity surprise you? How can we explain what happened?** Accept all reasonable answers.

- **Was it harder to "catch the teacher" when I was counting up from 1 to 5 or when I was counting backward from 5 to 1?** Likely answer: It was harder when you counted backward.

- **Can anyone figure out why this is harder?** Possible answer: It's harder to count backward and remember numbers at the same time.

4 Assess

Informal Assessment

Use the Student Assessment Record, **Assessment,** page 100, to record informal observations.

COMPUTING	ENGAGING
Count Up	**Catch the Teacher**
Did the child	Did the child
❏ respond accurately?	❏ pay attention to the contributions of others?
❏ respond quickly?	
❏ respond with confidence?	❏ contribute information and ideas?
❏ self-correct?	
	❏ improve on a strategy?
	❏ reflect on and check accuracy of work?

KEEP IT UP!

There's Strength in NUMBERS

No "lion"– Math is fun!

Lesson 5
Review

Objective

Children continue to explore number relationships and set size as they do the activity of their choice.

Program Materials

Warm Up
Large Number Cards

Engage
See Activity Cards for Materials

Additional Materials

Engage
See Activity Cards for Materials

Creating Context

Graphic organizers are excellent tools for helping English Learners develop higher-level thinking skills. For example, a Venn diagram can be used to show differences and similarities and can be completed using a combination of words and illustrations.

1 Warm Up   5

Concept Building   COMPUTING

Name That Numeral

Before beginning the **Free-Choice** activity, use **Name That Numeral** Variation: **Team Name That Numeral** with the whole class.

Purpose Team Name That Numeral challenges children to name numerals as they compete as teams.

Warm-Up Card 5

Monitoring Student Progress

| If . . . children are not ready to compete as teams, | Then . . . continue to use the original **Name That Numeral** activity. |

2 Engage   20

Skill Building   ENGAGING

Free-Choice Activity

For the last day of the week, allow children to choose an activity from this week or the previous weeks. These include the following:

• Object Land: **Watch, Listen, and Count**
• Picture Land: **Tapping and Clapping**
• Object Land: **Picnic Party**

Note which activity they select. Do they prefer easy or challenging activities? Do they prefer activities that involve movement? If you believe your children would benefit from extra practice on specific skills, choose an activity for them.

3 Reflect 10

Extended Response REASONING

Ask questions such as the following:

- **What did you like about playing** Catch the Teacher?
- **Was there anything about playing this game you didn't like?**
- **Did this game help you do something you couldn't do before? What did it help you do?**
- **What was easy when you were playing** Catch the Teacher?
- **What was hard when you were playing** Catch the Teacher?
- **What do you remember most about this game?**
- **What was your favorite part?**

4 Assess 10

A Gather Evidence

Formal Assessment

Have children complete the weekly test, **Assessment,** page 41. Use the Student Assessment Record, **Assessment,** page 100, to record formal assessment scores.

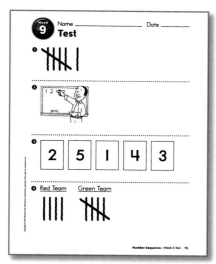

Assessment, p. 41

B Summarize Findings

Review the Student Assessment Record. Determine whether children have Minimal, Basic, or Secure understanding of the concepts presented in Week 9.

C Differentiate Instruction

Based on your observations, use these teaching strategies next week to follow up.

Minimal Understanding

- Repeat the Warm-Up and Engage activities to develop number awareness and counting skills.
- Use **Building Blocks** computer activities beginning with Build Stairs 1 to develop and reinforce counting and sequencing concepts.

Basic Understanding

- Repeat Engage activities in subsequent weeks to reinforce basic counting concepts.
- Use **Building Blocks** counting computer activities beginning with Build Stairs 3 to reinforce this week's concepts.

Secure Understanding

- Use Challenge variations of activities.
- Use computer activities to extend children's understanding of the Week 9 concepts.

Identifying Pattern and Quantity

Week at a Glance

This week children begin **Number Worlds,** Week 10, and continue to explore Picture Land.

Background

In Picture Land, numbers are represented as sets of dots. When a child is asked, "What number do you have?" he or she might count each dot to determine the set size or identify the amount by recognizing the pattern which marks that amount. Thinking about numbers as a product of counting and as a set that has a particular value will become intertwined in the child's sense of number.

How Children Learn

As children begin this week's lesson, they should begin to associate numbers with symbols such as dot sets and tally marks.

By the end of the week, children should have strengthened their recognition of dot-set patterns. This should ease their transition from object representations of numbers to set representations.

Skills Focus

- Count from 1 to 5, with appropriate finger displays
- Count from 1 to 10
- Identify pattern
- Identify quantity
- Compare quantities
- Identify numerals

Teaching for Understanding

As children engage in these activities, they match sets that are arranged in universal patterns but that are composed of familiar objects instead of dots. This builds their skills to prepare them for more abstract representations of numbers.

Observe closely while evaluating the Engage activities assigned for this week.

- Are children showing numbers with their fingers when asked?
- Are children able to match patterns?
- Are children able to find the set size and accurately compare it with other sets?

Math at Home

Give one copy of the Letter to Home, page A10, to each child. Complete the activity in class, and then encourage children to share it with their caregivers.

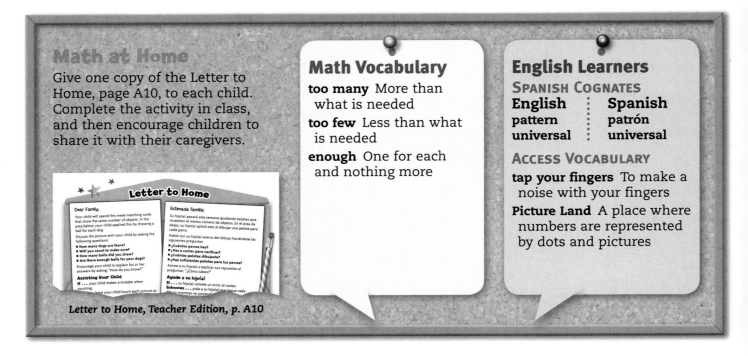

Letter to Home, Teacher Edition, p. A10

Math Vocabulary

too many More than what is needed

too few Less than what is needed

enough One for each and nothing more

English Learners

SPANISH COGNATES

English	Spanish
pattern	patrón
universal	universal

ACCESS VOCABULARY

tap your fingers To make a noise with your fingers

Picture Land A place where numbers are represented by dots and pictures

Week 10 Planner — Identifying Pattern and Quantity

PACING	LESSON	LEARNING GOALS	MATERIALS	TECHNOLOGY
DAY 1	**Warm Up** **Picture Land:** Count Up*	Children explore a visual example of how counting up increases set size.	No additional materials needed	**B**uilding **B**locks Matchmaker
	Engage **Picture Land:** Dog and Bone*	Children transition from counting objects to counting set representations.	**Program Materials** • Dog and Bone Cards • Dot Cube	ⓔ **MathTools** Set Tool
DAY 2	**Warm Up** **Picture Land:** Name that Numeral*	Children build number recognition skills.	**Program Materials** Large Number Cards	**B**uilding **B**locks Matchmaker
	Engage **Picture Land:** Dog and Bone*	Children gain additional practice identifying universal patterns of familiar objects.	**Program Materials** • Dog and Bone Cards • Dot Cube	
DAY 3	**Warm Up** **Picture Land:** Blastoff!*	Children learn that counting backward by one unit and subtracting one object from a set yield equivalent results.	No additional materials needed	**B**uilding **B**locks Matchmaker
	Engage **Picture Land:** Dog and Bone*	Children continue to practice pattern recognition and counting skills.	**Program Materials** • Dog and Bone Cards • Dot Cube	
DAY 4	**Warm Up** **Picture Land:** Blastoff!*	Children continue to build a foundation for mental subtraction.	No additional materials needed	**B**uilding **B**locks Matchmaker
	Engage **Picture Land:** Dog and Bone*	Children continue to build pattern recognition, counting, and comparison skills.	**Program Materials** • Dog and Bone Cards • Dot Cube	
DAY 5	**Warm Up** **Picture Land:** Blastoff!*	Children count down from 5 with their fingers.	No additional materials needed	**B**uilding **B**locks Review previous activities
	Review **Picture Land:** Free-Choice activity	Children review key number and sequencing concepts.	See materials listed on Activity Cards.	

* Includes Challenge Variations

Lesson 1

Objective

Children build counting skills, identify pattern or quantity on set representations, and verify equivalence by counting.

Program Materials

Engage

- Dog and Bone Cards
- Dot Cube

Additional Materials

No additional materials needed

Access Vocabulary

count with your fingers To hold up your fingers to show the number you are saying
bone Something a dog likes to chew

Creating Context

English Learners benefit from demonstrations of new instructions or concepts. When introducing **Dog and Bone,** demonstrate each step of the game so that all children understand the goal and the rules. Encourage children to check with each other in English or in their home language to make sure they know the rules before they begin to play.

---

1 Warm Up 5

Concept Building COMPUTING

Count Up

Before beginning **Dog and Bone,** use the **Count Up** activity to count up to 5 with the whole group.

Purpose **Count Up** gives children a visual example of how counting up increases set size.

Warm-Up Card 3

Monitoring Student Progress

If . . . children have trouble remembering the number sequence,

Then . . . have them practice counting to 5 with you throughout the day. Hold up the appropriate number of fingers as you count together.

2 Engage 30

Skill Building UNDERSTANDING

Dog and Bone

"Today we will practice matching numbers."

Follow the instructions on the Activity Card to play **Dog and Bone.** As children play, ask questions about what is happening in the activity.

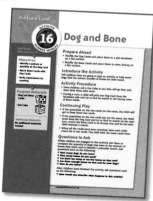

Activity Card 16

Purpose **Dog and Bone** helps children make a smooth transition from counting objects to counting set representations. The set representations are arranged in universal patterns, but show pictures of familiar objects instead of dots.

MathTools Use the Set Tool to demonstrate and explore matching quantities.

Monitoring Student Progress

| **If . . .** children have trouble matching the quantities, | **Then . . .** remind them to count the number of dogs and bones on each card. |

Teacher's Note Encourage children to use this week's vocabulary words as they engage in the activities, discuss math concepts, and make predictions.

Building Blocks For additional practice matching quantities, children should complete **Building Blocks** Matchmaker.

3 Reflect 10

Extended Response REASONING

Ask questions such as the following:

- **What is this activity about?** Possible answers: finding a Bone Card that has just enough bones for your Dog Card; finding a Bone Card that has the same number of bones as your Dog Card

- **How do you know there are enough bones for your dogs?** Possible answers: Count the dogs and the bones; look to see if the pictures are arranged the same way.

Informal Assessment

Use the Student Assessment Record, **Assessment,** page 100, to record informal observations.

COMPUTING	UNDERSTANDING
Count Up	**Dog and Bone**
Did the child	Did the child
❑ respond accurately?	❑ make important observations?
❑ respond quickly?	❑ extend or generalize learning?
❑ respond with confidence?	❑ provide insightful answers?
❑ self-correct?	❑ pose insightful questions?

Identifying Pattern and Quantity • Lesson 1 **137**

Lesson 2

Objective

Children will gain practice identifying patterns and determining set size.

Program Materials

Warm Up
Large Number Cards

Engage
- Dog and Bone Cards
- Dot Cube

Additional Materials

No additional materials needed

Access Vocabulary

patterns Objects that go in the same order over and over again

no matter what . . . Always stays the same

Creating Context

Help English Learners expand their English vocabularies by brainstorming a list of other things that go together, such as dogs and bones, birds and nests, or bikes and riders. Have children draw pictures of things that go together and share them with the whole group. Display the pictures on a bulletin board.

You hit the SPOT!

---

1 Warm Up 5

Concept Building COMPUTING

Name That Numeral

Before beginning **Dog and Bone,** use the **Name That Numeral** activity with the whole group.

Purpose Name That Numeral helps children refine their numeral recognition skills.

Warm-Up Card 5

Monitoring Student Progress

If . . . children do not recognize a number,

Then . . . discuss the lines and shapes of the numbers with the class.

---

2 Engage 30

Skill Building ENGAGING

Dog and Bone

"Today we will continue playing **Dog and Bone.**"

Follow the instructions on the Activity Card to play **Dog and Bone.** As children play, ask questions about what is happening in the activity.

Purpose Dog and Bone gives children additional practice identifying universal patterns of familiar objects. This activity will help them make a smooth transition to set representations.

Activity Card 16

Monitoring Student Progress

If . . . children have trouble matching the quantities,	Then . . . work only with cards that show quantities from 1–5, and gradually add cards that show quantities up to 10.

 Building Blocks For additional practice matching quantities, children should complete **Building Blocks** Matchmaker.

3 Reflect 10

Extended Response REASONING

Ask questions such as the following:

- **What do you know about the cards?** Possible answers: The number of dogs or bones on the card; how the dogs or bones are arranged on the card.

- **What patterns do you see?** Possible answers: The sets for each number are arranged the same way, no matter what the picture is; four is always arranged like a square.

4 Assess

Informal Assessment

Use the Student Assessment Record, **Assessment,** page 100, to record informal observations.

COMPUTING	ENGAGING
Name That Numeral	**Dog and Bone**
Did the child	Did the child
❏ respond accurately?	❏ pay attention to the contributions of others?
❏ respond quickly?	❏ contribute information and ideas?
❏ respond with confidence?	❏ improve on a strategy?
❏ self-correct?	❏ reflect on and check accuracy of work?

Lesson 3

Objective

Children continue to practice pattern recognition and counting skills.

Program Materials

Engage

- Dog and Bone Cards
- Dot Cube

Additional Materials

No additional materials needed

Access Vocabulary

Warm Up An activity to get ready for a lesson

Which one is bigger? When comparing two numbers, which one is more?

Creating Context

Some children may literally interpret the question, "which one is bigger?" and look for the dog that is largest in size. Help children understand that a set of objects is bigger when it contains more than another set. Make sure they understand that they are not comparing the size of the dogs, bones, or numerals. Encourage them to count the objects pictured on each card.

You're doing SWIMMINGLY

 1 **Warm Up** 5

Concept Building COMPUTING

Blastoff!

Before beginning **Dog and Bone,** use the **Blastoff!** activity with the whole group.

Purpose Blastoff! teaches that counting backward by one unit and subtracting one object from a set yield equivalent results.

Warm-Up Card 4

Monitoring Student Progress

If . . . children are ready,

Then . . . have them count down from 7 or 10.

2 **Engage** 30

Skill Building ENGAGING

Dog and Bone

"Today we will continue to make sure that the dogs on our cards have enough bones."

Follow the instructions on the Activity Card to play **Dog and Bone.** As children play, ask questions about what is happening in the activity.

Activity Card 16

Purpose Dog and Bone invites children to continue to practice pattern recognition and counting skills as they match cards with the same number of pictures.

Monitoring Student Progress

| If ... children have trouble matching the cards, | Then ... encourage them to focus on numbers and quantities. Provide extra practice throughout the day by having children count items in book illustrations or on magazine pages. For example, they can count the number of animals on a page or the number of cars in a picture of a city street. |

 Building Blocks For additional practice matching quantities, children should complete *Building Blocks* Matchmaker.

3 Reflect   10

Extended Response REASONING

Ask questions such as the following:

■ **How do you know how many dogs you have?** Possible answers: Count the dog pictures on the card; look at how the pictures are arranged on the card to see how many.

■ **If the numbers on the Dog Card and Bone Card aren't the same, which one is bigger? How do you know?** Possible answers: Because there are more pictures on one card; because 5 comes after 3 when we count.

4 Assess

Informal Assessment

Use the Student Assessment Record, **Assessment,** page 100, to record informal observations.

COMPUTING	ENGAGING
Blastoff!	**Dog and Bone**
Did the child	Did the child
❏ respond accurately?	❏ pay attention to the contributions of others?
❏ respond quickly?	
❏ respond with confidence?	❏ contribute information and ideas?
❏ self-correct?	
	❏ improve on a strategy?
	❏ reflect on and check accuracy of work?

Lesson 4

Objective

Children prepare for later work with numerals and patterns.

Program Materials

Engage
- Dog and Bone Cards
- Dot Cube

Additional Materials

No additional materials needed

Access Vocabulary

recount To count again

match numbers To pair numbers that show the same number of objects

Creating Context

When English Learners work in small groups or pairs with children who are more proficient in English, they have an opportunity to practice speaking fluently. Have children work together to create a story about the dogs and bones. Encourage children to retell their stories to give children practice speaking in English.

---

1 Warm Up 5

Concept Building COMPUTING

Blastoff!

Before beginning **Dog and Bone,** use the **Blastoff!** activity with the whole group.

Purpose **Blastoff!** builds children's foundation for mental subtraction as they count down with their fingers.

Warm-Up Card 4

Monitoring Student Progress

If . . . children cannot predict what number comes next,

Then . . . have them recount from 5 with their fingers.

---

2 Engage 30

Skill Building UNDERSTANDING

Dog and Bone

"Today we will play **Dog and Bone** again."

Follow the instructions on the Activity Card to play **Dog and Bone.** As children play, ask questions about what is happening in the activity.

Purpose **Dog and Bone** helps children continue to build pattern recognition and counting and comparison skills as they match sets arranged in universal patterns.

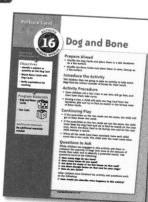

Activity Card 16

Monitoring Student Progress

If . . . children have trouble identifying how many objects are pictured on a card,

Then . . . encourage them to touch each picture as they count.

 Building Blocks For additional practice matching quantities, children should complete **Building Blocks** Matchmaker.

3 Reflect 10

Extended Response APPLYING

Ask questions such as the following:

- **What do you like about this game?** Accept all reasonable answers.
- **Can you think of other times at school or at home when you have to match amounts?** Accept all reasonable answers.

Informal Assessment

Use the Student Assessment Record, **Assessment,** page 100, to record informal observations.

COMPUTING	UNDERSTANDING
Blastoff!	**Dog and Bone**
Did the child	Did the child
❏ respond accurately?	❏ pay attention to the contributions of others?
❏ respond quickly?	
❏ respond with confidence?	❏ contribute information and ideas?
❏ self-correct?	
	❏ improve on a strategy?
	❏ reflect on and check accuracy of work?

Lesson 5
Review

Objective

Children continue to explore number and set relationships as they complete the activity of their choice.

Program Materials

Engage
See Activity Cards for Materials.

Additional Materials

Engage
See Activity Cards for Materials.

Access Vocabulary

What do you like best? What is your favorite part; what do you like most?
Picture Land A place where numbers are represented by dots and pictures

Creating Context

Have English Learners use finger displays to show you numbers that you name. Children with higher levels of English proficiency can be asked to name the displays they make.

You earned your stripes!

---

1 Warm Up   5

Concept Building COMPUTING

Blastoff!
Before beginning the **Free-Choice** activity, use the **Blastoff!** activity with the whole class.

Purpose **Blastoff!** teaches children to count down from 5 with their fingers.

Warm-Up Card 4

Monitoring Student Progress

| If . . . children have trouble counting with their fingers, | Then . . . have them practice short sequences after you. |

2 Engage   20

Skill Building ENGAGING

Free-Choice Activity

For the last day of the week, allow children to choose an activity from this week or previous weeks. These include:

- Picture Land: **Tapping and Clapping**
- Object Land: **Picnic Party**
- Picture Land: **Catch the Teacher**

Make a note of the activities children select. Do they prefer easy or challenging activities? If you believe your children would benefit from extra practice on specific skills, choose an activity for them.

3 Reflect   10

Extended Response   REASONING

Ask questions such as the following:

- **What did you like about playing** Dog and Bone?
- **Was there anything about playing this game you didn't like?**
- **Did this game help you do something you couldn't do before? What did it help you do?**
- **What was easy when you were playing** Dog and Bone?
- **What was hard when you were playing** Dog and Bone?
- **What do you remember most about this game?**
- **What was your favorite part?**

4 Assess   10

A Gather Evidence

Formal Assessment

Have children complete the weekly test on **Assessment,** page 43. Record formal assessment scores on the Student Assessment Record, **Assessment,** page 100.

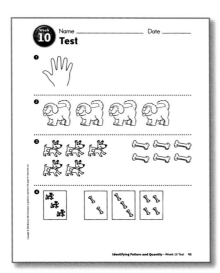

Assessment, p. 43

B Summarize Findings

Review the Student Assessment Records. Determine whether children have Minimal, Basic, or Secure understanding of the concepts presented in Week 10.

C Differentiate Instruction

Based on your observations, use these teaching strategies next week to follow up.

Minimal Understanding

- Repeat the Warm-Up and Engage activities to provide additional practice with number awareness and counting skills.
- Use **Building Blocks** computer activities beginning with Matchmaker to develop and reinforce counting and sequencing concepts.

Basic Understanding

- Repeat Engage activities in subsequent weeks to give children further practice identifying patterns and quantities.
- Use **Building Blocks** computer activities beginning with Matchmaker to reinforce this week's concepts.

Secure Understanding

- Use Challenge variations of **Dog and Bone.**
- Use computer activities to extend children's understanding of the Week 10 concepts.

Week 11

Matching Patterns and Quantities

Week at a Glance

This week, children begin **Number Worlds,** Week 11, and continue to explore Picture Land.

Background

In Picture Land, numbers are represented as a set of dots. When a child is asked, "What number do you have?" he or she might count each dot to determine the set size or identify the amount by recognizing the pattern that marks that amount. Thinking about numbers as a product of counting and as a set that has a particular value will become intertwined in the child's sense of number.

How Children Learn

As children begin this week's activities, they should have begun to build connections between the numbers they count and ways these numbers can be represented.

By the end of the week, children should have a better understanding of dot-set patterns and numerals.

Skills Focus

- Count from 1 to 5 and from 1 to 10 with appropriate finger displays
- Identify patterns or quantities
- Compare patterns and quantities
- Identify numerals

Teaching for Understanding

As children engage in these activities, they should become familiar with set representations of numbers. The challenge of finding matches should begin to build the connections needed for children to move to more advanced concepts, such as the way larger numbers are composed of two or more sets of smaller numbers.

Observe closely while evaluating the Engage activities assigned for this week.

- Are children using appropriate finger displays when counting?
- Are children able to identify the amount of dots on Dot Set Cards?
- Are children making accurate matches?
- Are children accurate when comparing quantities?

Math at Home

Give one copy of the Letter to Home, page A11, to each child. Complete the activity in class, and then encourage children to share it with their caregivers.

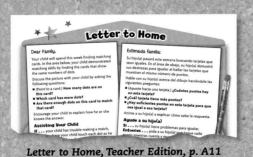

Letter to Home, Teacher Edition, p. A11

Math Vocabulary

numeral A symbol that represents a number

compare To think about how things are the same and how they are different

match To pair up two or more items that are alike

English Learners

SPANISH COGNATES

English	Spanish
concentration	concentración
pattern	patrón

ALTERNATE VOCABULARY

faceup Cards on a table with the dots or numerals showing

facedown Cards on a table with the dots or numerals down so they are not showing

Week 11 Planner — Matching Patterns and Quantities

PACING	LESSON	LEARNING GOALS	MATERIALS	TECHNOLOGY
DAY 1	**Warm Up** Picture Land: Count Up*	Children continue to practice the counting sequence.	**Program Materials** No additional materials needed	**Building Blocks** Number Snapshots 1
	Engage Picture Land: Concentration*	Children connect dot-set patterns with set size.	**Program Materials** • Small Dot Set Cards (1–10), 2 sets per group • Dot Cube, 1 per group	
DAY 2	**Warm Up** Picture Land: Count Up*	Children practice showing numbers with their fingers.	**Program Materials** No additional materials needed	**Building Blocks** Number Snapshots 1
	Engage Picture Land: Concentration*	Children build pattern recognition skills.	**Program Materials** • Small Dot Set Cards (1–10), 2 sets per group • Dot Cube, 1 per group	**MathTools** Coins and Money
DAY 3	**Warm Up** Picture Land: Blastoff!*	Children continue to learn to count backward.	**Program Materials** No additional materials needed	**Building Blocks** Number Snapshots 1
	Engage Picture Land: Concentration*	Children build skills in identifying and comparing quantities.	**Program Materials** • Small Dot Set Cards (1–10), 2 sets per group • Dot Cube, 1 per group	
DAY 4	**Warm Up** Picture Land: Name that Numeral*	Children recognize new numerals.	**Program Materials** Large Number Cards	**Building Blocks** Number Snapshots 1
	Engage Picture Land: Concentration Variation*	Children connect their knowledge of counting and patterns with numerals.	**Program Materials** • Small Dot Set Cards (1–10), 1 set per group • Small Number Cards (1–10), 1 set per group • Dot Cube, 1 per group	
DAY 5	**Warm Up** Picture Land: Blastoff!*	Children associate counting down with a decrease in set size.	**Program Materials** No additional materials needed	**Building Blocks** Review previous activities
	Review Free-Choice activity	Children build and reinforce skills learned this week and other weeks.	Materials will be selected from those used in previous weeks.	

* Includes Challenge Variations

Lesson 1

Objective

Children connect dot-set patterns with set size.

Program Materials

Engage

- Dot Set Cards (1–10), 2 sets per group
- Dot Cube, 1 per group

Additional Materials

No additional materials needed

Access Vocabulary

pile A group of objects that are on top of one another

Dot Set Card A card that shows a number of dots

Creating Context

When using cards, explain the concept of *faceup* and *facedown*. These idioms may not be familiar to children. The main, or most meaningful, side of the card is the face. A card that is faceup communicates specific information, while a facedown card gives little information.

Skill Building COMPUTING

Count Up

Before beginning **Concentration,** use the **Count Up** activity with the whole group.

Purpose **Count Up** gives children an opportunity to continue to practice the counting sequence.

Warm-Up Card 3

Monitoring Student Progress

If . . . children have trouble counting to 10,

Then . . . recount from 1 to 5 and gradually work up to 10.

2 **Engage** 30

Concept Building UNDERSTANDING

Concentration

"Today we will do an activity to see who can match the most cards."

Follow the instructions on the Activity Card to play **Concentration.** As children play, ask questions about what is happening in the activity.

Purpose **Concentration** helps children match dot-set patterns and builds children's connections between patterns and set size.

Activity Card 17

Monitoring Student Progress

If . . . a child has trouble matching cards,

Then . . . let the child count the dots on his or her card and tell the group the number. Next, let him or her count the dots on the card he or she believes is a match. Have the group remind the child of the number on the card from the pile. Ask whether the numbers are the same.

 Teacher's Note Encourage children to use this week's vocabulary words as they engage in the activities, discuss math concepts, and make predictions.

Building Blocks For additional practice matching quantities, children should complete **Building Blocks** Number Snapshots 1.

3 Reflect 10

Extended Response REASONING

Ask the following questions:

■ **How would you tell another child how to do this activity?** Possible answer: You pick one Dot Card from a pile and try to find the matching card. The person with the most matches wins.

■ **How did you find the card that matched your card?** Possible answer: I looked for a card that had dots in the same places.

4 Assess

Informal Assessment

Use the Student Assessment Record, **Assessment,** page 100, to record informal observations.

COMPUTING	UNDERSTANDING
Count Up	**Concentration**
Did the child	Did the child
❑ respond accurately?	❑ make important observations?
❑ respond quickly?	❑ extend or generalize learning?
❑ respond with confidence?	❑ provide insightful answers?
❑ self-correct?	❑ pose insightful questions?

Lesson 2

Objective

Children build pattern recognition skills.

Program Materials

Engage

- Dot Set Cards (1–10), 2 sets per group
- Dot Cube, 1 per group

Additional Materials

No additional materials needed

Access Vocabulary

match To find the card that looks the same
pay attention to To watch and listen carefully

Creating Context

Some English Learners may understand concepts but lack the English proficiency to quickly answer questions. Be sure to ask questions that will allow English Learners to participate. For example, include questions such as "Draw the dot pattern for 5" or "Point to the 4 card" so that students can show what they know without relying on oral descriptions.

---

1 Warm Up   5

Concept Building   COMPUTING

Count Up

Before beginning **Concentration,** use the **Count Up** activity with the whole group.

Purpose **Count Up** builds children's understanding of number representations as they practice showing numbers with their fingers while counting to 10.

Warm-Up Card 3

Monitoring Student Progress

| If . . . children have trouble showing numbers with their fingers, | Then . . . remind them to use their thumbs to hold down their fingers. |

---

2 Engage   30

Concept Building   ENGAGING

Concentration

"Today we will work in groups as we play **Concentration.**"

Follow the instructions on the Activity Card to play **Concentration.** As children play, ask questions about what is happening in the activity.

Purpose **Concentration** teaches children to understand patterns, relationships, and functions.

Activity Card 17

MathTools Use the Coins and Money tool to demonstrate and explore counting with pennies.

Monitoring Student Progress

If . . . children have trouble recognizing the number of dots on a card,

Then . . . have them count the dots, putting their fingers on each dot as they count.

 Building Blocks For additional practice matching quantities, children should complete **Building Blocks** Number Snapshots 1.

3 Reflect   10

Extended Response   APPLYING

Ask the following questions:

- **What patterns do you see on the Dot Set Cards?** Accept all reasonable answers.
- **Have you seen these patterns in other places? Where?** Possible answers: on the Dot Cubes; on the Dog and Bone Cards; on dominoes

4 Assess

Informal Assessment

Use the Student Assessment Record, **Assessment,** page 100, to record informal observations.

COMPUTING	ENGAGING
Count Up	**Concentration**
Did the child	Did the child
❏ respond accurately?	❏ pay attention to the contributions of others?
❏ respond quickly?	
❏ respond with confidence?	❏ contribute information and ideas?
❏ self-correct?	
	❏ improve on a strategy?
	❏ reflect on and check accuracy of work?

Lesson 3

Objective

Children further develop skills for identifying and comparing quantities.

Program Materials

Engage

- Dot Set Cards (1–10), 2 sets per group
- Dot Cube, 1 per group

Additional Materials

No additional materials needed

Access Vocabulary

warm up To get ready to learn
remind you of To make you think of

Creating Context

Explain to children that the word *concentration* comes from the verb *concentrate,* which means "to think hard about something." Then, ask them to tell you why **Concentration** is a good name for this matching game.

1 Warm Up 5

Skill Building COMPUTING

Blastoff!

Before beginning **Concentration**, use the **Blastoff!** activity with the whole group.

Purpose **Blastoff!** gives children an opportunity to practice and reinforce counting backward.

Warm-Up Card 4

Monitoring Student Progress

If . . . children are having trouble counting backward,

Then . . . have them repeat short sequences after you.

2 Engage 30

Concept Building REASONING

Concentration

"Today we will play **Concentration** again."

Follow the instructions on the Activity Card to play **Concentration**. As children play, ask questions about what is happening in the activity.

Purpose **Concentration** gives children opportunities to build on language skills as they compare quantities.

Activity Card 17

Monitoring Student Progress

If . . . children have trouble with larger numbers,

Then . . . use Dot Set Cards (1–5) and add more as children improve.

 Building Blocks For additional practice matching quantities, children should complete **Building Blocks** Number Snapshots 1.

3 Reflect 10

Extended Response APPLYING

Ask the following questions:

- **Does** Concentration **remind you of anything else you've done? What does it remind you of?** Accept all reasonable answers.

- **What else can you concentrate on?** Accept all reasonable answers.

4 Assess

Informal Assessment

Use the Student Assessment Record, **Assessment,** page 100, to record informal observations.

COMPUTING	REASONING
Blastoff! Did the child ❑ respond accurately? ❑ respond quickly? ❑ respond with confidence? ❑ self-correct?	**Concentration** Did the child ❑ provide a clear explanation? ❑ communicate reasons and strategies? ❑ choose appropriate strategies? ❑ argue logically?

Lesson 4

Objective

Children connect their knowledge of counting and patterns with numerals.

Program Materials

Warm Up
Large Number Cards

Engage
- Dot Set Cards (1–10), 1 set per group
- Number Cards (1–10), 1 set per group
- Dot Cube, 1 per group

Additional Materials

No additional materials needed

Access Vocabulary

numeral The written symbol that shows how many
Number Card A card that uses numerals instead of dots

Creating Context

English Learners often understand math concepts better when they can interact with other children who speak the same primary language. Pair together children who speak the same primary language so that they can check understanding with each other.

1 Warm Up 5

Skill Building COMPUTING

Name That Numeral
Before beginning **Concentration,** use the **Name That Numeral** activity with the whole group.

Purpose **Name That Numeral** reinforces numeral identification skills.

Warm-Up Card 5

Monitoring Student Progress

If . . . children have trouble remembering the name of the numeral,

Then . . . have another child prompt them.

2 Engage 30

Concept Building ENGAGING

Concentration

"Today we will match Dot Cards with the numbers that older children use."

Follow the instructions on the activity card to play **Concentration** Variation: **Matching Numeral to Dot Set.** As children play, ask questions about what is happening in the activity.

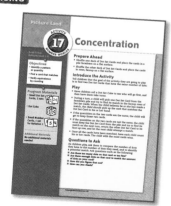

Activity Card 17

Purpose **Matching Numeral to Dot Set** helps children transition their knowledge of patterns and counting to numerals.

Monitoring Student Progress

If . . . children have trouble recognizing numerals,

Then . . . work as a group to count objects around the room, labeling them with the correct numeral.

 Teacher's Note Just as learning to say number words involved learning an arbitrary list with no patterns, learning to recognize numerals involves recognizing abstract symbols and finding ways to integrate this knowledge. This takes time and practice. Give children many opportunities to build these skills throughout the day.

Building Blocks For additional practice matching quantities, children should complete **Building Blocks** Number Snapshots 1.

3 Reflect   10

Extended Response   REASONING

Ask the following questions:

- **How would you explain this version of** Concentration **to someone who wasn't here?** Possible answer: You pick a Number Card and find a Dot Card that has the same number of dots.

- **Where else do you see numbers like the ones on the cards?** Accept all reasonable answers.

Informal Assessment

Use the Student Assessment Record, **Assessment,** page 100, to record informal observations.

COMPUTING	ENGAGING
Name That Numeral	**Concentration**
Did the child	Did the child
❏ respond accurately?	❏ pay attention to the contributions of others?
❏ respond quickly?	
❏ respond with confidence?	❏ contribute information and ideas?
❏ self-correct?	
	❏ improve on a strategy?
	❏ reflect on and check accuracy of work?

Lesson 5

Review

Objective

Children build and reinforce skills learned this week and other weeks by playing a game of their choice.

Program Materials

Engage
See Activity Cards for materials.

Additional Materials

Engage
See Activity Cards for materials.

Creating Context

Building on childrens' experiences with card games or other activities is helpful for English Learners. Ask children to tell you what it means to take turns and why it is important to take turns in a game.

1 Warm Up 5

Skill Practice COMPUTING

Blastoff!
Before beginning the **Free-Choice** activity, use the **Blastoff!** activity with the whole group.

Purpose **Blastoff!** helps children associate counting down with a decrease in set size.

Warm-Up Card 4

Monitoring Student Progress

If . . . children have trouble showing the numbers with their fingers,

Then . . . remind them to use their thumbs to hold the other fingers down.

2 Engage 20

Concept Building ENGAGING

Free-Choice Activity
For the last day of the week, allow children to choose an activity from the previous weeks. Some activities they may choose include the following:

- Object Land: **Matching Colors and Quantities**
- Object Land: **Matching Colors and Shapes**
- Picture Land: **Tapping and Clapping**
- Picture Land: **Dog and Bone**

Make a note of the activities children select. Do they prefer easy or challenging activities? If you believe your children would benefit from extra practice on specific skills, choose an activity for them.

3 Reflect
10

Extended Response REASONING
Ask the following questions:
- **What did you like about playing** Concentration?
- **Was there anything about playing this game you didn't like?**
- **Did this game help you do something you couldn't do before? What did it help you do?**
- **What was easy when you were playing** Concentration?
- **What was hard when you were playing** Concentration?
- **What do you remember most about this game?**
- **What was your favorite part?**

4 Assess
10

A Gather Evidence
Formal Assessment
Have children complete the weekly test, *Assessment,* page 45. Record formal assessment scores on the Student Assessment Record, *Assessment,* page 100.

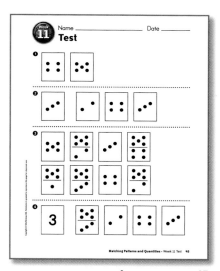

Assessment, p. 45

B Summarize Findings
Review the Student Assessment Records. Determine whether children have Minimal, Basic, or Secure understanding of the concepts presented in Week 11.

C Differentiate Instruction
Based on your observations, use these teaching strategies next week to follow up.

Minimal Understanding
- Repeat the Warm-Up and Engage activities to develop and reinforce concepts of number representation.
- Use **Building Blocks** computer activities beginning with Number Snapshots 1 to develop and reinforce concepts.

Basic Understanding
- Repeat Engage activities in subsequent weeks to reinforce basic concepts.
- Use **Building Blocks** computer activities beginning with Number Snapshots 1 to reinforce this week's concepts.

Secure Understanding
- Use Challenge variations of **Concentration.**
- Use computer activities to extend children's understanding of the Week 11 concepts.

Week 12 Introduction to the Number Line

Week at a Glance

This week, children begin **Number Worlds,** Week 12, and are introduced to Line Land.

Background

In Line Land, numbers are represented on a horizontal line using numerals. Children transition from the world of small, countable objects and patterns to a world of abstract numbers. Using number lines builds an understanding that one can move forward and backward through the number sequence. This will become the basis for later addition and subtraction.

How Children Learn

As children begin this week's activities, they should understand that numbers can be represented in different ways.

By the end of the week, children should begin to see numbers as steps along a path when they count. They should also be more familiar with numerals.

Skills Focus

- Count from 1 to 5 and from 1 to 10
- Visualize the number sequence as a line
- Predict the next number when counting up or back
- Associate counting with movement on a number line
- Associate position on a number line with distance traveled toward a goal

Teaching for Understanding

As children engage in these activities, they not only begin to build a foundation for mental addition and subtraction, but also build an understanding of relationships among numbers that gives them tools that will later help them begin counting from numbers other than 1.

Observe closely while evaluating the Engage activities assigned for this week.

- Are children making correct predictions about movement along the number line?
- Are children identifying positions on the number line?
- Are children able to see position on a number line as distance traveled toward a goal?

Math at Home

Give one copy of the Letter to Home, page A12, to each child. Complete the activity in class, and then encourage children to share it with their caregivers.

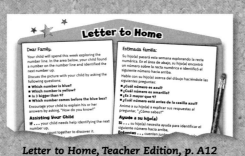

Letter to Home, Teacher Edition, p. A12

Math Vocabulary

number line A line with numbers on it used to solve math problems

far Not close to; a long way away

near Close to

English Learners

SPANISH COGNATES

English	Spanish
horizontal	horizontal
numerals	números
comparisons	comparaciónes

ALTERNATE VOCABULARY

pawn Something that shows where you are on a game board or line

position Where something is

Week 12 Planner   Introduction to the Number Line

PACING	LESSON	LEARNING GOALS	MATERIALS	TECHNOLOGY
DAY 1	**Warm Up** Line Land: Count Up*	Children learn to associate counting up with moving forward on a number line.	**Program Materials** Step-by-Step Number Line	**Building Blocks** Road Race Counting Game
	Engage Line Land: Teddy Bear Path*	Children begin to learn that the numbers they count can be seen as positions on a line.	**Program Materials** • Step-by-Step Number Line • Teddy Bear	
DAY 2	**Warm Up** Line Land: Count Up*	Children continue to build their understanding of the number line.	**Program Materials** Step-by-Step Number Line	**Building Blocks** Road Race Counting Game
	Engage Line Land: Teddy Bear Path*	Children associate movement along the number line with counting forward and backward.	**Program Materials** • Step-by-Step Number Line • Teddy Bear	
DAY 3	**Warm Up** Line Land: Count Up*	Children practice the number sequence and use a number line.	**Program Materials** Step-by-Step Number Line	**Building Blocks** Road Race Counting Game
	Engage Line Land: Teddy Bear Path Variation*	Children begin making comparisons between the positions of different numbers on a number line.	**Program Materials** Step-by-Step Number Line **Additional Materials** 2 stuffed animals for the Challenge	
DAY 4	**Warm Up** Line Land: Count Up*	Children build a foundation for mental addition.	**Program Materials** Step-by-Step Number Line	**Building Blocks** Road Race Counting Game
	Engage Line Land: Teddy Bear Path Variation*	Children compare positions to identify who is closer to a goal.	**Program Materials** Step-by-Step Number Line **Additional Materials** 2 stuffed animals	**e MathTools** Coins and Money
DAY 5	**Warm Up** Line Land: Count Up*	Children continue to associate counting up and moving forward along a line.	**Program Materials** Step-by-Step Number Line	**Building Blocks** Road Race Counting Game
	Review Free-Choice activity	Children review skills learned this week and in prior weeks.	Materials will be selected from those used in previous weeks.	

* Includes Challenge Variations

Lesson 1

Objective

Children learn that numbers can be seen as positions on a line.

Program Materials

Step-by-Step Number Line

Engage
Teddy Bear

Additional Materials

No additional materials needed.

Access Vocabulary

moving forward Going in the direction of the larger numbers

number line A line with numbers on it (recta numérica not línea numérica)

Creating Context

In mathematics, many key words come from Latin and Greek root words. Many of these words are nearly the same in English and in Spanish and other romance languages. Use a bilingual dictionary or glossary to find cognates that can be used to help explain concepts to children.

Easy as . . .
1 - 2 - 3

1 Warm Up 5

Skill Building COMPUTING
Count Up

Before beginning **Teddy Bear Path,** use the **Count Up** activity with the whole group.

Purpose Count Up teaches children to associate counting up with moving forward on a number line.

Warm-Up Card 6

Monitoring Student Progress

If . . . children are not participating, **Then . . .** work to include them by having them be the pawn.

2 Engage 30

Concept Building UNDERSTANDING
Teddy Bear Path

"Today we will use the Step-by-Step Number Line to learn more about counting."

Follow the instructions on the Activity Card to play **Teddy Bear Path.** As children play, ask questions about what is happening in the activity.

Purpose Teddy Bear Path helps children begin to understand how numbers relate to each other on a line.

Activity Card 22

Teacher's Note Allowing several children to answer each question you ask provides an opportunity for you to assess the understanding of different children in your class. By encouraging children to explain their reasoning, you are not only deepening the understanding of the child who is offering an explanation, you are also providing opportunities for children to benefit from the explanations offered by their peers.

Building Blocks For additional practice with number lines, children should complete **Building Blocks** Road Race Counting Game.

3 Reflect 10

Extended Response REASONING

Ask questions such as the following:

- **What do you notice about the Step-by-Step Number Line?** Accept all reasonable answers.
- **How would you explain this activity to someone who is not here?** Possible answer: One person is the pawn; we tell the pawn to move one step up or back and guess the number that the pawn will be on.

4 Assess

Informal Assessment

Use the Student Assessment Record, **Assessment,** page 100, to record informal observations.

COMPUTING	UNDERSTANDING
Count Up	**Teddy Bear Path**
Did the child	Did the child
❑ respond accurately?	❑ make important observations?
❑ respond quickly?	❑ extend or generalize learning?
❑ respond with confidence?	❑ provide insightful answers?
❑ self-correct?	❑ pose insightful questions?

Lesson 2

Objective

Children associate movement along the number line with counting forward and backward.

Program Materials

Step-by-Step Number Line

Engage
Teddy Bear

Additional Materials

No additional materials needed

Access Vocabulary

teddy bear A stuffed animal, or toy, shaped like a bear

What does this remind you of? What does this look like?

Creating Context

Many math concepts rely on comparison. In English, comparatives and superlatives are formed by adding the endings *-er* and *-est* to one- and two-syllable adjectives (e.g., *She is taller than her brother*). For adjectives of three or more syllables, *more* and *most* are used (e.g., *He is more intelligent than his dog*). Make a chart listing comparison adjectives, and then add to it throughout the weeks.

---

1 Warm Up 5

Concept Building COMPUTING

Count Up

Before beginning **Teddy Bear Path,** use the **Count Up** activity with the whole group.

Purpose Count Up continues to build children's understanding of the number line.

Warm-Up Card 6

Monitoring Student Progress

| If . . . children are not sure of the next number, | Then . . . have them choose a classmate to prompt them. |

2

Concept Building ENGAGING

Teddy Bear Path

"Today more of you will get to be the pawn as we work with the Step-by-Step Number Line."

Follow the instructions on the Activity Card to play **Teddy Bear Path.** As children play, ask questions about what is happening in the activity.

Activity Card 22

Purpose Teddy Bear Path associates numbers in the counting sequence with position on the number line. This helps children realize that numbers that come later in the sequence are bigger and farther away from 1 than numbers that come earlier in the sequence.

Monitoring Student Progress

| **If . . .** children are having trouble with larger numbers, | **Then . . .** fold the number line so that only numbers 1–5 are visible. |

 Teacher's Note Encourage children to use this week's vocabulary words as they engage in the activities, discuss math concepts, and make predictions.

Building Blocks For additional practice with number lines, children should complete **Building Blocks** Road Race Counting Game.

3 Reflect 10

Extended Response REASONING

Ask questions such as the following:

■ **Which numbers are far away from 1?** Possible answers: 5; 6; 7; 8; 9; 10

■ **Can we call these big numbers? How do we know they are big?** Possible answer: Yes, they are big because it takes a lot of steps to get there.

4 Assess

Informal Assessment

Use the Student Assessment Record, **Assessment,** page 100, to record informal observations.

COMPUTING	**ENGAGING**
Count Up	**Teddy Bear Path**
Did the child	Did the child
❏ respond accurately?	❏ pay attention to the contributions of others?
❏ respond quickly?	❏ contribute information and ideas?
❏ respond with confidence?	❏ improve on a strategy?
❏ self-correct?	❏ reflect on and check accuracy of work?

Lesson 3

Objective
Children begin making comparisons between the positions of different numbers on a number line.

Program Materials
Step-by-Step Number Line

Additional Materials
No additional materials needed

Access Vocabulary
pawn The student who moves on the Number Line
goal What you are trying to get to

Creating Context
Throughout this lesson, children will make comparisons (for example, the positions of the two pawns). Review the comparative and superlative forms of words children may use such as *near, nearer, nearest* and *far, farther, farthest.*

Math-a-saurus

1 Warm Up
5

Skill Building COMPUTING
Count Up
Before beginning **Teddy Bear Path,** use the **Count Up** activity with the whole group.

Purpose **Count Up** gives children practice with the number sequence and the number line.

Warm-Up Card 6

Monitoring Student Progress

If . . . children are having trouble counting,

Then . . . make sure you are giving them enough time to think of the next number.

2 Engage
30

Concept Building UNDERSTANDING
Teddy Bear Path
"Today we will have two people be pawns as we play with the number line."
Follow the instructions on the Activity Card to play **Teddy Bear Path** Variation: **Two Teddy Bear Paths.** As children play, ask questions about what is happening in the activity.

Activity Card 22

Purpose **Two Teddy Bear Paths** illustrates the relationships between numbers and builds children's understanding of numbers.

Monitoring Student Progress

| If . . . children are having trouble recognizing numerals, | Then . . . point out examples of how they are used around the classroom. |

 Teacher's Note This activity requires that children compare positions. To introduce these comparisons, have children count the number of steps to each position, starting at 1, to see which position requires more steps. Next, have them look at the numerals on which each pawn is standing to see if they match the number of steps counted. Finally, tell them that the number that is farther along the path and that takes more steps and energy to reach is a bigger number than the other number. We know this because it comes *after* the other number when we are counting and walking along the number line.

Building Blocks For additional practice with number lines, children should complete **Building Blocks** Road Race Counting Game.

3 Reflect 10

Extended Response REASONING

Ask questions such as the following:

- **How is this version of Step-by-Step Number Line different from what we did yesterday?** Possible answers: There are two lines and two people. We compare the two pawns to each other to see who is closer to the stuffed animal.

- **Which numbers are closer to 1?** Possible answers: 2; 3; 4

- **Can we call these small numbers or little numbers? How do we know they are little?** Possible answer: Yes; They are small numbers because it takes only a few steps to get there.

4 Assess

Informal Assessment

Use the Student Assessment Record, **Assessment**, page 100, to record informal observations.

COMPUTING	UNDERSTANDING
Count Up	**Teddy Bear Path**
Did the child	Did the child
❏ respond accurately?	❏ make important observations?
❏ respond quickly?	❏ extend or generalize learning
❏ respond with confidence?	❏ provide insightful answers?
❏ self-correct?	❏ pose insightful questions?

Lesson 4

Objective

Children compare two different positions on a number line to identify which one is closer to a goal.

Program Materials

Step-by-Step Number Line

Additional Materials

No additional materials needed

Access Vocabulary

compare To think about how things are the same and how they are different
ahead of Before; in front of

Creating Context

Work with children to create a word web focused on *goal*. Encourage children to name different kinds of goals and draw a quick picture of each kind. Goals may include things such as a soccer or football goal, a thermometer such as those used to track fundraising, or grades on a report card. Discuss what these goals have in common and how we work to meet goals.

1 Warm Up
5

Skill Building COMPUTING

Count Up
Before beginning **Teddy Bear Path,** use the **Count Up** activity with the whole group.

Purpose Count Up helps children build a foundation for mental addition.

Warm-Up Card 6

Monitoring Student Progress

If . . . children are having trouble counting,

Then . . . break the sequence into smaller groups and have children repeat after you.

2 Engage
30

Concept Building ENGAGING

Teddy Bear Path

"Today we will continue to play Teddy Bear Path."

Follow the instructions on the Activity Card to play **Teddy Bear Path** Variation: **Two Teddy Bear Paths.** As children play, ask questions about what is happening in the activity.

Purpose Two Teddy Bear Paths helps refine children's understanding of each number's position in the 1–10 sequence.

Activity Card 22

Monitoring Student Progress

| **If . . .** children can handle this activity with ease, | **Then . . .** use the **Two Teddy Bear Paths** Challenge. |

 For additional practice with number lines, children should complete **Building Blocks** Road Race Counting Game.

3 Reflect 10

Extended Response REASONING

Ask questions such as the following:

- **How do you know which pawn is on the bigger number?** Possible answer: The pawn is on the number it takes longer to count up or walk up to.

- **What facts do we know to help us decide which number is bigger?** Possible answers: We know that one is farther along the line; we know the order of the numbers when we count; we know that we hold up more fingers for one than for the other.

4 Assess

Informal Assessment

Use the Student Assessment Record, **Assessment**, page 100, to record informal observations.

COMPUTING	ENGAGING
Count Up	**Teddy Bear Path**
Did the child	Did the child
❏ respond accurately?	❏ pay attention to the contributions of others?
❏ respond quickly?	
❏ respond with confidence?	❏ contribute information and ideas?
❏ self-correct?	❏ improve on a strategy?
	❏ reflect on and check accuracy of work?

Lesson 5
Review

Objective

Children review skills learned this week and in prior weeks.

Program Materials

Warm Up
Step-by-Step Number Line

Engage
See Activity Cards for materials.

Additional Materials

Engage
See Activity Cards for materials.

Creating Context

To help children describe why they like certain **Number Worlds** activities, make a two-column chart labled *I Like* and *Because*. Have children give reasons they like something as you complete the chart.

✓ Cumulative Assessment

After Lesson 5 is complete, children should complete the Cumulative Assessment, **Assessment,** pages 88–89. Using the key on **Assessment,** page 87, identify incorrect responses. Reteach and review the suggested activities to reinforce concept understanding.

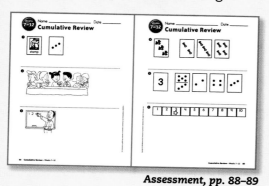

Assessment, pp. 88–89

1 Warm Up　　5

Skill Practice　COMPUTING

Count Up

Before beginning the **Free-Choice** activity, use the **Count Up** activity with the whole group.

Purpose　**Count Up** gives children practice with a number line so that they associate counting up with moving forward along the line.

Warm-Up Card 6

Monitoring Student Progress

If . . . children are having trouble remembering the sequence of numbers,

Then . . . model counting for them throughout the day.

2 Engage　　20

Concept Building　APPLYING

Free-Choice Activity

For the last day of the week, allow children to choose an activity from previous weeks. Some activities they may choose include the following:

- Object Land: **Matching Colors and Shapes**
- Object Land: **Picnic Party**
- Picture Land: **Dog and Bone**

Make a note of the activities children select. Do they prefer easy or challenging activities? If you believe your children would benefit from extra practice on specific skills, choose an activity for them.

3 Reflect   10

Extended Response   REASONING

Ask questions such as the following:

- **What did you like about playing** Teddy Bear Path?
- **Was there anything about playing this game you didn't like?**
- **Did this game help you do something you couldn't do before? What did it help you do?**
- **What was easy when you were playing** Teddy Bear Path?
- **What was hard when you were playing** Teddy Bear Path?
- **What do you remember most about this game?**
- **What was your favorite part?**

4 Assess   10

A Gather Evidence

Formal Assessment

Have children complete the weekly test, **Assessment,** page 47. Record formal assessment scores on the Student Assessment Record, **Assessment,** page 100.

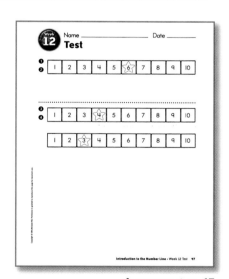

Assessment, p. 47

B Summarize Findings

Review the Student Assessment Records. Determine whether children have Minimal, Basic, or Secure understanding of the concepts presented in Week 12.

C Differentiate Instruction

Based on your observations, use these teaching strategies next week to follow up.

Minimal Understanding

- Repeat the Warm-Up and Engage activities to develop concepts of the number line.
- Use **Building Blocks** computer activities beginning with Road Race Counting Game to develop and reinforce numeration concepts.

Basic Understanding

- Repeat Engage activities in subsequent weeks to reinforce basic concepts.
- Use **Building Blocks** computer activities beginning with Road Race Counting Game to reinforce this week's concepts.

Secure Understanding

- Use Challenge variations of **Teddy Bear Path.**
- Use computer activities to extend children's understanding of the Week 12 concepts.

Position on the Number Line

Week at a Glance

This week, children begin **Number Worlds,** Week 13, and continue to explore Line Land.

Background

In Line Land, numbers are represented on a horizontal line using numerals. Children transition from the world of small, countable objects and patterns to a world of abstract numbers. Using number lines builds an understanding that one can move forward and backward through the number sequence. This will become the basis for addition and subtraction.

How Children Learn

As they begin this week's activities, children should be developing an understanding of the relationship between the number line and the counting sequence.

By the end of the week, children should have a better understanding of the relationship among set size, sequence, and the number line.

Skills Focus

- Learn that each numeral corresponds to a set that has one more than the numeral preceding it and one less than the numeral that follows it
- Count from 1 to 10
- Associate numerals with set size

Teaching for Understanding

As children engage in these activities, they make further connections between their existing knowledge of counting and set size and the new concept of the number line. This provides a foundation for mental addition and subtraction.

Observe closely while evaluating the Engage activities assigned for this week.

- Are children counting correctly?
- Are children identifying numerals?
- Are children able to answer questions about quantity?

Math at Home

Give one copy of the Letter to Home, page A13, to each child. Complete the activity in class, and then encourage children to share it with their caregivers.

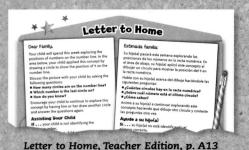

Letter to Home, Teacher Edition, p. A13

Math Vocabulary

position where something is

first number one on the number line

English Learners

SPANISH COGNATES

English	Spanish
explore	explorar
correspond	corresponde
sequence	secuencia
continue	continuar

ALTERNATE VOCABULARY

number line A horizontal line with the counting numbers listed in order

step-by-step A way of doing one thing at a time

Week 13 Planner | Position on the Number Line

PACING	LESSON	LEARNING GOALS	MATERIALS	TECHNOLOGY
DAY 1	**Warm Up** Line Land: Count Up*	Children associate numerals on the number line with the counting sequence.	**Program Materials** Step-by-Step Number Line	**Building Blocks** Road Race Counting Game
	Engage Line Land: Position on the Number Line*	Children explore the association between counting up by one and increasing set size by one.	**Program Materials** Step-by-Step Number Line	
DAY 2	**Warm Up** Line Land: Count Up*	Children continue to refine their counting skills.	**Program Materials** Step-by-Step Number Line	**Building Blocks** Road Race Counting Game
	Engage Line Land: Position on the Number Line*	Children learn the position of each number in the 1–10 sequence.	**Program Materials** Step-by-Step Number Line	
DAY 3	**Warm Up** Line Land: Count Up*	Children strengthen their understanding of the counting sequence.	**Program Materials** Step-by-Step Number Line	**Building Blocks** Road Race Counting Game
	Engage Line Land: Position on the Number Line*	Children build a foundation for mental addition.	**Program Materials** Step-by-Step Number Line	**MathTools** Set Tool
DAY 4	**Warm Up** Line Land: Count Up*	Children reinforce counting skills.	**Program Materials** Step-by-Step Number Line	**Building Blocks** Road Race Counting Game
	Engage Line Land: Position on the Number Line Variation*	Children build a foundation for mental subtraction.	**Program Materials** Step-by-Step Number Line	
DAY 5	**Warm Up** Line Land: Count Up*	Children learn that counting up by 1 and moving forward 1 give equivalent results.	**Program Materials** Step-by-Step Number Line	**Building Blocks** Road Race Counting Game
	Review Free-Choice activity	Children review and reinforce concepts learned this week and in prior weeks.	Materials will be selected from those used in previous weeks.	

* Includes Challenge Variations

Lesson 1

Objective

Children explore the association between counting up by 1 and increasing set size by 1.

Program Materials

Step-by-Step Number Line

Additional Materials

No additional materials needed

Access Vocabulary

number line A horizontal line with numbers listed in the order that you count them
stand next to . . . Stand at the side of . . .

Creating Context

Physical movement reinforces comprehension for English Learners. **Position on the Number Line** calls for children to physically line up on a number line so that each child represents a number. This is a great example of how children can better understand the concept of number lines visually and kinesthetically.

There's Strength in NUMBERS

---

1 Warm Up 5

Skill Building COMPUTING
Count Up

Before beginning **Position on the Number Line,** use the **Count Up** activity with the whole group.

Purpose **Count Up** helps children associate numerals on the number line with the counting sequence.

Warm-Up Card 6

Monitoring Student Progress

If . . . children have trouble remembering the number sequence,

Then . . . give them opportunities to model counting for one another throughout the day.

---

2 Engage 30

Concept Building UNDERSTANDING
Position on the Number Line

"Today we will use the number line to solve a problem."

Follow the instructions on the Activity Card to play **Position on the Number Line.** As children play, ask questions about what is happening in the activity.

Purpose **Position on the Number Line** provides an opportunity for children to discover that they can use the numbers on the number line to find how many children are standing on the number line.

Activity Card 23

Monitoring Student Progress

If . . . children cannot remember the name of the numeral they are standing on,

Then . . . review the numbers by counting up from 1.

Teacher's Note Encourage children to use this week's vocabulary words as they engage in the activities, discuss math concepts, and make predictions.

Building Blocks For additional practice with number lines, children should complete **Building Blocks** Road Race Counting Game.

3 Reflect 10

Extended Response REASONING

Ask the following questions:

- **How would you describe this activity to someone who isn't here?** Possible answer: One person lines up on each number in the number line. Then, we check to see what numbers they are on and how many people are on the line.
- **What do you notice about this activity?** Accept all reasonable answers.

4 Assess

Informal Assessment

Use the Student Assessment Record, **Assessment,** page 100, to record informal observations.

COMPUTING	UNDERSTANDING
Count Up	**Position on the Number Line**
Did the child	Did the child
❑ respond accurately?	❑ make important observations?
❑ respond quickly?	❑ extend or generalize learning?
❑ respond with confidence?	❑ provide insightful answers?
❑ self-correct?	❑ pose insightful questions?

Lesson 2

Objective

Children learn the position on the number line of each number in the 1–10 sequence and explore its relationship to set size.

Program Materials

Step-by-Step Number Line

Additional Materials

No additional materials needed

Access Vocabulary

forward In the direction in front of your body

backward In the direction of the back of your body

Creating Context

Many children enjoy learning to count in other languages. If children know how to count in another language, invite them to count in that language. You may wish to teach, or have the children teach, the counting sequence in these other languages.

Concept Building COMPUTING

Count Up

Before beginning **Position on the Number Line,** use the **Count Up** activity with the whole group.

Purpose **Count Up** allows children to continue to refine their counting skills.

Warm-Up Card 6

Monitoring Student Progress

| **If . . .** a child is not participating, | **Then . . .** try to get him or her involved by inviting the child to be the pawn. |

---

2 Engage 30

Concept Building ENGAGING

Position on the Number Line

"Today we will see how many people it takes to fill the number line."

Follow the instructions on the Activity Card to play **Position on the Number Line.** As children play, ask questions about what is happening in the activity.

Purpose **Position on the Number Line** helps children learn the position of each number in the 1–10 sequence.

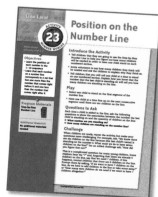

Activity Card 23

Monitoring Student Progress

| If . . . children are having trouble working with numbers over 5, | Then . . . fold the mat in half and use 1–5 before introducing the other numbers. |

 For additional practice with number lines, children should complete **Building Blocks** Road Race Counting Game.

3 Reflect 10

Extended Response REASONING

Ask the following questions:

- **What is the same each time you play the game?**
  Possible answer: The number of people on the line is the same number as the number the last person is standing on.

- **How did you figure that out?** Accept all reasonable answers.

4 Assess

Informal Assessment

Use the Student Assessment Record, **Assessment**, page 100, to record informal observations.

COMPUTING	ENGAGING
Count Up	**Position on the Number Line**
Did the child	Did the child
❏ respond accurately?	❏ pay attention to the contributions of others?
❏ respond quickly?	
❏ respond with confidence?	❏ contribute information and ideas?
❏ self-correct?	
	❏ improve on a strategy?
	❏ reflect on and check accuracy of work?

Lesson 3

Objective

Children continue to build a foundation for mental addition.

Program Materials

Step-by-Step Number Line

Additional Materials

No additional materials needed

Access Vocabulary

solve problems To think about something and find an answer

recount To count again

Creating Context

At this age, prepositions are new to most children, particularly to English Learners. Work with the Step-by-Step Number Line to emphasize some key location words such as *on, beside, in front of, behind,* or *in back of.*

1 Warm Up 5

Skill Building COMPUTING

Count Up

Before beginning **Position on the Number Line,** use the **Count Up** activity with the whole group.

Purpose Count Up strengthens children's understanding of the counting sequence.

Warm-Up Card 6

Monitoring Student Progress

If . . . children are having trouble predicting the next number,

Then . . . have them recount, starting from 1.

2 Engage 30

Concept Building REASONING

Position on the Number Line

"Today we will continue to play **Position on the Number Line.**"

Follow the instructions on the Activity Card to play **Position on the Number Line.** As children play, ask questions about what is happening in the activity.

Purpose Position on the Number Line helps children understand that the numbers they count can be seen along a line that will help them solve problems.

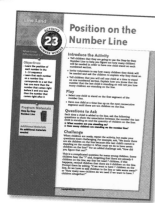

Activity Card 23

MathTools Use the Set Tool to demonstrate and explore counting and magnitude.

Monitoring Student Progress

If . . . children can handle this activity with ease,	Then . . . use **Position on the Number Line** Challenge variation.

Building Blocks For additional practice with number lines, children should complete **Building Blocks** Road Race Counting Game.

3 Reflect 10 ▶

Extended Response

Ask the following questions:

- **What is this problem about?** Possible answers: seeing how many children it takes to fill the line; seeing how many more children you need.
- **What does this remind you of?** Accept all reasonable answers.

4 Assess

Informal Assessment

Use the Student Assessment Record, **Assessment,** page 100, to record informal observations.

COMPUTING	REASONING
Count Up Did the child	**Position on the Number Line** Did the child
❏ respond accurately?	❏ provide a clear explanation?
❏ respond quickly?	❏ communicate reasons and strategies?
❏ respond with confidence?	❏ choose appropriate strategies?
❏ self-correct?	❏ argue logically?

Lesson 4

Objective

Children learn the correspondence between numbers on the number line and set size, and they build a foundation for mental subtraction.

Program Materials

Step-by-Step Number Line

Additional Materials

No additional materials needed

Access Vocabulary

set A group of objects that go together

"the number that comes right after it" The number that is the very next one when counting

Creating Context

One way to help children understand new concepts is to ask them questions that help them think about the process and the results. If you have English Learners at the early levels of proficiency, be sure to include questions with responses that require little language production, such as yes/no questions, one-word answers, or "point to the correct answer" questions.

---

1 Warm Up 5

Skill Building COMPUTING

Count Up

Before beginning **Position on the Number Line,** use the **Count Up** activity with the whole group.

Purpose **Count Up** helps children reinforce counting skills.

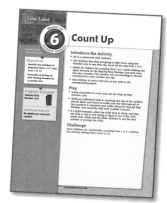

Warm-Up Card 6

Monitoring Student Progress

If . . . children have trouble counting to 10, **Then . . .** give them more practice counting to 5 and gradually work up to higher numbers.

---

2 Engage 30

Concept Building ENGAGING

Position on the Number Line

"Today we will fill the line with children and then have one child leave the line at a time until there are no children left."

Follow the instructions on the Activity Card to play **Position on the Number Line** Variation: **Count Down.** As children play, ask questions about what is happening in the activity.

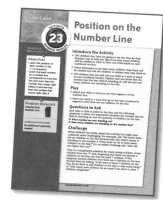

Activity Card 23

Purpose **Count Down** teaches children that each number in a number line corresponds to a set of a particular size.

Monitoring Student Progress

| If . . . children have trouble with this variation, | Then . . . return to playing the activity without this variation. |

 Building Blocks For additional practice with number lines, children should complete **Building Blocks** Road Race Counting Game.

3 Reflect 10

Extended Response REASONING

Ask the following questions:

- **What did you notice about the numbers we were saying and the numbers the new children were standing on?** Possible answer: They were getting bigger. (If children have trouble answering this question, have them play again to find the answer.)

- **What did you notice about the numbers we were saying as we were taking children off the line?** Possible answer: They were getting smaller.

4 Assess

Informal Assessment

Use the Student Assessment Record, **Assessment,** page 100, to record informal observations.

COMPUTING	ENGAGING
Count Up Did the child ❏ respond accurately? ❏ respond quickly? ❏ respond with confidence? ❏ self-correct?	**Position on the Number Line** Did the child ❏ pay attention to the contributions of others? ❏ contribute information and ideas? ❏ improve on a strategy? ❏ reflect on and check accuracy of work?

Lesson 5

Review

Objective

Children review and reinforce concepts learned this week and in prior weeks.

Program Materials

Warm Up
Step-by-Step Number Line

Engage
See Activity Cards for materials.

Additional Materials
See Activity Cards for materials.

Creating Context

Discuss with children multiple meanings of the word *free*. Children may know that *free* means "no cost," but may not know that *Free Choice* means that they can choose which activity they would like to do.

1 Warm Up 5

Skill Practice COMPUTING

Count Up
Before beginning the **Free-Choice** activity, use the **Count Up** activity with the whole group.

Purpose **Count Up** helps children learn that counting up by one unit and moving forward one on a number line give equivalent results.

Warm-Up Card 6

Monitoring Student Progress

If . . .	Then . . . model counting for
children have trouble with the counting sequence,	them throughout the day.

2 Engage 20

Concept Building APPLYING

Free-Choice activity

For the last day of the week, allow children to choose an activity from the previous weeks. Some activities they may choose include the following:

- Object Land: **Picnic Party**
- Picture Land: **Catch the Teacher**
- Line Land: **Teddy Bear Path**

Make a note of the activities children select. Do they prefer easy or challenging activities? If you believe your children would benefit from extra practice on specific skills, choose an activity for them.

3 Reflect 10

Extended Response REASONING

Ask the following questions:

- **What did you like about playing** Position on the Number Line?
- **Was there anything about playing this game you didn't like?**
- **Did this game help you do something you couldn't do before? What did it help you do?**
- **What was easy when you were playing** Position on the Number Line?
- **What was hard when you were playing** Position on the Number Line?
- **What do you remember most about this game?**
- **What was your favorite part?**

4 Assess 10

A Gather Evidence

Formal Assessment

Have children complete the weekly test, **Assessment,** page 49. Record formal assessment scores on the Student Assessment Record, **Assessment,** page 100.

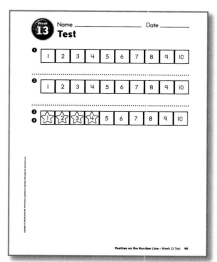

Assessment, p. 49

B Summarize Findings

Review the Student Assessment Records. Determine whether children have Minimal, Basic, or Secure understanding of the concepts presented in Week 13.

C Differentiate Instruction

Based on your observations, use these teaching strategies next week to follow up.

Minimal Understanding

- Repeat the Warm-Up and Engage activities to develop counting concepts.
- Use **Building Blocks** computer activities beginning with Road Race Counting Game to develop and reinforce numeration and addition concepts.

Basic Understanding

- Repeat Engage activities in subsequent weeks to reinforce basic concepts.
- Use **Building Blocks** computer activities beginning with Road Race Counting Game to reinforce this week's concepts.

Secure Understanding

- Use Challenge variations of **Position on the Number Line.**
- Use computer activities to extend children's understanding of the Week 13 concepts.

Sequence on the Number Line

Week at a Glance

This week, children begin **Number Worlds,** Week 14, and continue to explore Line Land.

Background

In Line Land, numbers are represented on a horizontal line using numerals. Children transition from the world of small countable objects and patterns to a world of abstract numbers. Using number lines builds an understanding that you can move forward and backward through the number sequence. This will become the basis for addition and subtraction.

How Children Learn

As they begin this week's activities, children should recognize dot-set patterns and know what number each pattern represents. They should also be familiar with number lines.

By the end of the week, children should be building connections among dot sets, numbers, and the number line.

Skills Focus

- Count up from 1 to 10 and back from 10 to 1
- Identify numerals
- Identify and compute set size
- Match set size with position on the number line

Teaching for Understanding

As children engage in these activities, they reinforce their understanding of numbers. This knowledge helps them learn that each number on a number line is associated with a set of a particular size and that set size increases by one with each successive number up in the sequence.

Observe closely while evaluating the Engage activities assigned for this week.

- Are children counting up and back correctly?
- Are children able to identify numerals?
- Are children making connections between set size and placement on the number line?

Math at Home

Give one copy of the Letter to Home, page A14, to each child. Complete the activity in class, and then encourage children to share it with their caregivers.

Letter to Home, Teacher Edition, p. A14

Math Vocabulary

first Number one on the number line

next The one after this one

English Learners

SPANISH COGNATES

English	Spanish
identify	identificar
position	posición
horizontal	horizontal
activities	actividades

ALTERNATE VOCABULARY

pattern Dots arranged a certain way

numeral The written symbol for a number

PACING	LESSON	LEARNING GOALS	MATERIALS	TECHNOLOGY
DAY 1	**Warm Up** Line Land: Blastoff!*	Children review the reverse counting sequence.	**Program Materials** Step-by-Step Number Line	**Building Blocks** Build Stairs 1
	Engage Line Land: Next Number Up*	Children associate set size with position on the number line.	**Program Materials** • Step-by-Step Number Line • Dot Set Cards	**MathTools** Number Line
DAY 2	**Warm Up** Line Land: Count Up*	Children continue to reinforce counting skills.	**Program Materials** Step-by-Step Number Line	**Building Blocks** Build Stairs 1
	Engage Line Land: Next Number Up*	Children identify set size and numerals.	**Program Materials** • Step-by-Step Number Line • Dot Set Cards	
DAY 3	**Warm Up** Line Land: Blastoff!*	Children reinforce their ability to count back.	**Program Materials** Step-by-Step Number Line	**Building Blocks** Build Stairs 1
	Engage Line Land: Next Number Up*	Children learn that set size increases by 1 with each position on the number line.	**Program Materials** • Step-by-Step Number Line • Number Cards	
DAY 4	**Warm Up** Picture Land: Name that Numeral*	Children reinforce numeral recognition.	**Program Materials** • Large Number Cards (1–5) • Large Number Cards (1–12) for the Challenge	**Building Blocks** Build Stairs 1
	Engage Line Land: Next Number Up Variation	Children use the reverse counting sequence to explore the number line.	**Program Materials** • Step-by-Step Number Line • Dot Set Cards	
DAY 5	**Warm Up** Line Land: Count Up*	Children associate counting up with moving forward on a number line.	**Program Materials** Step-by-Step Number Line	**Building Blocks** Build Stairs 1
	Review Free-Choice activity	Children review and reinforce skills and concepts learned in this and other weeks.	Materials will be selected from those used in previous weeks.	

* Includes Challenge Variations

Lesson 1

Objective

Children match sets to positions on a number line and build their understanding of the correspondence between set size and position on the number line.

Program Materials

Step-by-Step Number Line

Engage
Dot Set Cards

Additional Materials

No additional materials needed

Access Vocabulary

dots Round spots
pattern Dots arranged a certain way

Creating Context

English Learners may find it confusing to hear the many idiomatic expressions in English that use prepositions. The phrase "next number up" may be confusing to some English Learners. Explain to children that, in this phrase, *up* refers to the next number in the sequence.

1 Warm Up   5

Skill Building   COMPUTING

Blastoff!
Before beginning **Next Number Up,** use the **Blastoff!** activity with the whole group.

Purpose **Blastoff!** helps children review the reverse counting sequence and builds a foundation for subtraction.

Warm-Up Card 7

Monitoring Student Progress

| **If . . .** a child jumps up too early or too late, | **Then . . .** begin the countdown sequence again. |

2 Engage   30

Concept Building   UNDERSTANDING

Next Number Up
"Today we will use Dot Set Cards to tell us where we should line up on the number line."

Follow the instructions on the Activity Card to play **Next Number Up.** As children play, ask questions about what is happening in the activity.

Activity Card 24

Purpose **Next Number Up** helps children associate set size with positions on the number line.

ⓔMathTools Use the Number Line to demonstrate and explore counting.

Monitoring Student Progress

| **If . . .** children have trouble with the phrase *next number up,* | **Then . . .** practice the activity and review the numbers first. |

 **Teacher's Note** Encourage children to use this week's vocabulary words as they engage in the activities, discuss math concepts, and make predictions.

Building Blocks For additional practice visualizing magnitude, children should complete **Building Blocks** Build Stairs 1.

3 Reflect 10

Extended Response REASONING

Ask questions such as the following:

- **How do you know when you have the "next number up"?** Possible answer: The number of dots on my card means the same thing as the number on the number line.

- **Did anyone use a different way to figure out when to line up? What was it?** Accept all reasonable answers.

 ## 4 Assess

Informal Assessment

Use the Student Assessment Record, **Assessment,** page 100, to record informal observations.

COMPUTING	UNDERSTANDING
Blastoff! Did the child	**Next Number Up** Did the child
❑ respond accurately? ❑ respond quickly? ❑ respond with confidence? ❑ self-correct?	❑ make important observations? ❑ extend or generalize learning? ❑ provide insightful answers? ❑ pose insightful questions?

BOW-WOW!

Purr-fect

SQUEAKY CLEAN

Lesson 2

1 Warm Up

5

Concept Building COMPUTING

Count Up
Before beginning **Next Number Up,** use the **Count Up** activity with the whole group.

Purpose **Count Up** continues to reinforce children's counting skills.

Warm-Up Card 6

Objective
Children identify set size and match it to the corresponding numeral.

Program Materials
Step-by-Step Number Line

Engage
Dot Set Cards

Additional Materials
No additional materials needed

Access Vocabulary

line up To form a line, standing behind another child in line
mistake An error; a wrong answer

Creating Context
To help English Learners at early proficiency levels, remember to use questions during assessment that are not heavily dependent on language. Questions that require children to demonstrate the answer or to provide a short response are good ways to include early English Learners. These could include gestures such as thumbs up/thumbs down or pointing to the example.

Monitoring Student Progress

If . . . some children are having more trouble than others,

Then . . . use activities from earlier in the year to help reinforce the counting sequence.

2 Engage

30

Concept Building ENGAGING

Next Number Up

"Today we continue to line up on the number line using Dot Cards."

Follow the instructions on the Activity Card to play **Next Number Up.** As children play, ask questions about what is happening in the activity.

Purpose **Next Number Up** teaches children that set size increases by one with each position up the number line.

Activity Card 24

Monitoring Student Progress

If . . . children have trouble identifying the next number up,

Then . . . have them count from 1 until they reach the number after the last number on which a child is standing. They should then examine their Dot Set Cards to see if they have the number that was said.

 Building Blocks For additional practice visualizing magnitude, children should complete **Building Blocks** Build Stairs 1.

3 Reflect 10

Extended Response REASONING

Ask questions such as the following:

- **What does this remind you of?** Accept all reasonable answers.

- **How would you explain** Next Number Up **to someone who isn't here?** Possible answer: You get a Dot Set Card and figure out what number is on it. Then the teacher asks the person with the first number to line up on the number line. Everyone lines up so that the number on his or her card matches the numbers on the number line.

4 Assess

Informal Assessment

Use the Student Assessment Record, **Assessment,** page 100, to record informal observations.

COMPUTING	ENGAGING
Count Up	**Next Number Up**
Did the child	Did the child
❏ respond accurately?	❏ pay attention to the contributions of others?
❏ respond quickly?	
❏ respond with confidence?	❏ contribute information and ideas?
❏ self-correct?	
	❏ improve on a strategy?
	❏ reflect on and check accuracy of work?

Sequence on the Number Line • Lesson 2 **187**

Week 14

Lesson 3

Objective

Children build numeral recognition skills.

Program Materials

Step-by-Step Number Line

Engage
Number Cards

Additional Materials

No additional materials needed

Access Vocabulary

blastoff When a spaceship flies up to space
count back To count, going down by one
each time you say a number

Creating Context

In English, some nouns are collective nouns
— words that describe a collection of things.
English Learners may not be familiar with
these nouns. Tell children about collective
nouns such as a *bunch* of bananas or a *stack*
or *pile* of counters. Brainstorm a list of
collective nouns that children know.

off to a
Running Start!

1 Warm Up 5

Skill Building COMPUTING

Blastoff!
Before beginning **Next
Number Up,** use the
Blastoff! activity with
the whole group.

Purpose Blastoff!
reinforces children's
ability to count back.

Warm-Up Card 7

Monitoring Student Progress

If . . . children
have trouble
counting back
from 10,

Then . . . have them focus on
counting back from 5.

2 Engage 30

Concept Building REASONING

Next Number Up
"Today we will line up by
using Number Cards."
Follow the instructions on
the Activity Card to play
Next Number Up Challenge
variation. As children play,
ask questions about what
is happening in the activity.

**Purpose Next Number
Up** helps children build
number sequencing skills.

Activity Card 24

Monitoring Student Progress

If . . . children have trouble identifying the correct numerals,

Then . . . use the **Name That Numeral** activity to help them.

 Building Blocks For additional practice visualizing magnitude, children should complete **Building Blocks** Build Stairs 1.

---

3 Reflect 10

Extended Response REASONING

Ask questions such as the following:

- **How is this version of** Next Number Up **different from what we did earlier in the week?** Possible answer: This time we matched Number Cards instead of Dot Cards.

- **Which version was easier? Why?** Accept all reasonable answers.

---

4 Assess

Informal Assessment

Use the Student Assessment Record, **Assessment,** page 100, to record informal observations.

COMPUTING	REASONING
Blastoff!	**Next Number Up**
Did the child	Did the child
❏ respond accurately?	❏ provide a clear explanation?
❏ respond quickly?	
❏ respond with confidence?	❏ communicate reasons and strategies?
❏ self-correct?	❏ choose appropriate strategies?
	❏ argue logically?

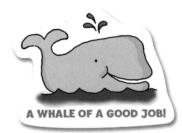

A WHALE OF A GOOD JOB!

FIN-tastic!

Lesson 4

Objective

Children use the reverse counting sequence to explore the number line.

Program Materials

- Step-by-Step Number Line
- Large Number Cards

Engage
Dot Set Cards

Additional Materials

No additional materials needed

Access Vocabulary

set of cards All of the cards needed to do the activity
dots Round spots

Creating Context

Prepositions are not in the beginning level of language proficiency. Help English Learners practice prepositions such as *before*, *after*, and *between* when they line up. Have children identify who is before them in line, who is after them, and who is between two other children.

Making Tracks

---

1 Warm Up 5

Skill Building COMPUTING

Name That Numeral
Before beginning **Next Number Up,** use the **Name That Numeral** activity with the whole group.

Purpose Name That Numeral reinforces numeral recognition.

Warm-Up Card 5

Monitoring Student Progress

| If . . . children are having trouble recognizing numerals, | Then . . . work with just a few numerals at a time. |

---

2 Engage 30

Concept Building ENGAGING

Next Number Up
"Today we will line up so that the people with higher numbers on their cards go first."

Follow the instructions on the activity card to play **Next Number Up** Variation: **Sequencing Down.** As children play, ask questions about what is happening in the activity.

Activity Card 24

Purpose Sequencing Down gives children the opportunity to use the reverse counting sequence while they explore the number line.

Monitoring Student Progress

| **If . . .** children have trouble counting down from 10, | **Then . . .** fold the Step-by-Step Number Line in half and count down from 5. |

 Building Blocks For additional practice visualizing magnitude, children should complete **Building Blocks** Build Stairs 1.

3 Reflect 10

Extended Response REASONING

Ask questions such as the following:

- **How can you check to see if you are standing on the correct spot?** Accept all reasonable answers.
- **Who has the biggest number?** The student who has the 10 card. **How do you know?** Possible answer: 10 has the most dots.
- **Who has the smallest number?** The student that has the 1 card. **How do you know?** Possible answer: 1 is first.

4 Assess

Informal Assessment

Use the Student Assessment Record, **Assessment,** page 100, to record informal observations.

COMPUTING	ENGAGING
Name That Numeral	**Next Number Up**
Did the child	Did the child
❏ respond accurately?	❏ pay attention to the contributions of others?
❏ respond quickly?	
❏ respond with confidence?	❏ contribute information and ideas?
❏ self-correct?	❏ improve on a strategy?
	❏ reflect on and check accuracy of work?

Lesson 5

Review

Objective

Children review and reinforce skills and concepts learned in this and other weeks.

Program Materials

Warm Up
Step-by-Step Number Line

Engage
See Activity Cards for materials.

Additional Materials

Engage
See Activity Cards for materials.

Creating Context

Role playing reinforces listening skills, vocabulary, and math concepts. Find other vocabulary that can be reinforced through physical movement and role play.

1 Warm Up 5

Skill Practice COMPUTING

Count Up

Before beginning the **Free-Choice** activity, use the **Count Up** activity with the whole group.

Purpose Count Up teaches children to associate counting up with moving forward on a number line.

Warm-Up Card 6

Monitoring Student Progress

| If . . . children can use the number line to count up from 1 to 10 with ease, | Then . . . use the **Blastoff!** activity and have them count back from 10 to 1. |

2 Engage 20

Concept Building APPLYING

Free-Choice Activity

For the last day of the week, allow children to choose an activity from the previous weeks. Some activities they may choose are the following:

- Picture Land: **Concentration**
- Line Land: **Position on the Step-by-Step Number Line**
- Line Land: **Teddy Bear Path**

Make a note of the activities children select. Do they prefer easy or challenging activities? If you believe your children would benefit from extra practice on specific skills, choose an activity for them.

3 Reflect 10

Extended Response REASONING

Ask questions such as the following:

- **What did you like about playing** Next Number Up?
- **Was there anything about playing this game you didn't like?**
- **Did this game help you do something you couldn't do before? What did it help you do?**
- **What was easy when you were playing** Next Number Up?
- **What was hard when you were playing** Next Number Up?
- **What do you remember most about this game?**
- **What was your favorite part?**

4 Assess 10

A Gather Evidence

Formal Assessment

Have children complete the weekly test, *Assessment,* page 51. Record formal assessment scores on the Student Assessment Record, *Assessment,* page 100.

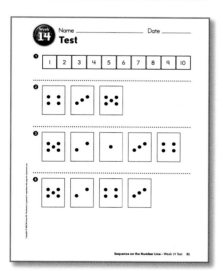

Assessment, p. 51

B Summarize Findings

Review the Student Assessment Records. Determine whether children have Minimal, Basic, or Secure understanding of the concepts presented in Week 14.

C Differentiate Instruction

Based on your observations, use these teaching strategies next week to follow up.

Minimal Understanding

- Repeat the Warm-Up and Engage activities to develop and reinforce concepts of counting objects.
- Use **Building Blocks** computer activities beginning with Build Stairs 1 to develop and reinforce numeration and addition concepts.

Basic Understanding

- Repeat Engage activities in subsequent weeks to reinforce basic concepts.
- Use **Building Blocks** computer activities beginning with Build Stairs 1 to reinforce this week's concepts.

Secure Understanding

- Use Challenge variations of **Next Number Up.**
- Use computer activities to extend children's understanding of the Week 14 concepts.

Using the Number Line

Week at a Glance

This week, children begin **Number Worlds,** Week 15, and continue to explore Line Land.

Background

In Line Land, numbers are represented on a horizontal line using numerals. Children transition from the world of small, countable objects and patterns to a world of abstract numbers. Using number lines builds an understanding that one can move forward and backward through the number sequence. This will become the basis for addition and subtraction.

How Children Learn

As they begin this week's activities, children should be beginning to understand the relationships among set size, counting, and the number line.

By the end of the week, children should be developing the skills to use small numbers alone to solve problems.

Skills Focus

- Identify and compute set size
- Associate set size with position on a number line
- Compare positions and amounts
- Associate increasing a set with moving forward on a number line
- Solve problems

Teaching for Understanding

As children engage in these activities, they continue to build an understanding of the number sequence. This lays a good foundation for number sense and the idea that numbers that come later in the sequence are always associated with larger quantities.

Observe closely while evaluating the Engage activities assigned for this week.

- Are children correctly identifying amounts?
- Are children able to recognize which positions are nearer to or farther from a goal?
- Are children using their knowledge to solve the problems presented to them?

Math at Home

Give one copy of the Letter to Home, page A15, to each child. Complete the activity in class, and then encourage children to share it with their caregivers.

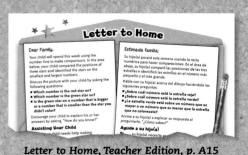

Letter to Home, Teacher Edition, p. A15

Math Vocabulary

more A bigger number or amount

few A small number of something

equal Having the same amount

prove To show that what you say is true

English Learners

SPANISH COGNATES

English	Spanish
compare	comparar
cube	cubo
equal	igual

ALTERNATE VOCABULARY

team A group of people who work together to get something done

pawn Something that shows where you are on a game board

Week 15 Planner — Using the Number Line

PACING	LESSON	LEARNING GOALS	MATERIALS	TECHNOLOGY
DAY 1	**Warm Up** **Line Land:** Count Up*	Children reinforce counting skills.	**Program Materials** Step-by-Step Number Line	**Building Blocks** Road Race Counting Game
	Engage **Line Land:** Number Line Game*	Children associate set size with position on the number line.	**Program Materials** • Number Line Game Board • Pawns, 1 per child • Counters, 10 per child • Dot Cube	
DAY 2	**Warm Up** **Line Land:** Blastoff!*	Children build a foundation for mental subtraction.	**Program Materials** Step-by-Step Number Line	**Building Blocks** Road Race Counting Game
	Engage **Line Land:** Number Line Team Game*	Children compare positions on a number line to identify more, less, and equal amounts.	**Program Materials** • Counters, 10 per team • Dot Cube **Prepare Ahead** 2 Step-by-Step Number Lines	
DAY 3	**Warm Up** **Line Land:** Count Up*	Children associate counting up by one with moving forward on a number line.	**Program Materials** Step-by-Step Number Line	**Building Blocks** Road Race Counting Game
	Engage **Line Land:** Number Line Game*	Children use number sense to solve problems.	**Program Materials** • Number Line Game Board • Pawns, 1 per child • Counters, 10 per child • Dot Cube	**ⓔ MathTools** Number Line
DAY 4	**Warm Up** **Line Land:** Blastoff!*	Children review the reverse counting sequence.	**Program Materials** Step-by-Step Number Line	**Building Blocks** Road Race Counting Game
	Engage **Line Land:** Number Line Team Game*	Children identify set size and associate it with position on a number line.	**Program Materials** • Counters, 10 per team • Dot Cube **Prepare Ahead** 2 Step-by-Step Number Lines	
DAY 5	**Warm Up** **Line Land:** Blastoff!	Children associate counting down with moving backward on a number line.	**Program Materials** Step-by-Step Number Line	**Building Blocks** Review previous activities
	Review Free-Choice activity	Children review and reinforce skills and concepts.	Materials will be selected from those used in previous weeks.	

* Includes Challenge Variations

Lesson 1

Objective

Children associate set size with position on the number line.

Program Materials

Warm Up
Step-by-Step Number Line

Engage
- Number Line Game Board
- Pawns, 1 per child
- Counters, 10 per child
- Dot Cube

Additional Materials
No additional materials needed

Access Vocabulary

remind To make you think of something
game board A piece of cardboard or thick paper with a drawing on it that is used to play a game

Creating Context

Playing games is an excellent way to give English Learners practice listening to and speaking English. The natural repetition of procedural and counting language replaces tedious drill with authentic, active experience. Wait time is built into the process, and the low-stakes environment makes the use of games an enjoyable and efficient learning tool in math.

---

1 Warm Up 5

Skill Building COMPUTING

Count Up
Before beginning **Number Line Game,** use the **Count Up** activity with the whole group.

Purpose Count Up helps children reinforce counting skills.

Warm-Up Card 6

Monitoring Student Progress

| If . . . children have trouble remembering the number sequence, | Then . . . give them additional opportunities to practice during the day. |

2 Engage 30

Concept Building UNDERSTANDING

Number Line Game
"Today we will play a board game to practice counting."

Follow the instructions on the Activity Card to play **Number Line Game.** As children play, ask questions about what is happening in the activity.

Purpose Number Line Game teaches children that set size can be associated with positions on the number line.

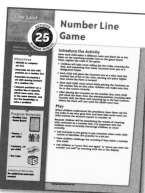

Activity Card 25

Monitoring Student Progress

If . . . children have trouble moving their Pawns to the correct spot,	**Then . . .** remind them to count aloud and touch their Pawns to each Counter as they count.

 Building Blocks For additional practice using the number line, children should complete **Building Blocks** Road Race Counting Game.

3 Reflect   10

Extended Response APPLYING

Ask the following questions:

■ **What does this game remind you of?** Possible answers: other board games, the Step-by-Step Number Line.

■ **How would you explain this game to someone who is not here?** Possible answer: You roll the Cube and count the number of dots. You ask for that number of Counters. You put each Counter on a space on the game board and move your Pawn to the last Counter. You take turns until someone gets to 10 and wins.

4 Assess

Informal Assessment

Use the Student Assessment Record, **Assessment,** page 100, to record informal observations.

COMPUTING	**UNDERSTANDING**
Count Up	**Number Line Game**
Did the child	Did the child
❏ respond accurately?	❏ make important observations?
❏ respond quickly?	❏ extend or generalize learning?
❏ respond with confidence?	❏ provide insightful answers?
❏ self-correct?	❏ pose insightful questions?

Lesson 2

Objective

Children compare positions on a number line to see who has the most, the least, or the same amount.

Program Materials

Warm Up
Step-by-Step Number Line

Engage
- 20 Counters, 10 yellow and 10 red
- Dot Cube

Prepare Ahead

Engage
2 Step-by-Step Number Lines

Access Vocabulary

Cube roller The person who rolls the Dot Cube for a team

Counter handler The person who holds the Counters and counts them out

Creating Context

Teaching children to raise their hands—to take turns or to be called on by the teacher—is not a universal practice in all cultures. Some children are taught to sit quietly as a sign of respect. Help all children learn this common signal used in United States schools.

FIN-tastic!

---

1 Warm Up 5

Concept Building COMPUTING

Blastoff!
Before beginning **Number Line Team Game,** use the **Blastoff!** activity with the whole group.

Purpose Blastoff! helps children build a foundation for mental subtraction.

Warm-Up Card 7

Monitoring Student Progress

If . . . children cannot remember the reverse counting sequence,

Then . . . review it with them in small chunks before having them do a longer sequence.

---

2 Engage 30

Concept Building ENGAGING

Number Line Team Game
"Today we will play **Number Line Team Game** using the Step-by-Step Number Line."

Follow the instructions on the Activity Card to play **Number Line Team Game.** As children play, ask questions about what is happening in the activity.

Activity Card 26

Purpose Number Line Team Game helps children focus on the quantities associated with position on a number line and learn to compare positions and identify which have more, less, and equal amounts.

Monitoring Student Progress

| If . . . children have trouble comparing more than two positions or amounts at a time, | Then . . . have them first compare their Counters with only one other child's Counters. |

Building Blocks For additional practice using the number line, children should complete **Building Blocks** Road Race Counting Game.

3 Reflect 10

Extended Response REASONING

Ask the following questions:

- **How do you know who has more Counters?**
  Possible answer: The person who is closest to 10 on the number line has more.
- **How did you figure that out?** Accept all reasonable answers.

4 Assess

Informal Assessment

Use the Student Assessment Record, **Assessment,** page 100, to record informal observations.

COMPUTING	ENGAGING
Blastoff! Did the child	**Number Line Team Game** Did the child
❑ respond accurately?	❑ pay attention to the contributions of others?
❑ respond quickly?	
❑ respond with confidence?	❑ contribute information and ideas?
❑ self-correct?	
	❑ improve on a strategy?
	❑ reflect on and check accuracy of work?

Lesson 3

Objective

Children solve problems using the number line and counting skills.

Program Materials

Warm Up
Step-by-Step Number Line

Engage
- Number Line Game Board
- Pawns, 1 per child
- Counters, 10 per child
- Dot Cube

Additional Materials

No additional materials needed

Access Vocabulary

Dot Cube A square object that, when rolled, shows a pattern of dots that tells you how many

winner's square The space on the game board that you are trying to get to first; the 10 space

Creating Context

Be sensitive to the fact that some children may not have experience with board games. Discuss the routines used in board games such as determining who starts the game, taking turns, and knowing how the game is won.

Hot Dog!

1 Warm Up 5

Skill Building COMPUTING

Count Up

Before beginning **Number Line Game,** use the **Count Up** activity with the whole group.

Purpose **Count Up** teaches children to associate counting up by one with moving forward on a number line.

Warm-Up Card 6

Monitoring Student Progress

If . . . children have trouble predicting the next number up,

Then . . . have them count up from 1.

2 Engage 30

Concept Building REASONING

Number Line Game

"Today we will continue playing **Number Line Game.**"

Follow the instructions on the Activity Card to play **Number Line Game.** As children play, ask questions about what is happening in the activity.

Purpose **Number Line Game** gives children an opportunity to use number sense to solve problems.

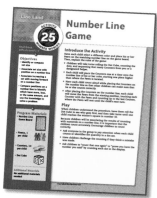

Activity Card 25

MathTools Use the Number Line Tool to demonstrate and explore counting on the number line.

Monitoring Student Progress

| **If . . .** children have trouble predicting the total number of Counters they will have when they add the new set, | **Then . . .** have them count all of the Counters, starting with the first set and continuing to the new set. |

Teacher's Note When children are comfortable with the game, begin introducing some of the ideas in the Challenges.

Building Blocks For additional practice using the number line, children should complete **Building Blocks** Road Race Counting Game.

3 Reflect 10

Extended Response REASONING

Ask the following questions:

- **When you are rolling the Dot Cube, which numbers do you want to roll? Why?** Possible answer: A 4 or a 5 because they're bigger numbers and I can get to the end faster with big numbers.

- **How can you check to see if you've moved the correct number of spaces?** Possible answer: Count the dots on the Cube and recount the number of spaces I moved.

4 Assess

Informal Assessment

Use the Student Assessment Record, **Assessment,** page 100, to record informal observations.

COMPUTING	**REASONING**
Count Up	**Number Line Game**
Did the child	Did the child
❑ respond accurately?	❑ provide a clear explanation?
❑ respond quickly?	❑ communicate reasons and strategies?
❑ respond with confidence?	❑ choose appropriate strategies?
❑ self-correct?	❑ argue logically?

Lesson 4

Objective

Children compare positions on a number line to identify which positions have more, less, or the same amount.

Program Materials

Warm Up
Step-by-Step Number Line

Engage
- 20 Counters, 10 yellow and 10 red
- Dot Cube

Prepare Ahead
Engage
2 Step-by-Step Number Lines

Access Vocabulary

turn A chance or time to do something
prove To show that what you say is true

Creating Context

To reinforce the concepts of *near* and *far*, have children work in pairs to generate a list of places that are near to or far from school. Make a chart with columns labeled *Near* and *Far* to record the answers, and have children illustrate or add a symbol for each location.

---

1 Warm Up 5

Skill Building COMPUTING
Blastoff!
Before beginning **Number Line Team Game,** use the **Blastoff!** activity with the whole group.

Purpose **Blastoff!** provides an opportunity for children to review the reverse counting sequence.

Warm-Up Card 7

Monitoring Student Progress

If . . . children are not yet ready to count down from 10 to 1,

Then . . . fold the Step-by-Step Number Line in half so that only numbers 1–5 are visible.

---

2 Engage 30

Concept Building ENGAGING
Number Line Team Game

"Today we will play **Number Line Team Game** with the whole class."

Follow the instructions on the Activity Card to play **Number Line Team Game.** As children play, ask questions about what is happening in the activity.

Activity Card 26

Purpose **Number Line Team Game** helps children associate set size and position on a number line.

Monitoring Student Progress

If . . . children can play this game with ease and answer the basic questions,	**Then . . .** use one of the Challenge activities.

 Building Blocks For additional practice using the number line, children should complete **Building Blocks** Road Race Counting Game.

3 Reflect 10

Extended Response **REASONING**

Ask the following questions:

■ **How is playing with teams different?** Accept all reasonable answers.

■ **Is playing with teams harder? Why?** Accept all reasonable answers.

■ **How is playing with teams the same?** Accept all reasonable answers.

4 Assess

Informal Assessment

Use the Student Assessment Record, **Assessment,** page 100, to record informal observations.

COMPUTING	ENGAGING
Blastoff!	**Number Line Team Game**
Did the child	Did the child
❏ respond accurately?	❏ pay attention to the contributions of others?
❏ respond quickly?	
❏ respond with confidence?	❏ contribute information and ideas?
❏ self-correct?	❏ improve on a strategy?
	❏ reflect on and check accuracy of work?

Week 15

Lesson 5

Review

Objective

Children review and reinforce skills and concepts learned in this and previous weeks.

Program Materials

Warm Up
Step-by-Step Number Line

Engage
See Activity Cards for materials.

Additional Materials

Engage
See Activity Cards for materials.

Creating Context

Discuss with children different meanings of the word *like*. Make a two-column chart on the board. Label one column *I Like* and the other column *Because*. Have children give reasons they like something as you complete the chart on the board. This will help them describe why they like certain *Number Worlds* activities.

1 Warm Up · 5

Skill Practice COMPUTING
Blastoff!
Before beginning the **Free-Choice** activity, use the **Blastoff!** activity with the whole group.

Purpose Blastoff! helps children associate counting down with moving backward on a number line.

Warm-Up Card 7

Monitoring Student Progress

If . . . children have trouble predicting the next number down,

Then . . . ask a classmate to help.

2 Engage · 20

Concept Building APPLYING
Free-Choice Activity
For the last day of the week, allow children to choose an activity from previous weeks. Some activities they may choose include the following:
- Picture Land: **Catch the Teacher**
- Line Land: **Position on the Step-by-Step Number Line**
- Line Land: **Next Number Up**

Make a note of the activities children select. Do they prefer easy or challenging activities? If you believe your children would benefit from extra practice on specific skills, choose an activity for them.

3 Reflect 10

Extended Response REASONING

Ask the following questions:

- **What did you like about playing** Number Line Game?
- **Was there anything about playing this game you didn't like?**
- **Did this game help you do something you couldn't do before? What did it help you do?**
- **What was easy when you were playing** Number Line Team Game?
- **What was hard when you were playing** Number Line Team Game?
- **What do you remember most about this game?**
- **What was your favorite part?**

4 Assess 10

A Gather Evidence

Formal Assessment

Have children complete the weekly test, **Assessment,** page 53. Record formal assessment scores on the Student Assessment Record, **Assessment,** page 100.

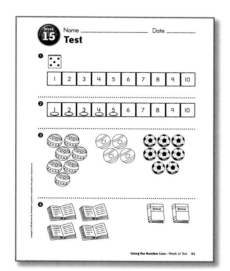

Assessment, p. 53

B Summarize Findings

Review the Student Assessment Records. Determine whether children have Minimal, Basic, or Secure understanding of the concepts presented in Week 15.

C Differentiate Instruction

Based on your observations, use these teaching strategies next week to follow up.

Minimal Understanding

- Repeat the Warm-Up and Engage activities to develop counting concepts.
- Use **Building Blocks** computer activities beginning with Road Race Counting Game to develop and reinforce numeration and addition concepts.

Basic Understanding

- Repeat Engage activities in subsequent weeks to reinforce basic concepts.
- Use **Building Blocks** computer activities beginning with Road Race Counting Game to reinforce this week's concepts.

Secure Understanding

- Use Challenge variations of **Number Line Game**
- Use computer activities to extend children's understanding of the Week 15 concepts.

Week 16 — Moving Up and Down

Week at a Glance

This week, children begin **Number Worlds,** Week 16, and are introduced to Sky Land.

Background

In Sky Land, numbers are represented along a vertical scale such as a thermometer. This gives children an opportunity to develop the language of height as it is used in mathematics. Children learn that this language can be used interchangeably with the language used in Line Land, helping them consolidate the knowledge from both Lands.

How Children Learn

As they begin this week's activities, children should know the counting sequence through 10 and have an understanding of set size and the number line.

By the end of the week children should build connections between the number line and scale measures of numbers that move up and down.

Skills Focus

- Count up from 1 to 10 and down from 10 to 1
- Compare numbers and relative position
- Associate set size with points on a vertical scale
- Solve problems

Teaching for Understanding

As children engage in these activities, they learn to associate set size with position on a scale. This knowledge helps them realize that numbers that are higher up on the scale are associated with larger quantities than numbers that are lower down.

Observe closely while evaluating the assigned tasks this week.

- Are children correctly counting up to and down from 10?
- Are children using vocabulary related to height when they make comparisons?
- Are children realizing that set size is the same in all the **Number Worlds** they have experienced?

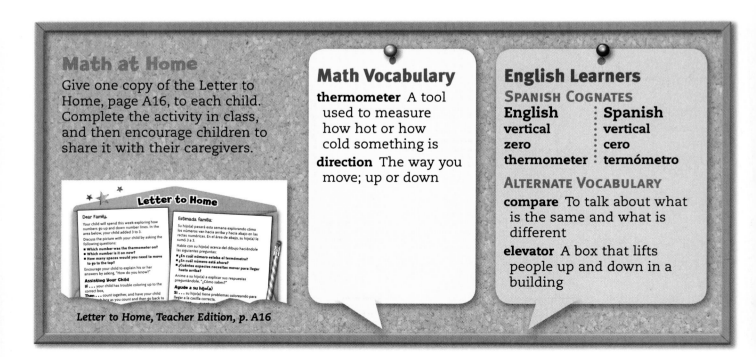

Math at Home

Give one copy of the Letter to Home, page A16, to each child. Complete the activity in class, and then encourage children to share it with their caregivers.

Letter to Home, Teacher Edition, p. A16

Math Vocabulary

thermometer A tool used to measure how hot or how cold something is

direction The way you move; up or down

English Learners

SPANISH COGNATES

English	Spanish
vertical	vertical
zero	cero
thermometer	termómetro

ALTERNATE VOCABULARY

compare To talk about what is the same and what is different

elevator A box that lifts people up and down in a building

Week 16 Planner — Moving Up and Down

PACING	LESSON	LEARNING GOALS	MATERIALS	TECHNOLOGY
DAY 1	**Warm Up** Sky Land: Count Up*	Children review the counting sequence.	**Program Materials** Classroom Thermometer	**Building Blocks** Build Stairs 2
	Engage Sky Land: Going Up*	Children associate increasing quantity with moving up a vertical scale.	**Program Materials** • Elevator Game Boards, 1 per child • Pawns, 1 per child • Dot Cube	**e MathTools** Graphing Tool
DAY 2	**Warm Up** Sky Land: Blastoff!*	Children review counting down.	**Program Materials** Classroom Thermometer	**Building Blocks** Build Stairs 2
	Engage Sky Land: Going Up*	Children associate set size with points on a scale.	**Program Materials** • Elevator Game Boards, 1 per child • Pawns, 1 per child • Dot Cube	
DAY 3	**Warm Up** Sky Land: Count Up*	Children reinforce counting skills.	**Program Materials** Classroom Thermometer	**Building Blocks** Build Stairs 2
	Engage Sky Land: Going Down*	Children associate counting down with decreases in the height of a scale measure.	**Program Materials** • Elevator Game Boards, 1 per child • Pawns, 1 per child • Dot Cube	
DAY 4	**Warm Up** Sky Land: Blastoff!*	Children learn applications for the reverse counting sequence.	**Program Materials** Classroom Thermometer	**Building Blocks** Build Stairs 2
	Engage Sky Land: Going Down*	Children use their knowledge of counting and scales to solve problems.	**Program Materials** • Elevator Game Boards, 1 per child • Pawns, 1 per child • Dot Cube	
DAY 5	**Warm Up** Sky Land: Count Up Variation*	Children learn to begin counting in sequence from numbers other than 1.	**Program Materials** Classroom Thermometer	**Building Blocks** Review previous activities
	Review Free-Choice activity	Children review and reinforce concepts and skills previously learned.	Materials will be selected from those used in previous weeks.	

* Includes Challenge Variations

Lesson 1

Objective

Children associate increasing quantity with moving up a vertical scale.

Program Materials

Warm Up
Class Thermometer

Engage
- Elevator Game Boards, 1 per child
- Pawns, 1 per child
- Dot Cube

Additional Materials
No additional materials needed

Access Vocabulary

thermometer A tool used to measure temperature
pawn An object used to show where a player is on a game board
take turns To wait for each of the other players to go before rolling and moving again

Creating Context

It may help English Learners to review some of the vocabulary associated with elevators, such as *lobby*, *ground floor*, *elevator car*, *up*, and *down*. Have children role play riding in an elevator to help them learn the vocabulary.

1 Warm Up 5

Skill Building COMPUTING

Count Up
Before beginning **Going Up**, use the **Count Up** activity with the whole group.

Purpose **Count Up** reinforces children's knowledge of the counting sequence.

Warm-Up Card 9

Monitoring Student Progress

If . . . children have trouble when each one says a number individually,

Then . . . try doing the activity with the whole class counting together several times first.

2 Engage 30

Concept Building UNDERSTANDING

Going Up

"Today we will do an activity to see who can get to the top of a building the fastest."

Follow the instructions on the Activity Card to play **Going Up**. As children play, ask questions about what is happening in the activity.

Purpose **Going Up** teaches children to associate set size with points on a vertical scale.

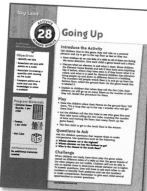

Activity Card 28

MathTools Use the Graphing Tool to demonstrate and explore bar graphs.

Monitoring Student Progress

If . . . children have trouble moving to the correct space,

Then . . . let them use Counters like they do in **Number Line Game.**

 Teacher's Note Sky Land activities are also designed to introduce children to bar graphs. Use bar graphs in the classroom to keep track of items and children's progress so that children are exposed to these representations and begin to understand them.

Building Blocks For additional practice counting, children should complete **Building Blocks** Build Stairs 2.

3 Reflect 10

Extended Response REASONING

Ask the following questions:

■ **Who got to the top first? How can we explain this? Was there anything special about the numbers this player rolled?** Possible answers: She got a 4 or a 5 and I only got 1; she got to go higher.

■ **What does this remind you of?** Accept all reasonable answers.

4 Assess

Informal Assessment

Use the Student Assessment Record, **Assessment,** page 100, to record informal observations.

COMPUTING	UNDERSTANDING
Count Up	**Going Up**
Did the child	Did the child
❑ respond accurately?	❑ make important observations?
❑ respond quickly?	
❑ respond with confidence?	❑ extend or generalize learning?
❑ self-correct?	❑ provide insightful answers?
	❑ pose insightful questions?

Lesson 2

Objective

Children compare points on a scale to identify which have more, less, or the same amount.

Program Materials

Warm Up
Class Thermometer

Engage
- Elevator Game Boards, 1 per child
- Pawns, 1 per child
- Dot Cube

Additional Materials

No additional materials needed

Access Vocabulary

"who is winning" The player closest to finishing the game
solve problems To think of and figure out the answer

Creating Context

Demonstrations can help introduce new concepts, especially those that are more complicated, to English Learners. To explain scales, bring in a measuring cup and fill it with water. Show children the scale on the side of the cup. Pour some liquid out of the cup and show the children that you move down the scale when there is less liquid in the cup.

Concept Building COMPUTING
Blastoff!
Before beginning **Going Up**, use the **Blastoff!** activity with the whole group.

Purpose Blastoff! lets children practice predicting the next number down when counting down.

Warm-Up Card 10

Monitoring Student Progress

If . . . children have trouble understanding the Class Thermometer,

Then . . . discuss and compare the thermometer with the Step-by-Step Number Line.

2 **Engage** 30

Concept Building ENGAGING
Going Up

"Today we will compare to see whose pawn is higher and whose is lower while playing **Going Up**."

Follow the instructions on the Activity Card to play **Going Up**. As children play, ask questions about what is happening in the activity.

Activity Card 28

Purpose Going Up teaches children to compare points on a scale and to use this knowledge to solve a problem.

Monitoring Student Progress

If . . . children have trouble identifying whose pawn is highest or closest to the 10th floor,

Then . . . review the concepts *high–low* and *close to–far away* with the children.

Teacher's Note Encourage children to use this week's vocabulary words as they engage in the activities, discuss math concepts, and make predictions.

Building Blocks For additional practice counting, children should complete **Building Blocks** Build Stairs 2.

3 Reflect 10

Extended Response REASONING

Ask the following questions:

- **How do you know who is winning?** Possible answer: The person who is winning is higher up on the board than everyone else.

- **Is there another way to tell who is winning? What is it?** Possible answer: Count to see who has a bigger number.

4 Assess

Informal Assessment

Use the Student Assessment Record, **Assessment,** page 100, to record informal observations.

COMPUTING	ENGAGING
Blastoff!	**Going Up**
Did the child	Did the child
❑ respond accurately?	❑ pay attention to the contributions of others?
❑ respond quickly?	
❑ respond with confidence?	❑ contribute information and ideas?
❑ self-correct?	
	❑ improve on a strategy?
	❑ reflect on and check accuracy of work?

Lesson 3

Objective
Children associate decreasing quantity with moving down the scale.

Program Materials
Warm Up
Class Thermometer

Engage
- Elevator Game Boards, 1 per child
- Pawns, 1 per child
- Dot Cube

Additional Materials
No additional materials needed

Access Vocabulary
higher Bigger or taller
lower Smaller or shorter

Creating Context
English Learners may need multiple exposures to new words and phrases before they can commit them to memory and use them regularly. Remember that phrases such as *count up, count down, count forward,* and *count backward* can be confusing. Try to use gestures with these phrases to reinforce the directions you give.

---

1 Warm Up 5

Skill Building COMPUTING
Count Up!
Before beginning **Going Down,** use the **Count Up** activity with the whole group.

Purpose **Count Up** helps children review the counting sequence.

Warm-Up Card 9

Monitoring Student Progress

If . . . children have trouble remembering the next number, **Then . . .** encourage them to listen and count silently while others count and to count with their fingers to help them remember.

2 Engage 30

Concept Building ENGAGING
Going Down
"Today we will see who can get to the bottom of the Elevator Game Board first."

Follow the instructions on the Activity Card to play **Going Down.** As children play, ask questions about what is happening in the activity.

Purpose **Going Down** teaches children to associate decreases in quantity with moving down the scale.

Activity Card 29

Monitoring Student Progress

If . . . children are having trouble making comparisons, **Then . . .** have them compare other heights throughout the day.

 Building Blocks For additional practice counting, children should complete **Building Blocks** Build Stairs 2.

3 Reflect 10

Extended Response REASONING

Ask the following questions:

- **Did anything about this activity surprise you? Why?** Accept all reasonable answers.
- **What did you find interesting as you played?** Accept all reasonable answers.

4 Assess

Informal Assessment

Use the Student Assessment Record, **Assessment,** page 100, to record informal observations.

COMPUTING	ENGAGING
Count Up	**Going Down**
Did the child	Did the child
❑ respond accurately?	❑ provide a clear explanation?
❑ respond quickly?	
❑ respond with confidence?	❑ communicate reasons and strategies?
❑ self-correct?	❑ choose appropriate strategies?
	❑ argue logically?

Lesson 4

Objective

Children use their knowledge of counting and scale measures to solve problems.

Program Materials

Warm Up
Class Thermometer

Engage
- Elevator Game Boards, 1 per child
- Pawns, 1 per child
- Dot Cube

Additional Materials
No additional materials needed

Access Vocabulary

going down Moving down by counting backward the number of dots on the Dot Cube

bottom The lowest place

Creating Context

You can help enhance English Learners' comprehension and vocabulary building by associating counting with physical objects or points. These activities include many opportunities for children to practice counting skills using the points on a vertical scale to help keep track.

KEEP IT UP!

1 Warm Up 5

Skill Building COMPUTING

Blastoff!

Before beginning **Going Down,** use **Blastoff!** with the whole group.

Purpose **Blastoff!** helps children learn applications for the reverse counting sequence.

Warm-Up Card 10

Monitoring Student Progress

If . . . children find it difficult to remember the next number down,

Then . . . encourage them to listen and count silently while others count and to count with their fingers to help them remember.

2 Engage 30

Concept Building ENGAGING

Going Down

"Today we will compare numbers to see who is winning."

Follow the instructions on the Activity Card to play **Going Down.** As children play, ask questions about what is happening in the activity.

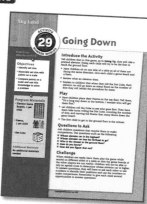

Activity Card 29

Purpose **Going Down** helps children learn problem-solving skills as they compare to see which amounts are higher or lower.

Monitoring Student Progress

| **If . . .** children cannot tell whose car is farthest from or nearest to the bottom of the board, | **Then . . .** review and demonstrate what *farthest* and *nearest* mean. |

 Building Blocks For additional practice counting, children should complete *Building Blocks* Build Stairs 2.

3 Reflect 10 ▶

Extended Response REASONING

Ask the following questions:

- **Can you think of any other times you might compare numbers or positions?** Accept all reasonable answers.
- **How would you make those comparisons?** Accept all reasonable answers.

Informal Assessment

Use the Student Assessment Record, **Assessment,** page 100, to record informal observations.

COMPUTING	ENGAGING
Blastoff!	**Going Down**
Did the child	Did the child
❏ respond accurately?	❏ pay attention to the contributions of others?
❏ respond quickly?	
❏ respond with confidence?	❏ contribute information and ideas?
❏ self-correct?	
	❏ improve on a strategy?
	❏ reflect on and check accuracy of work?

Lesson 5
Review

Objective

Children review and reinforce concepts and skills previously learned

Program Materials

Warm Up
Class Thermometer

Engage
See Activity Cards for materials.

Additional Materials

See Activity Cards for materials.

Creating Context

Pointing and winking are not considered polite in all cultures. It is important to explain to English Learners that when doing this activity, pointing and winking are signals and are not rude.

No "lion"— Math is fun!

You earned Your stripes!

Skill Practice COMPUTING

Count Up

Before beginning the **Free-Choice** activity, use **Count Up** Variation: **Pointing and Winking** with the whole group.

Purpose **Pointing and Winking** helps children learn to begin counting in sequence from numbers other than 1.

Warm-Up Card 9

Monitoring Student Progress

If . . . children have trouble giving the next number in sequence,

Then . . . say longer strings of numbers before asking for the next number.

---

2 Engage 20

Concept Building APPLYING

Free-Choice Activity

For the last day of the week, allow children to choose an activity from the previous weeks. Some activities they may choose include the following:

- Line Land: **Teddy Bear Path**
- Line Land: **Number Line Game**
- Line Land: **Position on the Step-by-Step Number Line**
- Sky Land: **Going Up**
- Sky Land: **Going Down**

Make a note of the activities children select. Do they prefer easy or challenging activities? If you believe your children would benefit from extra practice on specific skills, choose an activity for them.

3 Reflect · 10

Extended Response REASONING

Ask the following questions:

- **What did you like about playing** Going Up?
- **Was there anything about playing this game you didn't like?**
- **Did this game help you do something you couldn't do before? What did it help you do?**
- **What was easy when you were playing** Going Down?
- **What was hard when you were playing** Going Down?
- **What do you remember most about this game?**
- **What was your favorite part?**

4 Assess · 10

A Gather Evidence

Formal Assessment

Have students complete the weekly test on **Assessment**, page 55. Record formal assessment scores on the Student Assessment Record, **Assessment**, page 100.

Assessment, p. 55

B Summarize Findings

Review the Student Assessment Records. Determine whether students have Minimal, Basic, or Secure understanding of the concepts presented in Week 16.

C Differentiate Instruction

Based on your observations, use these teaching strategies next week to follow up.

Minimal Understanding

- Repeat the Warm-Up and Engage activities to develop and reinforce counting concepts.
- Use **Building Blocks** computer activities beginning with Build Stairs 1 to develop and reinforce numeration and addition concepts.

Basic Understanding

- Repeat Engage activities in subsequent weeks to reinforce basic concepts.
- Use **Building Blocks** computer activities beginning with Build Stairs 2 to reinforce this week's concepts.

Secure Understanding

- Use Challenge variations of this week's activities.
- Use computer activities to extend children's understanding of the Week 16 concepts.

Week 17 Counting Rotations

Week at a Glance

This week, children begin **Number Worlds,** Week 17, and are introduced to Circle Land.

Background

In Circle Land, children continue to explore numbers by examining dials. Children develop spatial intuitions in Circle Land that become the foundation for understanding many concepts in mathematics that deal with circular motion, such as pie charts, time, and number bases.

How Children Learn

As they begin this week's activities, children should understand that numbers can be viewed on a line to help make comparisons.

By the end of the week, they should be expanding that understanding to include the circular path on a dial.

Skills Focus

- Count from 1 to 9
- Associate change in quantity with movement around a dial
- Compute set size
- Compare points on a dial
- Solve problems

Teaching for Understanding

As children engage in these activities, they learn that changes in quantity can be represented as movement around a dial, not just along a line. Children must keep track not just of distance around a dial but also of revolutions around a dial. Because zero is both the starting point and the finishing point in Circle Land, children begin to make connections that will help them understand place value in later grades.

Observe closely while evaluating the Engage activities assigned for this week.

- Are children counting from 1 to 10 correctly?
- Are children making comparisons that identify more, less, and the same amount?
- Are children using the information they discover to solve problems and answer questions?

Math at Home

Give one copy of the Letter to Home, page A17, to each child. Complete the activity in class, and then encourage children to share it with their caregivers.

Letter to Home, Teacher Edition, p. A17

Math Vocabulary

dial A circle that shows numbers using a pointer

most More than anyone else

around Going in a circle

clockwise Moving in the direction that the hands of a clock move, as numbers go up

English Learners

SPANISH COGNATES

English	Spanish
count	contar
direction	dirección
revolutions	revoluciones
rotations	rotaciones

ALTERNATE VOCABULARY

skating party A party held at an ice-skating rink, a place where people go to ice-skate

PACING	LESSON	LEARNING GOALS	MATERIALS	TECHNOLOGY
DAY 1	**Warm Up** Circle Land: Count Up*	Children learn that numbers can be used for measurement as well as counting.	**Program Materials** Classroom Dial	**Building Blocks** Pizza Pizzazz 2
	Engage Circle Land: Skating Party	Children associate set size with a point on the dial.	**Program Materials** • Skating Party Game Board • Pawns, 1 per child • Dot Cube • 20 Award Cards	**e MathTools** Stopwatch
DAY 2	**Warm Up** Circle Land: Count Up*	Children review the number sequence from 1 to 5.	**Program Materials** Classroom Dial	**Building Blocks** Pizza Pizzazz 2
	Engage Circle Land: Skating Party	Children associate increasing quantity with moving clockwise around a dial.	**Program Materials** • Skating Party Game Board • Pawns, 1 per child • Dot Cube • 20 Award Cards	
DAY 3	**Warm Up** Circle Land: Count Up*	Children associate counting up with movement around a dial.	**Program Materials** Classroom Dial	**Building Blocks** Pizza Pizzazz 2
	Engage Circle Land: Skating Party	Children compare quantities and points on a dial.	**Program Materials** • Skating Party Game Board • Pawns, 1 per child • Dot Cube • 20 Award Cards	
DAY 4	**Warm Up** Circle Land: Blastoff!*	Children reinforce counting and number concepts.	**Program Materials** Classroom Dial	**Building Blocks** Pizza Pizzazz 2
	Engage Circle Land: Skating Party	Children use knowledge of quantity to solve problems.	**Program Materials** • Skating Party Game Board • Pawns, 1 per child • Dot Cube • 20 Award Cards	
DAY 5	**Warm Up** Circle Land: Blastoff*	Children see that counting down moves counterclockwise toward 0.	**Program Materials** Classroom Dial	**Building Blocks** Review previous activities
	Review Free-Choice activity	Children review and reinforce concepts learned this week and in previous weeks.	Materials will be selected from those used in previous weeks.	

* Includes Challenge Variations

Lesson 1

Objective

Children identify and compute set size and associate it with a point on the dial.

Program Materials

Warm Up
Classroom Dial

Engage
- Skating Party Game Board
- Pawns, 1 per child
- Dot Cube
- 20 Award Cards

Additional Materials

No additional materials needed

Access Vocabulary

hand A long pointer on a clock or dial
dial A circle with numbers on it in counting order

Creating Context

Many young children have not experienced skating at a skating rink. Explain to children that at a skating rink, everyone must skate in the same direction so that they don't bump into each other. If possible, find a picture or video that shows a skating rink to help build background. Ask children if any of them have ever gone to a skating rink and what it was like.

---

1 Warm Up 5

Skill Building COMPUTING

Count Up

Before beginning **Skating Party,** use the **Count Up** activity with the whole group.

Purpose **Count Up** shows children that numbers can be used for measurement as well as counting.

Warm-Up Card 11

Monitoring Student Progress

If . . . children have trouble counting individually,

Then . . . have them count as a group several times while volunteers move the hand on the dial.

---

2 Engage 30

Concept Building UNDERSTANDING

Skating Party

"Today we will see who can go around the Skating Party Game Board the most times."

Follow the instructions on the Activity Card to play **Skating Party.** As children play, ask questions about what is happening in the activity.

Activity Card 34

Purpose **Skating Party** provides a concrete example to help children associate increasing quantity and counting with moving clockwise around a dial.

MathTools Use the Stopwatch Tool to demonstrate and explore real-world uses of dials.

Monitoring Student Progress

| If . . . children are having trouble finding out who wins, | Then . . . remind them to count their Award Cards and see who has the most. |

 Teacher's Note Feel free to change the time limit to fit the needs of your class. Make the time limit long enough so each child will be able to go around the rink a few times, but short enough to keep the activity exciting.

Building Blocks For additional practice counting, children should complete **Building Blocks** Pizza Pizzazz 2 (1–5).

3 Reflect 10

Extended Response REASONING

Ask the following questions:

- **What does this remind you of?** Accept all reasonable answers.

- **How would you explain how to play** Skating Party **to someone who isn't here?** Possible answer: You roll the Dot Cube and move the number of dots on the Cube. When you pass 0, you get an Award Card. You play until the teacher says to stop, and then everybody counts their Award Cards to see who went around the most.

4 Assess

Informal Assessment

Use the Student Assessment Record, **Assessment,** page 100, to record informal observations.

COMPUTING	UNDERSTANDING
Count Up	**Skating Party**
Did the child	Did the child
❏ respond accurately?	❏ make important observations?
❏ respond quickly?	❏ extend or generalize learning?
❏ respond with confidence?	❏ provide insightful answers?
❏ self-correct?	❏ pose insightful questions?

Lesson 2

Objective

Children associate increasing quantity with moving around a dial.

Program Materials

Warm Up
Classroom Dial

Engage
- Skating Party Game Board
- Pawns, 1 per child
- Dot Cube
- 20 Award Cards

Additional Materials

No additional materials needed

Access Vocabulary

award A prize, points
call "time" To say that time is up, the game is over, and stop playing

Creating Context

English Learners may think that skating *rink* is skating *ring* because of the shape of the path and because the final sound in rink may not be clearly articulated. Every language has its own set of phonemes that may or may not overlap with English phonemes. If children speak a primary language that does not have /k/ in the final position, they may need practice listening to and pronouncing minimal pairs such as rink/ring, pink/ping, wink/wing.

---

1 Warm Up 5

Concept Building COMPUTING
Count Up

Before beginning **Skating Party,** use the **Count Up** activity with the whole group.

Purpose **Count Up** helps children review the number sequence from 1 to 5.

Warm-Up Card 11

Monitoring Student Progress

If . . . children are comfortable counting from 1 to 5,

Then . . . try the Challenge activity.

---

2 Engage 30

Concept Building ENGAGING
Skating Party

"Today we will compare Award Cards to see who wins."

Follow the instructions on the Activity Card to play **Skating Party.** As children play, ask questions about what is happening in the activity.

Purpose **Skating Party** helps children realize that a dial is another device for representing quantity.

Activity Card 34

Monitoring Student Progress

If . . . children are not moving the correct number of spaces,

Then . . . remind them to count aloud and listen for mistakes when others count.

 Teacher's Note Encourage children to use this week's vocabulary words as they engage in the activities, discuss math concepts, and make predictions.

Building Blocks For additional practice counting, children should complete **Building Blocks** Pizza Pizzazz 2 (1–5).

3 Reflect 10

Extended Response REASONING

Ask the following questions:

- **How did you figure out who won?** Possible answer: We all counted our Award Cards, and the person with the most won.

- **What has to happen for you to get an Award Card?** Possible answer: You have to go all the way around the rink, starting at 0 and ending up back at 0. (Encourage children to articulate the complete explanation instead of saying "Go past 0.")

4 Assess

Informal Assessment

Use the Student Assessment Record, **Assessment,** page 100, to record informal observations.

COMPUTING	ENGAGING
Count Up	**Skating Party**
Did the child	Did the child
❑ respond accurately?	❑ pay attention to the contributions of others?
❑ respond quickly?	
❑ respond with confidence?	❑ contribute information and ideas?
❑ self-correct?	❑ improve on a strategy?
	❑ reflect on and check accuracy of work?

Lesson 3

Objective

Children compare points on a dial to see who has more, less, or the same amount.

Program Materials

Warm Up
Classroom Dial

Engage
- Skating Party Game Board
- Pawns, 1 per child
- Dot Cube
- 20 Award Cards

Additional Materials
No additional materials needed

Access Vocabulary

Award Card A card you get when you go all the way around the game board
more Having a bigger number than another amount
less Fewer; not as many

Creating Context

Although it is important to ask beginning English Learners questions that have a lower language demand, this does not mean that questions should have a lower cognitive load. Ask English Learners to demonstrate their understanding with manipulatives, models, illustrations, and gestures. Check children's understanding by asking them to show, demonstrate, or point to the answer.

EXCELLENT

---

1 Warm Up   5

Skill Building   COMPUTING
Count Up

Before beginning **Skating Party,** use the **Count Up** activity with the whole group.

Purpose Count Up helps children associate counting up with movement around a dial.

Warm-Up Card 11

Monitoring Student Progress

If . . . children have trouble understanding the dial,

Then . . . discuss objects that remind them of a dial.

---

2 Engage   30

Concept Building   REASONING
Skating Party

"Today you should watch carefully to try to figure out why some people have more cards than others."

Follow the instructions on the Activity Card to play **Skating Party.** As children play, ask questions about what is happening in the activity.

Activity Card 34

Purpose Skating Party asks children to compare quantities and points on a dial to solve problems.

Monitoring Student Progress

If . . . a child counts his or her starting space when moving the pawn,

Then . . . tell the child that you do not start counting until you have moved one space. Reinforce this rule as often as necessary.

 Building Blocks For additional practice counting, children should complete **Building Blocks** Pizza Pizzazz 2 (1–5).

3 Reflect 10

Extended Response REASONING

Ask the following questions:

■ **How do you know when you've finished going around the Game Board?** Possible answer: You go past 0.

■ **Why do some people have more Award Cards than other people?** Possible answers: because they go faster; because they roll bigger numbers

4 Assess

Informal Assessment

Use the Student Assessment Record, **Assessment,** page 100, to record informal observations.

COMPUTING	REASONING
Count Up	**Skating Party**
Did the child	Did the child
❏ respond accurately?	❏ provide a clear explanation?
❏ respond quickly?	❏ communicate reasons and strategies?
❏ respond with confidence?	❏ choose appropriate strategies?
❏ self-correct?	❏ argue logically?

Lesson 4

Objective

Children use their knowledge of quantity to solve problems.

Program Materials

Warm Up
Classroom Dial

Engage
- Skating Party Game Board
- Pawns, 1 per child
- Dot Cube
- 20 Award Cards

Additional Materials

No additional materials needed

Access Vocabulary

"time is up" When there is no more time to play

figure out To think about a problem and find an answer

Creating Context

There are many expressions in English that include the word *time*, such as *in time*, *on time*, *time out*, *time up*, and *overtime*. Make sure children understand these expressions. For example, explain that *time up* does not really indicate location; it means stop. Gesturing with your hands can help English Learners understand this expression.

Hot Dog!

1 Warm Up 5

Skill Building COMPUTING

Blastoff!
Before beginning **Skating Party,** use the **Blastoff!** activity with the whole group.

Purpose Blastoff! reinforces counting and number concepts.

Warm-Up Card 12

Monitoring Student Progress

If . . . a child cannot remember the next number in sequence,

Then . . . make sure you give enough time for the child to think of the next number before asking another child to help.

2 Engage 30

Concept Building ENGAGING

Skating Party

"Today we will keep playing **Skating Party.**"

Follow the instructions on the Activity Card to play **Skating Party.** As children play, ask questions about what is happening in the activity.

Purpose Skating Party teaches children to use their knowledge of quantity to solve problems.

Activity Card 34

Monitoring Student Progress

If . . . a child is reluctant to participate in an activity,

Then . . . decide whether the reason is a social or a skill issue. Work on inclusive strategies if the child feels left out. Work individually on developing skills if the child can't complete the task successfully.

 For additional practice counting, children should complete **Building Blocks** Pizza Pizzazz 2 (1–5).

3 Reflect 10

Extended Response REASONING

Ask the following questions:

- **How do you know how many spaces to move?** Possible answer: I count the dots on the Dot Cube.
- **How can you check your answer?** Accept all reasonable answers.

4 Assess

Informal Assessment

Use the Student Assessment Record, **Assessment**, page 100, to record informal observations.

COMPUTING	ENGAGING
Blastoff!	**Skating Party**
Did the child	Did the child
❏ respond accurately?	❏ pay attention to the contributions of others?
❏ respond quickly?	
❏ respond with confidence?	❏ contribute information and ideas?
❏ self-correct?	❏ improve on a strategy?
	❏ reflect on and check accuracy of work?

Stand-up JOB

Whooo did a great job? YOU!

High Flyer!

Lesson 5

Review

Objective

Children review and reinforce concepts learned this week and in previous weeks.

Program Materials

Warm Up
Classroom Dial

Engage
See Activity Cards for materials.

Additional Materials

Engage
See Activity Cards for materials.

Creating Context

Young learners may need practice and guidance when taking turns playing the games. Help children learn the ground rules for board games. Explain how to decide who goes first, rotating turns in order, and why it is important that everyone has his or her turn.

STEADY JOB!

1 Warm Up 5

Skill Practice COMPUTING

Blastoff!
Before beginning the **Free-Choice** activity, use the **Blastoff!** activity with the whole group.

Purpose **Blastoff!** builds a foundation for subtraction and demonstrates that counting down moves toward 0.

Warm-Up Card 12

Monitoring Student Progress

If . . . children cannot predict which number comes next, **Then . . .** have them recount from 5.

2 Engage 20

Concept Building APPLYING

Free-Choice Activity
For the last day of the week, allow the children to choose an activity from the previous weeks. Some activities they may choose include the following:
- Object Land: **Matching Colors and Quantities**
- Line Land: **Number Line Game**
- Sky Land: **Going Up**
- Sky Land: **Going Down**

Make a note of the activities children select. Do they prefer easy or challenging activities? If you believe your children would benefit from extra practice on specific skills, choose an activity for them.

3 Reflect 10

Extended Response REASONING

Ask the following questions:

- **What did you like about playing** Skating Party?
- **Was there anything about playing this game you didn't like?**
- **Did this game help you do something you couldn't do before? What did it help you do?**
- **What was easy when you were playing** Skating Party?
- **What was hard when you were playing** Skating party?
- **What do you remember most about this game?**
- **What was your favorite part?**

4 Assess 10

A Gather Evidence
Formal Assessment

Have children complete the weekly test, *Assessment,* page 57. Record formal assessment scores on the Student Assessment Record, *Assessment,* page 100.

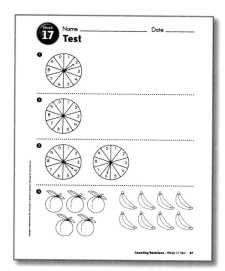

Assessment, p. 57

B Summarize Findings

Review the Student Assessment Records. Determine whether children have Minimal, Basic, or Secure understanding of the concepts presented in Week 17.

C Differentiate Instruction

Based on your observations, use these teaching strategies next week to follow up.

Minimal Understanding

- Repeat the Warm-Up and Engage activities to develop concepts of counting objects.
- Use **Building Blocks** computer activities beginning with Pizza Pizzazz 2 to develop and reinforce numeration concepts.

Basic Understanding

- Repeat Engage activities in subsequent weeks to reinforce basic counting concepts.
- Use **Building Blocks** computer activities beginning with Pizza Pizzazz 2 to reinforce this week's concepts.

Secure Understanding

- Use Challenge variations of activities.
- Use computer activities to extend children's understanding of the Week 17 concepts.

Adding to a Set

Week at a Glance

This week, children begin **Number Worlds,** Week 18, and return to Object Land and Picture Land.

Background

In Object Land, numbers are represented as groups of objects. This is the first way numbers were represented historically, and this is the first way children naturally learn about numbers. In Object Land, children work with tangible objects and with pictures of objects. In Picture Land, children work with abstract representations of numbers.

How Children Learn

As they begin this week's activities, children should be able to identify set size, especially by counting up from one.

By the end of the week, children should understand that adding 1 object to a set increases the set size to the next number up as in counting up, and they should be able to use this knowledge to make predictions.

Skills Focus

- Count from 1 to 5 and 1 to 10
- Count from 5 to 1 and 10 to 1
- Identify set size
- Recognize that when 1 object is added to a set, set size increases by 1
- Predict set size if 1 is added to the set

Teaching for Understanding

As children engage in these activities, they will build connections between an increase in quantity in the real world and the +1 symbol that represents this increase in the world of formal mathematics.

Observe closely while evaluating the Engage activities assigned for this week.

- Are children able to count and identify set size?
- Are children recognizing that the size of a set increases by 1 when 1 object is added?
- Are children able to make predictions about set size when 1 object is added to a set?

Math at Home

Give one copy of the Letter to Home, page A18, to each child. Complete the activity in class, and then encourage children to share it with their caregivers.

Letter to Home, Teacher Edition, p. A18

Math Vocabulary

symbol A letter, number, or picture that has a special meaning or stands for something else

plus Symbol that means "add"

predict To say what you think will happen

English Learners

SPANISH COGNATES

English	Spanish
activity	actividad
numbers	números
Counters	contadores

ALTERNATE VOCABULARY

check your answer To look at your answer carefully to see if it is correct

set A group of objects or numbers that go together

PACING	LESSON	LEARNING GOALS	MATERIALS	TECHNOLOGY
DAY 1	**Warm Up** **Picture Land:** Count Up*	Children review the counting sequence.	No additional materials needed.	**Building Blocks** Build Stairs 1
	Engage **Object Land:** Plus Pup*	Children determine set size.	**Program Materials** • Plus Pup Card • 10 Counters **Additional Materials** Paper bag	**ⓔ MathTools** Number Stairs
DAY 2	**Warm Up** **Picture Land:** Blastoff!*	Children reinforce the reverse counting sequence.	No additional materials needed.	**Building Blocks** Build Stairs 1
	Engage **Object Land:** Plus Pup*	Children recognize that adding 1 object to a set increases set size by 1.	**Program Materials** • Plus Pup Card • 10 Counters **Additional Materials** Paper bag	
DAY 3	**Warm Up** **Picture Land:** Count Up*	Children reinforce that when they count up by 1 unit or add 1 object to a set and recount, the results are equivalent.	No additional materials needed.	**Building Blocks** Build Stairs 1
	Engage **Object Land:** Plus Pup*	Children develop a mental image of the addition operation.	**Program Materials** • Plus Pup Card • 10 Counters **Additional Materials** Paper bag	
DAY 4	**Warm Up** **Picture Land:** Blastoff!*	Children associate counting down with a decrease in set size.	No additional materials needed.	**Building Blocks** Build Stairs 1
	Engage **Object Land:** Plus Pup Variation*	Children make predictions about set size when an object is added to the set.	**Program Materials** • Plus Pup Card • 10 Counters **Additional Materials** Paper bag	
DAY 5	**Warm Up** **Picture Land:** Blastoff!*	Children build a foundation for mental subtraction.	No additional materials needed.	**Building Blocks** Build Stairs 1
	Review Free-Choice activity	Children review and reinforce concepts learned this week and in previous weeks.	Materials will be selected from those used in previous weeks.	

* Includes Challenge Variations

Lesson 1

Objective

Children count objects to determine set size.

Program Materials

Engage

- Plus Pup Card
- 10 Counters

Additional Materials

Engage
Paper bag

Access Vocabulary

plus sign A symbol that means "add"
pup A young dog; puppy

Creating Context

Young children enjoy learning through stories. In this lesson, the character Plus Pup is introduced to help children remember to add one when they see a plus sign. Using puppets or other characters encourages English Learners to speak because the focus is on the character, not directly on them.

BEARY NICE

1 Warm Up 5

Skill Building COMPUTING

Count Up

Before beginning **Plus Pup,** use the **Count Up** activity with the whole group.

Purpose **Count Up** helps children review the counting sequence.

Warm-Up Card 3

Monitoring Student Progress

If . . . children have trouble remembering how to show numbers with their fingers,

Then . . . spend time reviewing with them and model finger displays throughout the day.

2 Engage 30

Concept Building UNDERSTANDING

Plus Pup

"Today we will meet Plus Pup."

Follow the instructions on the Activity Card to play **Plus Pup.** As children play, ask questions about what is happening in the activity.

Purpose **Plus Pup** teaches children to determine set size.

Activity Card 8

MathTools Use the Number Stairs to demonstrate and explore adding 1 to a set.

Monitoring Student Progress

If . . . children have trouble remembering the number of Counters in the bag, **Then . . .** invite them to count with appropriate finger displays as you count.

Teacher's Note Encourage children to use this week's vocabulary words as they engage in the activities, discuss math concepts, and make predictions.

Building Blocks For additional practice adding, children should complete **Building Blocks** Build Stairs 1.

3 Reflect 10

Extended Response REASONING

Ask the following questions:

- **How do you know how many Counters are in the bag?** Possible answers: We count them. You counted them, and I remembered the last number you said.
- **What happens when Plus Pup comes?** The number of Counters goes up by 1.

4 Assess

Informal Assessment

Use the Student Assessment Record, **Assessment,** page 100, to record informal observations.

COMPUTING	UNDERSTANDING
Count Up	**Plus Pup**
Did the child	Did the child
❏ respond accurately?	❏ make important observations?
❏ respond quickly?	
❏ respond with confidence?	❏ extend or generalize learning?
❏ self-correct?	❏ provide insightful answers?
	❏ pose insightful questions?

off to a Running Start!

SQUEAKY CLEAN

Purr-fect

Lesson 2

Objective

Children learn that adding 1 object to a set increases set size by 1.

Program Materials

Engage
- Plus Pup Card
- 10 Counters

Additional Materials

Engage
Paper bag

Access Vocabulary

biggest number The last number we said when counting
predict To say what you think will happen

Creating Context

An excellent way for English Learners to remember new structures is to use rhythm and chant.

There are _____ in the bag,
now the game's begun.
Plus Pup comes to visit,
and he adds one.
The bag has _____ . Wow!
To be sure, let's count now.

---

1 Warm Up

5

Concept Building COMPUTING

Blastoff!
Before beginning **Plus Pup**, use the **Blastoff!** activity with the whole group.

Purpose **Blastoff!** reinforces the reverse counting sequence.

Warm-Up Card 4

Monitoring Student Progress

If . . . children jump up too soon,

Then . . . remind them to remain seated until they count down to 1 and say "Blastoff!"

---

2 Engage

30

Concept Building ENGAGING

Plus Pup
"Today Plus Pup will help us learn about adding."

Follow the instructions on the Activity Card to play **Plus Pup**. As children play, ask questions about what is happening in the activity.

Activity Card 8

Purpose **Plus Pup** helps children build a foundation for addition by teaching them that adding 1 object to a set increases set size by 1.

Monitoring Student Progress

| **If . . .** children are having trouble predicting how many Counters are in the bag after Plus Pup adds 1 Counter, | **Then . . .** ask how many Counters were in the bag before Plus Pup added 1. Then ask children what number comes next when counting. Tell them, "That number will tell you how many Counters are in the bag now." |

 Building Blocks For additional practice adding, children should complete **Building Blocks** Build Stairs 1.

 3 Reflect 10

Extended Response REASONING

Ask the following questions:

- **What was the biggest number we counted to before Plus Pup came?** Possible answer: 4
- **What was the biggest number we counted to after Plus Pup visited?** Possible answer: 5
- **Is the number we count to higher before or after Plus Pup visits?** after he visits

 4 Assess

Informal Assessment

Use the Student Assessment Record, **Assessment,** page 100, to record informal observations.

COMPUTING	**ENGAGING**
Blastoff!	**Plus Pup**
Did the child	Did the child
❏ respond accurately?	❏ pay attention to the contributions of others?
❏ respond quickly?	
❏ respond with confidence?	❏ contribute information and ideas?
❏ self-correct?	
	❏ improve on a strategy?
	❏ reflect on and check accuracy of work?

Lesson 3

Objective

Children develop a mental image of the addition operation that serves as a conceptual bridge between concrete objects and formal mathematics.

Program Materials

Engage
- Plus Pup Card
- 10 Counters

Additional Materials

Engage
Paper bag

Access Vocabulary

I wonder I am curious, I want to know
surprise Something you do not expect

Creating Context

It may be helpful to provide a template so English Learners at early proficiency levels can more easily describe what happens when playing **Plus Pup.** For example, "I predict there are _____ Counters in the bag. Plus Pup put one more in the bag of _____ Counters. Now there will be _____ Counters."

Pretty Good!

---

1 Warm Up 5

Skill Building COMPUTING
Count Up

Before beginning **Plus Pup,** use the **Count Up** activity with the whole group.

Purpose **Count Up** reinforces that when children count up by 1 unit or add 1 object to a set and recount, they achieve the same results.

Warm-Up Card 3

Monitoring Student Progress

If . . . children have trouble showing numbers with their fingers,

Then . . . remind them to use their thumbs to hold their other fingers down.

---

2 Engage 30

Concept Building REASONING
Plus Pup

"Today we will play **Plus Pup** again."

Follow the instructions on the Activity Card to play **Plus Pup.** As children play, ask questions about what is happening in the activity.

Purpose **Plus Pup** helps children develop a meaningful mental image of the addition operation.

Activity Card 8

Monitoring Student Progress

If . . . children are comfortable predicting how many Counters are in the bag after Plus Pup adds 1 Counter to quantities up to 4,

Then . . . use the Challenge activity.

 Building Blocks For additional practice adding, children should complete **Building Blocks** Build Stairs 1.

3 Reflect 10

Extended Response REASONING

Ask the following questions:

■ **Does anything about this activity surprise you? Why?** Answers will vary.

■ **What do you like about** Plus Pup? Answers will vary.

4 Assess

Informal Assessment

Use the Student Assessment Record, **Assessment,** page 100, to record informal observations.

COMPUTING	REASONING
Count Up	**Plus Pup**
Did the child	Did the child
❏ respond accurately?	❏ provide a clear explanation?
❏ respond quickly?	❏ communicate reasons and strategies?
❏ respond with confidence?	❏ choose appropriate strategies?
❏ self-correct?	❏ argue logically?

Lesson 4

Objective

Children make predictions about set size when an object is added to the set.

Program Materials

Engage
- Plus Pup Card
- 10 Counters

Additional Materials

Engage
Paper bag

Access Vocabulary

check your answer To think about your answer and make sure it is correct
empty To take everything out

Creating Context

To build vocabulary, have children compare **Plus Pup** and **Blastoff!** Make a two-column chart with the headings *Alike* and *Different*. Brainstorm ways the two games are similar and ways they are different. To encourage participation of English Learners at early proficiency levels, ask yes/no questions.

Bonus!

---

1 Warm Up
5

Skill Building COMPUTING

Blastoff!
Before beginning **Plus Pup**, use the **Blastoff!** activity with the whole group.

Purpose **Blastoff!** builds children's understanding of the association between counting down and a decrease in set size.

Warm-Up Card 4

Monitoring Student Progress

If . . . children are comfortable counting down from 5,

Then . . . begin counting down from higher numbers.

2 Engage
30

Concept Building ENGAGING

Plus Pup
"Today we will continue to play **Plus Pup**."

Follow the instructions on the Activity Card to play **Plus Pup** Variation: **Plus Pup Predictions**. As children play, ask questions about what is happening in the activity.

Activity Card 8

Purpose **Plus Pup Predictions** teaches children about making predictions and checking to see if they are correct.

Monitoring Student Progress

| If . . . children are guessing more than predicting, | Then . . . model the process for making predictions by explaining how children can think about what they already know. |

 Building Blocks For additional practice adding, children should complete **Building Blocks** Build Stairs 1.

3 Reflect 10

Extended Response REASONING

Ask the following questions:

■ **How would you explain this to someone who has never done it before?** Possible answer: We count the number of Counters in the bag. Then Plus Pup comes and puts a Counter into the bag. We think about and predict the number of Counters in the bag. Then we count again to see if we are correct.

■ **What patterns do you see?** Possible answers: We count on higher after Plus Pup comes. There are always more Counters after Plus Pup comes.

■ **How much higher do you count? How many more Counters are there after Plus Pup comes?** Possible answers: We count up by 1 number. There is always one more Counter than we started with.

4 Assess

Informal Assessment

Use the Student Assessment Record, **Assessment,** page 100, to record informal observations.

COMPUTING	ENGAGING
Blastoff!	**Plus Pup**
Did the child	Did the child
❏ respond accurately?	❏ pay attention to the contributions of others?
❏ respond quickly?	
❏ respond with confidence?	❏ contribute information and ideas?
❏ self-correct?	
	❏ improve on a strategy?
	❏ reflect on and check accuracy of work?

Lesson 5
Review

Objective

Children review and reinforce concepts learned this week and in previous weeks.

Program Materials

Engage
See Activity Cards for materials.

Additional Materials

Engage
See Activity Cards for materials.

Creating Context

English Learners may benefit from practicing patterns for forming comparative adjectives and superlatives. Help students complete the following chart:

| big | bigger | biggest |
| small | smaller | smallest |

✓ Cumulative Assessment

After Lesson 5 is complete, students should complete the Cumulative Assessment, **Assessment,** pages 91–92. Using the key on **Assessment,** page 90, identify incorrect responses. Reteach and review the suggested activities to reinforce concept understanding.

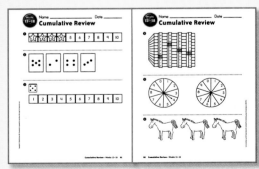

Assessment, pp. 91–92

---

1 Warm Up 5

Skill Practice COMPUTING
Blastoff!
Before beginning the **Free-Choice** activity, use the **Blastoff!** activity with the whole group.

Purpose **Blastoff!** helps children build a foundation for mental subtraction.

Warm-Up Card 4

Monitoring Student Progress

If . . . children need additional practice counting down,

Then . . . use the versions of **Blastoff!** more often.

2 Engage 20

Concept Building APPLYING
Free-Choice Activity
For the last day of the week, allow children to choose an activity from the previous weeks. Some activities they may choose include the following:

- Object Land: **Picnic Party**
- Line Land: **Step-by-Step Number Line**
- Circle Land: **Skating Party**

Make a note of the activities children select. Do they prefer easy or challenging activities? If you believe your children would benefit from extra practice on specific skills, choose an activity for them.

3 Reflect 10

Extended Response REASONING

Ask the following questions:

- **What did you like about playing** Plus Pup?
- **Was there anything about playing this game you didn't like?**
- **Did this game help you do something you couldn't do before? What did it help you do?**
- **What was easy when you were playing** Plus Pup?
- **What was hard when you were playing** Plus Pup?
- **What do you remember most about this game?**
- **What was your favorite part?**

4 Assess 10

A Gather Evidence

Formal Assessment

Have children complete the weekly test, **Assessment,** page 59. Record formal assessment scores on the Student Assessment Record, **Assessment,** page 100.

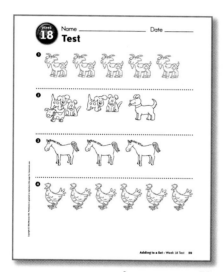

Assessment, p. 59

B Summarize Findings

Review the Student Assessment Records. Determine whether children have Minimal, Basic, or Secure understanding of the concepts presented in Week 18.

C Differentiate Instruction

Based on your observations, use these teaching strategies next week to follow up.

Minimal Understanding

- Repeat the Warm-Up and Engage activities to develop concepts of counting objects.
- Use **Building Blocks** computer activities beginning with Build Stairs 1 to develop and reinforce numeration and addition concepts.

Basic Understanding

- Repeat Engage activities in subsequent weeks to reinforce basic counting concepts.
- Use **Building Blocks** computer activities beginning with Build Stairs 1 to reinforce this week's concepts.

Secure Understanding

- Use Challenge variations of **Plus Pup.**
- Use computer activities to extend children's understanding of the Week 18 concepts.

A WHALE OF A GOOD JOB!

Grrrr8!

Week 19 Subtracting from a Set

Week at a Glance

This week, children begin **Number Worlds,** Week 19, and continue to explore Object Land and Sky Land.

Background

In Object Land, numbers are represented as groups of objects. This is the first way numbers were represented historically, and this is the first way children naturally learn about numbers. In Object Land, children work with real, tangible objects and with pictures of objects. In Sky Land, children work with numbers on a vertical scale such as a thermometer.

How Children Learn

As they begin this week's activities, children should be able to identify set size and should know the reverse counting sequence.

By the end of the week, children should have begun combining their knowledge of counting and set size into an early understanding of basic subtraction.

Skills Focus

- Count up from 1 to 10 and down from 10 to 1
- Calculate set size
- Predict the effect of adding or removing an object from a set
- Recognize that when 1 object is removed, set size decreases by 1

Teaching for Understanding

As children engage in these activities, they build connections between a decrease in quantity in the real world and the −1 symbol that represents this decrease in formal mathematics.

Observe closely while evaluating the Engage activities assigned for this week.

- Are children counting correctly?
- Are children accurately predicting set size after objects are removed?
- Are children recognizing that removing an object from a set decreases the set size by 1?

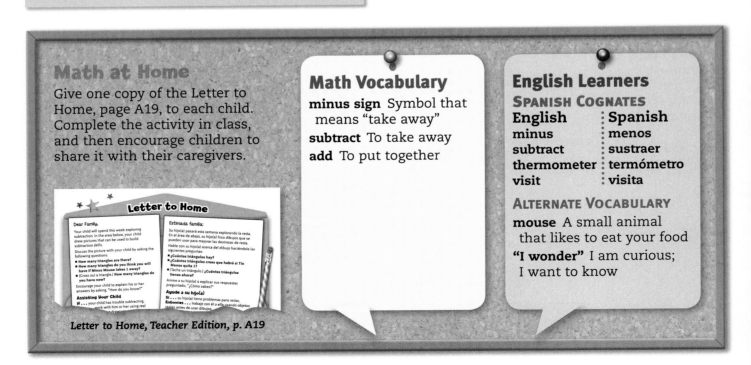

Math at Home

Give one copy of the Letter to Home, page A19, to each child. Complete the activity in class, and then encourage children to share it with their caregivers.

Letter to Home, Teacher Edition, p. A19

Math Vocabulary

minus sign Symbol that means "take away"

subtract To take away

add To put together

English Learners

SPANISH COGNATES

English	Spanish
minus	menos
subtract	sustraer
thermometer	termómetro
visit	visita

ALTERNATE VOCABULARY

mouse A small animal that likes to eat your food

"I wonder" I am curious; I want to know

Week 19 Planner Subtracting from a Set

PACING	LESSON	LEARNING GOALS	MATERIALS	TECHNOLOGY
DAY 1	**Warm Up** **Sky Land:** Count Up*	Children practice counting from 1 to 10.	**Program Materials** Classroom Thermometer	**Building Blocks** Before and After Math
	Engage **Object Land:** Minus Mouse*	Children learn that when 1 object is taken away from a set, the size of the set decreases by 1.	**Program Materials** • Minus Mouse Card • 10 Counters **Additional Materials** Paper bag	**e MathTools** Number Line
DAY 2	**Warm Up** **Sky Land:** Blastoff!*	Children practice the reverse counting sequence.	**Program Materials** Classroom Thermometer	**Building Blocks** Before and After Math
	Engage **Object Land:** Minus Mouse*	Children count a set of objects and identify set size.	**Program Materials** • Minus Mouse Card • 10 Counters **Additional Materials** Paper bag	
DAY 3	**Warm Up** **Sky Land:** Count Up Variation*	Children practice counting and stopping at specified numbers.	**Program Materials** Classroom Thermometer	**Building Blocks** Before and After Math
	Engage **Object Land:** Minus Mouse Variation 1*	Children make predictions about set size when an object is removed from the set.	**Program Materials** • Minus Mouse Card • 10 Counters **Additional Materials** Paper bag	
DAY 4	**Warm Up** **Sky Land:** Blastoff! Variation 1*	Children identify the next number down in sequence.	**Program Materials** Classroom Thermometer	**Building Blocks** Before and After Math
	Engage **Object Land:** Minus Mouse Variation 2	Children practice skills used for basic addition and subtraction.	**Program Materials** • Minus Mouse Card • Plus Pup Card • 10 Counters **Additional Materials** Paper bag	
DAY 5	**Warm Up** **Sky Land:** Blastoff! Variation 2*	Children associate counting down with decreases in height.	No additional materials needed	**Building Blocks** Review previous activities
	Review Free-Choice activity	Children review and reinforce concepts and skills learned this week and in previous weeks.	Materials will be selected from those used in previous weeks.	

* Includes Challenge Variations

Lesson 1

Objective

Children learn that when one object is taken away from a set, the size of the set decreases by 1.

Program Materials

Warm Up
Classroom Thermometer

Engage
- Minus Mouse Card
- 10 Counters

Additional Materials

Engage
Paper bag

Access Vocabulary

Counters Small, plastic circles used to practice counting

minus sign Symbol that means "take away"

Creating Context

English Learners may not yet be able to read and write, but they can gesture or draw pictures to show answers. Ask children to show how Minus Mouse takes away 1. They can draw or use finger displays to demonstrate the answer.

---

1 Warm Up 5

Skill Building COMPUTING

Count Up

Before beginning **Minus Mouse,** use the **Count Up** activity with the whole group.

Purpose **Count Up** gives children practice counting from 1 to 10.

Warm-Up Card 9

Monitoring Student Progress

If . . . children have trouble counting up to 10,

Then . . . have them count up to 5 and gradually work up to 10.

---

2 Engage 30

Concept Building UNDERSTANDING

Minus Mouse

"Today you will meet another character who will help you learn about math."

Follow the instructions on the Activity Card to play **Minus Mouse.** As children play, ask questions about what is happening in the activity.

Purpose **Minus Mouse** teaches children the basic subtraction principle that when 1 object is taken away from a set, the size of the set decreases by 1.

Activity Card 9

e MathTools Use the Number Line Tool to demonstrate and explore subtracting by 1.

Monitoring Student Progress

If . . . children do not recognize the change in set size,

Then . . . begin with smaller sets to make the change more apparent.

Teacher's Note Encourage children to use this week's vocabulary words as they engage in the activities, discuss math concepts, and make precitions.

Building Blocks For additional practice adding and subtracting by 1, children should complete **Building Blocks** Before and After Math.

3 Reflect 10 ▶

Extended Response REASONING

Ask the following questions:

- **What happens when Minus Mouse comes to visit?** Possible answer: He takes away a Counter.

- **Do you notice any patterns?** Possible answer: Every time Minus Mouse comes, the number of Counters in the bag goes down by 1.

4 Assess

Informal Assessment

Use the Student Assessment Record, **Assessment,** page 100, to record informal observations.

COMPUTING	UNDERSTANDING
Count Up	**Minus Mouse**
Did the child	Did the child
❑ respond accurately?	❑ make important observations?
❑ respond quickly?	
❑ respond with confidence?	❑ extend or generalize learning?
❑ self-correct?	❑ provide insightful answers?
	❑ pose insightful questions?

Lesson 2

Objective

Children count a set of objects and identify set size.

Program Materials

Warm Up
Classroom Thermometer

Engage
- Minus Mouse Card
- 10 Counters

Additional Materials

Engage
Paper bag

Access Vocabulary

empty the bag Take all the Counters out of the bag

return them to the bag Put the Counters back into the bag

Creating Context

Have English Learners work with partners or in small groups to make up some simple subtraction stories about Minus Mouse. Help them use subtraction phrases such as *take away*, *less than*, and *fewer*.

1 Warm Up 5

Concept Building COMPUTING

Blastoff!

Before beginning **Minus Mouse**, use the **Blastoff!** activity with the whole group.

Purpose **Blastoff!** gives children practice with the reverse counting sequence.

Warm-Up Card 10

Monitoring Student Progress

If . . . children have trouble remembering the next number down,

Then . . . have them repeat after you as you say the numbers in smaller sets.

2 Engage 30

Concept Building ENGAGING

Minus Mouse

"Today Minus Mouse will visit again."

Follow the instructions on the Activity Card to play **Minus Mouse.** As children play, ask questions about what is happening in the activity.

Purpose **Minus Mouse** reinforces identifying set size and counting skills.

Activity Card 9

Monitoring Student Progress

If . . . children have trouble remembering the number of Counters in the bag,

Then . . . invite them to count with appropriate finger displays as you count.

 Building Blocks For additional practice adding and subtracting by 1, children should complete **Building Blocks** Before and After Math.

3 Reflect 10

Extended Response REASONING

Ask the following questions:

- **How would you explain this activity to someone who has never done it before?** Possible answer: The teacher puts Counters into a bag. We take them out and count them. Then Minus Mouse comes and takes a Counter. We try to figure out how many Counters are left. Then we take them out of the bag and count to see if we are correct.

- **What does this remind you of?** Answers will vary. If children say **Plus Pup,** affirm the similarity and ask them how this game differs from **Plus Pup.**

4 Assess

Informal Assessment

Use the Student Assessment Record, **Assessment,** page 100, to record informal observations.

COMPUTING	ENGAGING
Blastoff!	**Minus Mouse**
Did the child	Did the child
❑ respond accurately?	❑ pay attention to the contributions of others?
❑ respond quickly?	
❑ respond with confidence?	❑ contribute information and ideas?
❑ self-correct?	
	❑ improve on a strategy?
	❑ reflect on and check accuracy of work?

Week 19

Lesson 3

Objective

Children make predictions about what happens to a set when an object is taken away and explain their reasoning.

Program Materials

Warm Up
Classroom Thermometer

Engage
- Minus Mouse Card
- 10 Counters

Additional Materials

Engage
Paper bag

Access Vocabulary

wink To close and open one eye quickly
". . . after Minus Mouse leaves" When he goes away

Creating Context

Some young children may find it difficult to wink. You may wish to have children blink instead. Ask children to brainstorm other signals that can be used to indicate when to start and stop counting.

---

1 Warm Up 5

Skill Building COMPUTING

Count Up

Before beginning **Minus Mouse,** use **Count Up** Variation: **Pointing and Winking** with the whole group.

Purpose Pointing and Winking gives children practice counting and stopping at specified numbers.

Warm-Up Card 9

Monitoring Student Progress

If . . . children have trouble completing the sequence individually,

Then . . . let the class complete the sequence after you stop counting.

2 Engage 30

Concept Building REASONING

Minus Mouse

"Today we will try to predict what happens after Minus Mouse leaves."

Follow the instructions on the Activity Card to play **Minus Mouse** Variation 1: **Minus Mouse Predictions.** As children play, ask questions about what is happening in the activity.

Purpose Minus Mouse Predictions builds children's language skills by having them explain their reasoning when making predictions.

Activity Card 9

Monitoring Student Progress

| **If . . .** children cannot explain why they made a prediction, | **Then . . .** call on other children to say how they figured it out. If necessary, model a thought process for them. |

Building Blocks For additional practice adding and subtracting by 1, children should complete **Building Blocks** Before and After Math.

3 Reflect 10

Extended Response REASONING

Ask the following questions:

- **What information did you use to predict how many Counters were in the bag?** Possible answers: the number of Counters in the bag before Minus Mouse came; that the number always goes down 1 when Minus Mouse comes.

- **How can you check your answer?** Possible answer: count the Counters in the bag; show the first number with my fingers, take away 1 finger, and count the fingers that are still up.

4 Assess

Informal Assessment

Use the Student Assessment Record, **Assessment,** page 100, to record informal observations.

COMPUTING	**REASONING**
Count Up	**Minus Mouse**
Did the child	Did the child
❏ respond accurately?	❏ provide a clear explanation?
❏ respond quickly?	❏ communicate reasons and strategies?
❏ respond with confidence?	❏ choose appropriate strategies?
❏ self-correct?	❏ argue logically?

Week 19

Lesson 4

Objective

Children practice skills used for basic addition and subtraction.

Program Materials

Warm Up
Classroom Thermometer

Engage
- Minus Mouse Card
- Plus Pup Card
- 10 Counters

Additional Materials

Engage
Paper bag

Access Vocabulary

chilly day A very cold, windy day
surprise Something you do not expect

Creating Context

Graphic organizers help English Learners build comprehension. Draw a Venn diagram and put the Plus Pup Card above one side and the Minus Mouse Card above the other. Ask children to describe the characters and what they do. Place shared characteristics in the overlapping area.

---

1 Warm Up 5

Skill Building COMPUTING

Blastoff!

Before beginning **Minus Mouse,** use **Blastoff! Variation 1: It's a Chilly Day** with the whole group.

Purpose It's a Chilly Day helps children understand the real-world application of mathematics.

Warm-Up Card 10

Monitoring Student Progress

If . . . a child makes a mistake, | **Then . . .** have that child begin the sequence again.

---

2 Engage 30

Concept Building ENGAGING

Minus Mouse

"Today we will work with Minus Mouse and Plus Pup."

Follow the instructions on the Activity Card to play **Minus Mouse** Variation 2: **Minus Mouse Meets Plus Pup.** As children play, ask questions about what is happening in the activity.

Activity Card 9

Purpose Minus Mouse Meets Plus Pup teaches children to predict how many there will be if 1 is added to or taken away from a set.

Monitoring Student Progress

| **If . . .** children are confused by using both Minus Mouse and Plus Pup in the same activity, | **Then . . .** wait to combine the two or switch only after several visits. |

 Building Blocks For additional practice adding and subtracting by 1, children should complete **Building Blocks** Before and After Math.

3 Reflect 10

Extended Response REASONING

Ask the following questions:

■ **How did you figure out what to do when we picked a card and one of these characters came to visit? How did you know whether to add 1 Counter or to take away 1 Counter?** Accept all reasonable answers.

■ **Is it easier to predict the number of Counters after Plus Pup visits or after Minus Mouse visits? Why?** Accept all reasonable answers.

4 Assess

Informal Assessment

Use the Student Assessment Record, **Assessment,** page 100, to record informal observations.

COMPUTING	**ENGAGING**
Blastoff!	**Minus Mouse**
Did the child	Did the child
❏ respond accurately?	❏ pay attention to the contributions of others?
❏ respond quickly?	
❏ respond with confidence?	❏ contribute information and ideas?
❏ self-correct?	
	❏ improve on a strategy?
	❏ reflect on and check accuracy of work?

Lesson 5
Review

Objective

Children review and reinforce concepts and skills learned in this week and in previous weeks.

Program Materials

Engage
See Activity Cards for materials.

Additional Materials

Engage
See Activity Cards for materials.

Creating Context

Incredible Shrinking People has children get smaller as they count down. Explain to children that this is the opposite of what happens during **Blastoff!** Brainstorm a list of other opposites and act out the words to help English Learners build their vocabularies.

1 Warm Up 5

Skill Practice COMPUTING
Blastoff!
Before beginning the **Free-Choice** activity, use **Blastoff!** Variation 2: **The Incredible Shrinking People** with the whole group.

Purpose Incredible Shrinking People helps children associate counting down with decreases in height.

Warm-Up Card 10

Monitoring Student Progress

If . . . children are not combining counting down with decreases in body height,

Then . . . show them where their bodies should be when they reach 5 in the countdown sequence: roughly a sitting position if they imagine a chair behind them.

2 Engage 20

Concept Building APPLYING
Free-Choice Activity
For the last day of the week, allow children to choose an activity from previous weeks. Some activities they may choose include the following:
- Picture Land: **Catch the Teacher**
- Sky Land: **Going Down**
- Object Land: **Plus Pup**

Make a note of the activities children select. Do they prefer easy or challenging activities? If you believe your children would benefit from extra practice on specific skills, choose an activity for them.

Reflect 10

Exended Response REASONING

Ask the following questions:

- **What did you like about playing** Minus Mouse?
- **Was there anything about playing this game you didn't like?**
- **Did this game help you do something you couldn't do before? What did it help you do?**
- **What was easy when you were playing** Minus Mouse?
- **What was hard when you were playing** Minus Mouse?
- **What do you remember most about this game?**
- **What was your favorite part?**

Assess 10

A Gather Evidence

Formal Assessment

Have children complete the weekly test, **Assessment,** page 61. Record formal assessment scores on the Student Assessment Record, **Assessment,** page 100.

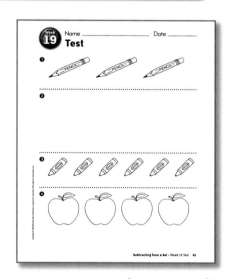

Assessment, p. 61

B Summarize Findings

Review the Student Assessment Records. Determine whether children have Minimal, Basic, or Secure understanding of the concepts presented in Week 19.

C Differentiate Instruction

Based on your observations, use these teaching strategies next week to follow up.

Minimal Understanding

- Repeat the Warm-Up and Engage activities to develop concepts of counting objects.
- Use **Building Blocks** computer activities beginning with Dino Shop 1 to develop and reinforce numeration concepts.

Basic Understanding

- Repeat Engage activities in subsequent weeks to reinforce basic counting concepts.
- Use **Building Blocks** computer activities beginning with Before and After Math to reinforce this week's concepts.

Secure Understanding

- Use Challenge variations of **Minus Mouse.**
- Use computer activities to extend children's understanding of the Week 19 concepts.

Comparing Sets

Week at a Glance

This week, children begin **Number Worlds,** Week 20, and continue to explore Object Land and Line Land.

Background

In Object Land, numbers are represented as groups of objects. This is the first way numbers were represented historically, and this is the first way children naturally learn about numbers. In Object Land, children work with real, tangible objects and with pictures of objects. In Line Land children explore number lines.

How Children Learn

As they begin this week's activities, children should be able to calculate set size both visually and by counting.

By the end of the week, children should recognize that counting and finding the set size with numbers is more accurate than a simple visual assessment.

Skills Focus

- Compare sets of objects
- Count up from 1 to 5 and from 1 to 10
- Count down from 5 to 1 and from 10 to 1
- Recognize that numbers provide consistency for comparisons
- Put sets in order based on set size

Teaching for Understanding

As children engage in these activities, the visual and spatial cues young children use to make quantity comparisons are gradually withdrawn, leading children to understand that counting is a more precise and reliable way to assess relative magnitude when other cues are available and is the only way when there are no other cues.

Observe closely while evaluating the Engage activities assigned for this week.

- Are children counting correctly?
- Are children accurately identifying and comparing set sizes?

Math at Home

Give one copy of the Letter to Home, page A20, to each child. Complete the activity in class, and then encourage children to share it with their caregivers.

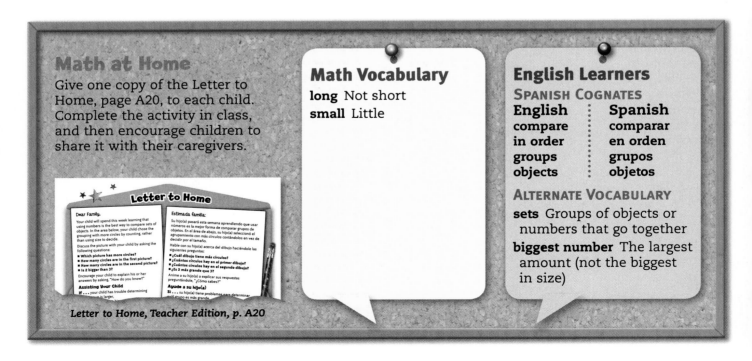

Letter to Home, Teacher Edition, p. A20

Math Vocabulary

long Not short
small Little

English Learners

SPANISH COGNATES

English	Spanish
compare	comparar
in order	en orden
groups	grupos
objects	objetos

ALTERNATE VOCABULARY

sets Groups of objects or numbers that go together

biggest number The largest amount (not the biggest in size)

Week 20 Planner | Comparing Sets

PACING	LESSON	LEARNING GOALS	MATERIALS	TECHNOLOGY
DAY 1	**Warm Up** Line Land: Count Up*	Children associate moving forward on a number line with counting.	**Program Materials** Step-by-Step Number Line	**Building Blocks** Comparisons
	Engage Object Land: Let's Compare*	Children compare sets of objects.	**Program Materials** Counters, 3–7 per child	
DAY 2	**Warm Up** Line Land: Blastoff!*	Children count down from 5 to 1 and from 10 to 1.	**Program Materials** Step-by-Step Number Line	**Building Blocks** Comparisons
	Engage Object Land: Let's Compare Variation 1	Children compare sets of objects without visual clues to predict which is longer.	**Additional Materials** Paper-clip chains, 5 chains with 3–7 paper clips in each chain	**e MathTools** Set Tool
DAY 3	**Warm Up** Line Land: Count Up*	Children reinforce the counting sequence.	**Program Materials** Step-by-Step Number Line	**Building Blocks** Comparisons
	Engage Object Land: Let's Compare Variation 2	Children learn how to compare sets of objects when visual cues are deceptive.	**Additional Materials** Rubber bands, 3–9 per child	
DAY 4	**Warm Up** Line Land: Blastoff!*	Children associate counting down with moving backward on a number line.	**Program Materials** Step-by-Step Number Line	**Building Blocks** Comparisons
	Engage Object Land: Let's Compare Variation 3	Children learn to deal with conflicting clues when comparing sets.	**Program Materials** • 10 Counters • 6 Dot Cubes • 7 Pawns **Additional Materials** • 8 beads • 9 paper clips	
DAY 5	**Warm Up** Line Land: Blastoff!*	Children build a foundation for mental subtraction.	**Program Materials** Step-by-Step Number Line	**Building Blocks** Review previous activities
	Review Free-Choice activity	Children review and reinforce concepts and skills learned in this and previous weeks.	Materials will be selected from those used in previous weeks.	

* Includes Challenge Variations

Week 20

Lesson 1

Objective

Children compare sets of objects to identify which set is longer or shorter, bigger or smaller.

Program Materials

Warm Up
Step-by-Step Number Line

Engage
Counters, 3–7 per child

Additional Materials
No additional materials needed

Access Vocabulary

row A straight, horizontal line of objects
compare To say how things are similar and how they are different

Creating Context

Discuss the concepts of *most* and *fewest* with children. Ask them to point to the part of the classroom with the most books, the most jackets, the fewest children, and so on. Discuss which child has the most pencils in his or her desk, who has the fewest crayons, and so on.

FIN-tastic!

1 Warm Up   5

Skill Building `COMPUTING`
Count Up

Before beginning **Let's Compare,** use the **Count Up** activity with the whole group.

Purpose **Count Up** reinforces children's association between moving forward on a number line and counting up.

Warm-Up Card 6

Monitoring Student Progress

If . . . a child makes a mistake,

Then . . . ask the other children for help.

2 Engage   30

Concept Building `UNDERSTANDING`
Let's Compare

"Today we are going to see who has more Counters."

Follow the instructions on the Activity Card to play **Let's Compare.** As children play, ask questions about what is happening in the activity.

Purpose **Let's Compare** helps children learn how best to compare sets of objects.

Activity Card 10

Monitoring Student Progress

If . . . a child continues to identify the bigger row as the one that takes up the most space,

Then . . . try asking which set has more Counters.

Teacher's Note Encourage children to use this week's vocabulary words as they engage in the activities, discuss math concepts, and make predictions.

Building Blocks For additional practice with comparisons, children should complete **Building Blocks** Comparisons.

3 Reflect   10

Extended Response   REASONING

Ask the following questions:

■ **What did you do in this activity?** Possible answer: We compared to see who had the biggest line of counters.

■ **How could you tell who had the biggest line of Counters?** Accept all reasonable answers. Allow several children to answer before you summarize the strategies they offer.

4 Assess

Informal Assessment

Use the Student Assessment Record, **Assessment,** page 100, to record informal observations.

COMPUTING	UNDERSTANDING
Count Up	**Let's Compare**
Did the child	Did the child
❑ respond accurately?	❑ make important observations?
❑ respond quickly?	❑ extend or generalize learning?
❑ respond with confidence?	❑ provide insightful answers?
❑ self-correct?	❑ pose insightful questions?

Lesson 2

Objective
Children compare sets of objects without visual clues to predict which is longer.

Program Materials
Warm Up
Step-by-Step Number Line

Additional Materials
Engage
Paper-clip chains, 5 chains varying in length from 3–7 paper clips

Access Vocabulary
predict To say what you think will happen
paper-clip chains Metal objects hooked together to make a long string

Creating Context
This week the activities focus on comparing. Work with the class to brainstorm words that describe sets that are big and sets that are small. Make lists of words to display in the classroom. Refer to the lists when making comparisons and when helping children make comparisons.

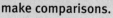
High Flyer!

1 Warm Up 5

Concept Building [COMPUTING]
Blastoff!
Before beginning **Let's Compare,** use the **Blastoff!** activity with the whole group.

Purpose **Blastoff!** helps children count down from 5 to 1 and from 10 to 1.

Warm-Up Card 7

Monitoring Student Progress

If . . . children have trouble counting down from 10, **Then . . .** fold the Step-by-Step Number Line in half and count down from 5.

2 Engage 30

Concept Building [ENGAGING]
Let's Compare
"Today we will compare paper-clip chains."

Follow the instructions on the Activity Card to play **Let's Compare** Variation 1: **Compare Paper Clips.** As children play, ask questions about what is happening in the activity.

Activity Card 10

Purpose **Compare Paper Clips** teaches children to compare sets of objects and predict which is longer when presented with conflicting visual clues.

MathTools Use the Set Tool to demonstrate and explore counting and comparing.

Monitoring Student Progress

If . . . no one suggests counting the paper clips to see which set is larger,	**Then . . .** mention that the paper clips can be counted.

 Building Blocks For additional practice with comparisons, children should complete **Building Blocks** Comparisons.

3 Reflect 10

Extended Response REASONING

Ask the following questions:

- **What did you find interesting about this problem?** Accept all reasonable answers.
- **What would happen if the paper clips were different colors?** The answer would be the same.

Informal Assessment

Use the Student Assessment Record, **Assessment,** page 100, to record informal observations

COMPUTING	ENGAGING
Blastoff!	**Let's Compare**
Did the child	Did the child
❏ respond accurately?	❏ pay attention to the contributions of others?
❏ respond quickly?	
❏ respond with confidence?	❏ contribute information and ideas?
❏ self-correct?	
	❏ improve on a strategy?
	❏ reflect on and check accuracy of work?

Lesson 3

Objective

Children learn that numbers provide the most reliable evidence when comparing sets of objects with deceptive visual cues.

Program Materials

Warm Up
Step-by-Step Number Line

Additional Materials

Engage
Rubber bands, 3–9 per child

Access Vocabulary

bunch A lot of something
rubber bands Elastic circles that are used to hold things together

Creating Context

Have English Learners work in pairs or small groups to describe each set. Provide sentence frames such as "The set of rubber bands is _____ than the set of paper clips." Refer to the list created in Lesson 2 to find words to fill in the blank.

Math-a-saurus

---

1 Warm Up 5

Skill Building COMPUTING

Count Up

Before beginning **Let's Compare,** use the **Count Up** activity with the whole group.

Purpose Count Up reinforces the counting sequence.

Warm-Up Card 6

Monitoring Student Progress

If . . . children are not participating,

Then . . . make sure they all have the opportunity to use the Step-by-Step Number Line.

2 Engage 30

Concept Building REASONING

Let's Compare

"Today we will compare to see who has more rubber bands."

Follow the instructions on the Activity Card to play **Let's Compare** Variation 2: **Compare Rubber Bands.** As children play, ask questions about what is happening in the activity.

Purpose Compare Rubber Bands teaches children how to compare sets of objects when visual cues are deceptive. Because the visible amount in each group may be deceiving, this activity encourages children to rely on using numbers to decide which of two groups is bigger.

Activity Card 10

Monitoring Student Progress

| **If . . .** children are having trouble finding the number of rubber bands in their sets, | **Then . . .** encourage children to arrange the rubber bands so that they are easier to count. |

 Building Blocks For additional practice with comparisons, children should complete **Building Blocks** Comparisons.

 3 **Reflect** 10

Extended Response REASONING

Ask the following questions:

- **Is it harder to compare rubber bands or paper clips?** Probable answer: rubber bands **Why?** Accept all reasonable answers.
- **How did you figure out who had the biggest bunch of rubber bands?** We counted and compared.

 4 **Assess**

Informal Assessment

Use the Student Assessment Record, **Assessment,** page 100, to record informal observations.

COMPUTING	**REASONING**
Count Up	**Let's Compare**
Did the child	Did the child
❏ respond accurately?	❏ provide a clear explanation?
❏ respond quickly?	❏ communicate reasons and strategies?
❏ respond with confidence?	❏ choose appropriate strategies?
❏ self-correct?	❏ argue logically?

Lesson 4

Objective

The visual and spatial clues that children may have been using to make comparisons are removed so children must count to make comparisons.

Program Materials

Warm Up
Step-by-Step Number Line

Engage
- 10 Counters
- 6 Dot Cubes
- 7 Pawns

Additional Materials

Engage
- 8 beads
- 9 paper clips

Access Vocabulary

pawn An object that marks where you are on a game board

bead A small object used in jewelry such as necklaces

Creating Context

Compare Little Things uses a number of small, common classroom objects. Help English Learners build vocabulary by using the names of objects, such as rubber bands, repeatedly during the lesson. The repetition and visual cues help to cement the new vocabulary for English Learners.

---

1 Warm Up 5

Skill Building COMPUTING

Blastoff!
Before beginning **Let's Compare,** use the **Blastoff!** activity with the whole group.

Purpose **Blastoff!** builds children's ability to associate counting down with moving backward on a number line.

Warm-Up Card 7

Monitoring Student Progress

If . . . children are skipping numbers as they walk down the Step-by-Step Number Line,

Then . . . remind them to move down one number at a time.

2 Engage 30

Concept Building ENGAGING

Let's Compare
"Today you will compare to see who has the largest amount, which means the biggest number of things."

Follow the instructions on the Activity Card to play **Let's Compare** Variation 3: **Compare Little Things.** As children play, ask questions about what is happening in the activity.

Activity Card 10

Purpose **Compare Little Things** helps children deal with conflicting cues when comparing sets.

Monitoring Student Progress

If . . . children have trouble comparing when there are five different objects,

Then . . . try using different groupings of two types of objects.

 Building Blocks For additional practice with comparisons, children should complete **Building Blocks** Comparisons.

3 Reflect

10

Extended Response REASONING

Ask the following questions:

- **How was what you did today like what you did yesterday?** Possible answer: We compared to find the biggest group.

- **How was what you did today different from what you did yesterday?** Possible answer: There were lots of different things to compare.

- **How do you know which group of objects is the biggest?** Possible answer: We count and compare.

4 Assess

Informal Assessment

Use the Student Assessment Record, **Assessment,** page 100, to record informal observations.

COMPUTING	ENGAGING
Blastoff!	**Let's Compare**
Did the child	Did the child
❏ respond accurately?	❏ pay attention to the contributions of others?
❏ respond quickly?	
❏ respond with confidence?	❏ contribute information and ideas?
❏ self-correct?	
	❏ improve on a strategy?
	❏ reflect on and check accuracy of work?

Lesson 5
Review

Objective

Children review and reinforce concepts and skills learned this week and in previous weeks.

Program Materials

Warm Up
Step-by-Step Number Line

Engage
See Activity Cards for materials.

Additional Materials

Engage
See Activity Cards for materials.

Creating Context

In addition to assessing and observing children using math concepts, complete an observation of children's language proficiency with academic English for mathematics. If your designated English Language Proficiency Test has an observation checklist, use it to determine which English forms, functions, and vocabulary are still needed for academic language proficiency.

1 Warm Up 5

Skill Practice COMPUTING
Blastoff!
Before beginning the **Free-Choice** activity, use the **Blastoff!** activity with the whole group.

Purpose **Blastoff!** helps children build a foundation for mental subtraction.

Warm-Up Card 7

Monitoring Student Progress

If . . . anyone jumps up too soon,

Then . . . have children repeat the sequence.

2 Engage 20

Concept Building APPLYING
Free-Choice Activity

For the last day of the week, allow children to choose an activity from previous weeks. Some activities they may choose include the following:

- Object Land: **Using the Sorting Mats**
- Picture Land: **Dog and Bone**
- Line Land: **Position on the Step-by-Step Number Line**

Make a note of the activities children select. Do they prefer easy or challenging activities? If you believe your children would benefit from extra practice on specific skills, choose an activity for them.

3 Reflect
10

Extended Response REASONING

Ask the following questions:

- **What did you like about playing** Let's Compare?
- **Was there anything about playing this game you didn't like?**
- **Did this game help you do something you couldn't do before? What did it help you do?**
- **What was easy when you were playing** Let's Compare?
- **What was hard when you were playing** Let's Compare?
- **What do you remember most about this game?**
- **What was your favorite part?**

4 Assess
10

A Gather Evidence

Formal Assessment

Have children complete the weekly test, *Assessment,* page 63. Record formal assessment scores on the Student Assessment Record, *Assessment,* page 100.

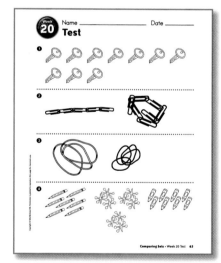

Assessment, p. 63

B Summarize Findings

Review the Student Assessment Records. Determine whether children have Minimal, Basic, or Secure understanding of the concepts presented in Week 20.

C Differentiate Instruction

Based on your observations, use these teaching strategies next week to follow up.

Minimal Understanding

- Repeat the Warm-Up and Engage activities in subsequent weeks to develop concepts of counting objects.
- Use **Building Blocks** computer activities beginning with Comparisons to develop and reinforce numeration concepts.

Basic Understanding

- Repeat Engage activities in subsequent weeks to reinforce basic counting concepts.
- Use **Building Blocks** computer activities beginning with Comparisons to reinforce this week's concepts.

Secure Understanding

- Use Challenge variations of **Let's Compare.**
- Use computer activities to extend children's understanding of the Week 20 concepts.

Week 21 Using Graphs

Week at a Glance

This week, children begin *Number Worlds,* Week 21, and continue to explore Sky Land and Line Land.

Background

In Sky Land, numbers are represented along a vertical scale such as a thermometer. This gives children an opportunity to develop the language of height as it is used in mathematics. Children learn that this language can be used interchangeably with the language used in Line Land to describe the number line, which helps them consolidate the knowledge from both Lands.

How Children Learn

As they begin this week's activities, children should be familiar with scale measures and the number line.

By the end of the week, children should have experience with basic graphing skills and should be better able to order sets.

Skills Focus

- Identify set size
- Record data on a bar graph
- Associate increasing quantity with increasing height on a scale measure
- Use graphs to compare quantities
- Order sets by size and position them on the number line

Teaching for Understanding

As they engage in these activities, children learn about the relationship between set size and scale height by representing objects on a bar graph.

Observe closely while evaluating the Engage activities assigned for this week.

- Are children correctly identifying set size?
- Are children able to match set sizes to the corresponding places on the number line?
- Are children connecting the information on the bar graph with the objects being counted?

Math at Home

Give one copy of the Letter to Home, page A21, to each child. Complete the activity in class, and then encourage children to share it with their caregivers.

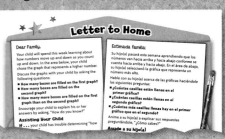

Letter to Home, Teacher Edition, p. A21

Math Vocabulary

target Something that you try to hit

bar graph A way of showing numbers

compare To look at two or more objects to see how they are similar

English Learners

ALTERNATE VOCABULARY

thermometer A tool used to measure temperature

prediction A guess based on information and experience

PACING	LESSON	LEARNING GOALS	MATERIALS	TECHNOLOGY
DAY 1	**Warm Up** **Line Land:** Line Up*	Children identify or compute set size.	**Program Materials** • Dot Set Cards (1–5), 1 per child • Dot Set Cards (1–10), 1 per child for the Challenge	**Building Blocks** Build Stairs 2
	Engage **Sky Land:** Beanbag Toss	Children associate increasing quantity with increasing the height of a scale measure.	**Additional Materials** • Beanbags, 1 per child • 18-in. square of paper • Markers, 2 in different colors	**MathTools** Graphing Tool
DAY 2	**Warm Up** **Sky Land:** Count Up*	Children reinforce counting skills.	**Program Materials** Classroom Thermometer	**Building Blocks** Build Stairs 2
	Engage **Sky Land:** Beanbag Toss	Children use a graph to compare quantities.	**Additional Materials** • Beanbags, 1 per child • 18-in. square of paper • Markers, 2 in different colors	
DAY 3	**Warm Up** **Line Land:** Line Up*	Children associate set size with positions on the number line.	**Program Materials** • Dot Set Cards (1–5), 1 per child • Dot Set Cards (1–10), 1 per child for the Challenge	**Building Blocks** Build Stairs 2
	Engage **Sky Land:** Beanbag Toss	Children identify and compare set size.	**Additional Materials** • Beanbags, 1 per child • 18-in. square of paper • Markers, 2 in different colors	
DAY 4	**Warm Up** **Sky Land:** Blastoff!*	Children review the reverse counting sequence.	**Program Materials** Classroom Thermometer	**Building Blocks** Build Stairs 2
	Engage **Sky Land:** Beanbag Toss	Children record quantity as units on a bar graph.	**Additional Materials** • Beanbags, 1 per child • 18-in. square of paper • Markers, 2 in different colors	
DAY 5	**Warm Up** **Line Land:** Line Up Variation*	Children recognize that a number line increases in length as each successive numeral is added.	**Program Materials** • Number Cards (1–5), 1 per child • Number Cards (1–10), 1 per child for the Challenge	**Building Blocks** Review previous activities
	Review Free-Choice activity	Children will review concepts and skills learned in this and other weeks.	Materials will be selected from those used in previous weeks.	

* Includes Challenge Variations

Lesson 1

Objective

Children associate increasing quantity with increasing the height of a scale measure.

Program Materials

Warm Up
- Dot Set Cards (1–5), 1 per child
- Dot Set Cards (1–10), 1 per child for the Challenge

Additional Materials

Engage
- Beanbags, 1 per child
- 18-in. square piece of paper or fabric
- Markers, 2 in different colors

Access Vocabulary

keeping track of numbers Paying attention to the numbers; remembering the numbers
beanbag A small, square bag made of cloth that has small beans inside

Creating Context

English Learners benefit from the use of visual displays, pictures, and graphs. In this lesson, tossing a beanbag adds a physical element to the graphing activity. This helps children build comprehension and associate increasing quantity with increasing the height of a scale measure. Consider giving children other visuals to show the numbers in this lesson and other lessons.

1 Warm Up 5

Skill Building COMPUTING

Line Up
Before beginning **Beanbag Toss,** use the **Line Up** activity with the whole group.

Purpose **Line Up** helps children identify or compute set size.

Warm-Up Card 8

Monitoring Student Progress

If . . . children do not identify their number correctly,

Then . . . remind them to count each dot once to find the answer.

2 Engage 30

Concept Building UNDERSTANDING

Beanbag Toss
"Today we will explore a new way of keeping track of numbers."

Follow the instructions on the Activity Card to play **Beanbag Toss.** As children play, ask questions about what is happening in the activity.

Purpose **Beanbag Toss** teaches children to associate increasing quantity with increasing the height of a scale measure.

Activity Card 30

MathTools Use the Graphing Tool to demonstrate and explore bar graphs.

Monitoring Student Progress

If . . . children have trouble keeping track of the numbers on the graphs,	Then . . . ask for a volunteer to help count the number of units that have been filled in.

 **Beanbag Toss** also provides an opportunity for children to improve their gross motor skills. If you do not have beanbags, you can substitute Counters.

Building Blocks For additional practice counting, children should complete **Building Blocks** Build Stairs 2.

3 Reflect 10

Extended Response REASONING

Ask the following questions:

- **What do you notice about the graph on the board?** Accept all reasonable answers.

- **Does it remind you of anything else? What does it remind you of? Why?** Possible answer: Classroom Thermometer; it goes up and down.

4 Assess

Informal Assessment

Use the Student Assessment Record, **Assessment**, page 100, to record informal observations.

COMPUTING	UNDERSTANDING
Line Up	**Beanbag Toss**
Did the child	Did the child
❏ respond accurately?	❏ make important observations?
❏ respond quickly?	❏ extend or generalize learning?
❏ respond with confidence?	❏ provide insightful answers?
❏ self-correct?	❏ pose insightful questions?

Lesson 2

Objective
Children use a graph to compare quantities.

Program Materials
Warm Up
Classroom Thermometer

Additional Materials
Engage
- Beanbags, 1 per child
- 18-in. square piece of paper or fabric
- Markers, 2 in different colors

Access Vocabulary
target Something that you try to hit with the beanbag
compare To look at two or more objects to see how they are similar

Creating Context
Children will compare quantities. English Learners may need to review words used to make comparisons. Explain a series of comparatives and superlatives by setting up three sets of objects and then pointing to the objects to demonstrate *big, bigger,* and *biggest* and *small, smaller,* and *smallest.*

---

1 Warm Up 5

Concept Building COMPUTING
Count Up
Before beginning **Beanbag Toss,** use the **Count Up** activity with the whole group.

Purpose **Count Up** reinforces children's counting skills.

Warm-Up Card 9

Monitoring Student Progress

If . . . some children are struggling with skills their classmates have mastered,

Then . . . give them opportunities to use activities from earlier in the year, especially during Free-Choice activity time.

2 Engage 30

Concept Building ENGAGING
Beanbag Toss
"Today we will see how many more beanbags go onto or off the target."

Follow the instructions on the Activity Card to play **Beanbag Toss.** As children play, ask questions about what is happening in the activity.

Purpose **Beanbag Toss** teaches children the basics of using graphs to compare quantities.

Activity Card 30

Monitoring Student Progress

If . . . children are overwhelmed by the numbers being compared,

Then . . . have smaller groups toss the beanbags and do multiple rounds of comparisons.

 **Teacher's Note** Encourage children to use this week's vocabulary words as they engage in the activities, discuss math concepts, and make predictions.

Building Blocks For additional practice counting, children should complete **Building Blocks** Build Stairs 2.

3 Reflect 10

Extended Response REASONING

Ask the following questions:

■ **Is it easier to see if there are more beanbags in the target by looking at the target or at the graph? Why?** Answers will vary.

■ **How do you know if there are more beanbags in the target or outside the target?** Possible answer: The graph with more beanbags is taller.

Informal Assessment

Use the Student Assessment Record, **Assessment,** page 100, to record informal observations.

COMPUTING	ENGAGING
Count Up Did the child	**Beanbag Toss** Did the child
❏ respond accurately?	❏ pay attention to the contributions of others?
❏ respond quickly?	
❏ respond with confidence?	❏ contribute information and ideas?
❏ self-correct?	
	❏ improve on a strategy?
	❏ reflect on and check accuracy of work?

Week 21

Lesson 3

Objective

Children identify and compare set size.

Program Materials

Warm Up
- Dot Set Cards (1–5), 1 per child
- Dot Set Cards (1–10), 1 per child for the Challenge

Additional Materials

Engage
- Beanbags, 1 per child
- 18-in. square piece of paper or fabric
- Markers, 2 in different colors

Access Vocabulary

pattern A predictable, repeating design made up of numbers, colors, or pictures
set A group of objects or numbers that go together

Creating Context

A hands-on approach to demonstrating new concepts is especially helpful for English Learners because they do not need to rely exclusively on language. Use manipulatives to create graphs so English Learners can practice the academic vocabulary and see a concrete representation of a graph.

---

1 Warm Up 5

Skill Building COMPUTING

Line Up

Before beginning **Beanbag Toss,** use the **Line Up** activity with the whole group.

Purpose **Line Up** builds children's associations between set size and position on the number line.

Warm-Up Card 8

Monitoring Student Progress

If . . . children need more practice with sequencing,

Then . . . use Line Up throughout the day when leaving the classroom.

---

2 Engage 30

Concept Building REASONING

Beanbag Toss

"Today we will use graphs to help us compare numbers."

Follow the instructions on the Activity Card to play **Beanbag Toss.** As children play, ask questions about what is happening in the activity.

Purpose **Beanbag Toss** teaches children new ways to identify and compare set size.

Activity Card 30

Monitoring Student Progress

| **If . . .** children need to review using numbers to make comparisons, | **Then . . .** have them play some of the variations of **Let's Compare.** |

Building Blocks For additional practice counting, children should complete **Building Blocks** Build Stairs 2.

3 Reflect 10

Extended Response REASONING

Ask the following questions:

■ **What patterns do you see? How would you describe this pattern?** The colors on the graphs are a pattern. They go back and forth.

■ **What does the pattern help you see?** It makes it easier to count.

FIN-tastic!

4 Assess

Informal Assessment

Use the Student Assessment Record, **Assessment,** page 100, to record informal observations.

COMPUTING	REASONING
Line Up Did the child ❑ respond accurately? ❑ respond quickly? ❑ respond with confidence? ❑ self-correct?	**Beanbag Toss** Did the child ❑ provide a clear explanation? ❑ communicate reasons and strategies? ❑ choose appropriate strategies? ❑ argue logically?

A WHALE OF A GOOD JOB!

Lesson 4

1 Warm Up 5

Skill Building COMPUTING
Blastoff!
Before beginning **Beanbag Toss,** use the **Blastoff!** activity with the whole group.

Purpose **Blastoff!** provides an opportunity to review the reverse counting sequence.

Warm-Up Card 10

Objective
Children record quantities as units on a bar graph.

Program Materials
Warm Up
Classroom Thermometer

Additional Materials
Engage
• Beanbags, 1 per child
• 18-in. square piece of paper or fabric
• Markers, 2 in different colors

Monitoring Student Progress

| **If . . .** children are having trouble remembering the reverse counting sequence, | **Then . . .** use the **Blastoff!** activity more often. |

Access Vocabulary
toss A gentle throw
take turns To wait for each person to go before you go again

Creating Context
Help English Learners understand that counting as a group can help a team do something in unison, such as lifting an injured person. Ask children to think of instances where counting in unison helps them work together.

2 Engage 30

Concept Building ENGAGING
Beanbag Toss
"Today we will continue to play **Beanbag Toss.**"

Follow the instructions on the Activity Card to play **Beanbag Toss.** As children play, ask questions about what is happening in the activity.

Purpose **Beanbag Toss** helps children compare quantities by recording them as units on a bar graph.

Activity Card 30

Monitoring Student Progress

If . . . children do not understand how to figure out "how many more" there are in one column than in the other column,

Then . . . work with them to explain that they need to see how many extra beanbags are in the "larger" column.

Building Blocks For additional practice counting, children should complete **Building Blocks** Build Stairs 2.

3 Reflect 10

Extended Response REASONING

Ask the following questions:

- **What does this graph tell us?** Accept all reasonable answers.
- **When you look at the graph, how can you figure out if there is more in the *On* column or more in the *Off* column?** Possible answers: by counting; by looking to see which graph is higher up
- **Can you think of another way to figure this out?** Accept all reasonable answers.

Informal Assessment

Use the Student Assessment Record, **Assessment,** page 100, to record informal observations.

COMPUTING	ENGAGING
Blastoff!	**Beanbag Toss**
Did the child	Did the child
❏ respond accurately?	❏ pay attention to the contributions of others?
❏ respond quickly?	
❏ respond with confidence?	❏ contribute information and ideas?
❏ self-correct?	
	❏ improve on a strategy?
	❏ reflect on and check accuracy of work?

off to a Running Start!

Week 21 · Using Graphs

Lesson 5
Review

Objective

Children will review concepts and skills learned in this and other weeks.

Program Materials

Warm Up
- Number Cards (1–5), 1 per child
- Number Cards (1–10), 1 per child for the Challenge

Engage
See Activity Cards for materials.

Additional Materials

Engage
See Activity Cards for materials.

Creating Context

Young children are just beginning to develop their understanding of prepositions such as *on, in, off,* and *up.* Young English Learners may find it difficult to distinguish between these terms. When introducing new phrases, use gestures to show children the meaning of prepositions. Using gestures makes new material clearer and more memorable.

Your Work Hogs the Spotlight

1 Warm Up 5

Skill Practice COMPUTING

Line Up
Before beginning the **Free-Choice** activity, use **Line Up** Version 1: **Number Card Line Up** with the whole group.

Purpose **Number Card Line Up** helps children recognize that a number line increases in length as each successive numeral is added.

Warm-Up Card 8

Monitoring Student Progress

If . . . children have trouble identifying numerals, | **Then . . .** discuss the features of each numeral to help them remember.

2 Engage 20

Concept Building APPLYING

Free-Choice Activity

For the last day of the week, allow children to choose an activity from previous weeks. Some activities they may choose include the following:
- Line Land: **The Number Line Game**
- Sky Land: **Going Up**
- Object Land: **Let's Compare**

Make a note of the activities children select. Do they prefer easy or challenging activities? If you believe your children would benefit from extra practice on specific skills, choose an activity for them.

3 Reflect 10

Extended Response REASONING

Ask the following questions:

- **What did you like best about playing** Beanbag Toss?
- **Was there anything about playing this game you didn't like?**
- **Did this game help you do something you couldn't do before? What did it help you do?**
- **What was easy when you were playing** Beanbag Toss?
- **What was hard when you were playing** Beanbag Toss?
- **What do you remember most about this game?**
- **What was your favorite part?**

4 Assess 10

A Gather Evidence

Formal Assessment

Have children complete the weekly test, **Assessment,** page 65. Record formal assessment scores on the Student Assessment Record, **Assessment,** page 100.

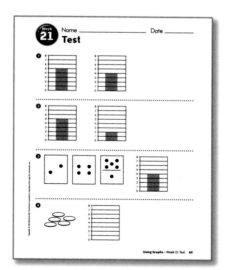

Assessment, p. 65

B Summarize Findings

Review the Student Assessment Records. Determine whether children have Minimal, Basic, or Secure understanding of the concepts presented in Week 21.

C Differentiate Instruction

Based on your observations, use these teaching strategies next week to follow up.

Minimal Understanding

- Repeat the Warm-Up and Engage activities to develop concepts of counting objects.
- Use **Building Blocks** computer activities beginning with Build Stairs 1 to develop and reinforce numeration concepts.

Basic Understanding

- Repeat Engage activities in subsequent weeks to reinforce basic counting concepts.
- Use **Building Blocks** computer activities beginning with Build Stairs 2 to reinforce this week's concepts.

Secure Understanding

- Use Challenge variations of this week's activities.
- Use computer activities to extend children's understanding of the Week 21 concepts.

Week 22 More Subtraction

Week at a Glance

This week, children begin **Number Worlds,** Week 22, and continue to explore Object Land and Picture Land.

Background

In Object Land, numbers are represented as groups of objects. This is the first way numbers were represented, and this is the first way children learn about numbers. In Object Land, children work with tangible objects and with pictures of objects. In Picture Land, they work with patterns to learn more abstract representations of numbers.

How Children Learn

As they begin this week's activities, children should have begun to develop an intuitive understanding of basic subtraction.

By the end of the week, children should have developed further proficiency in counting and comparing quantities and strengthened their understanding of basic subtraction.

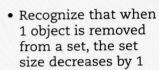

Skills Focus

- Count up from 1 to 10 and back from 10 to 1
- Calculate set size
- Recognize that when 1 object is removed from a set, the set size decreases by 1

Teaching for Understanding

As children engage in these activities, they review the concepts learned in earlier activities and continue to build on that knowledge.

Observe closely while evaluating the Engage activities assigned for this week.

- Are children correctly counting the objects in each set?
- Are children recognizing that when a set is reduced by 1 object, the set size goes down by 1?

Math at Home

Give one copy of the Letter to Home, page A22, to each child. Complete the activity in class, and then encourage children to share it with their caregivers.

Letter to Home, Teacher Edition, p. A22

Math Vocabulary

predict To say what you think will happen

tally marks Straight lines grouped in sets of five

figure out To solve a problem by thinking about it, using information and your experience

English Learners

SPANISH COGNATES

English	Spanish
subtraction	sustracción
minus	menos
comparison	comparación
count	contar

ALTERNATE VOCABULARY

compare To look at two or more objects to see how they are similar and different

Week 22 Planner | More Subtraction

PACING	LESSON	LEARNING GOALS	MATERIALS	TECHNOLOGY
DAY 1	**Warm Up** **Picture Land:** Count Up*	Children review the counting sequence.	No additional materials needed	**Building Blocks** Pizza Pizzazz 4
	Engage **Object Land:** Mouse in the Cookie Jar*	Children develop proficiency in counting and comparing quantities.	**Program Materials** Counters, 5 per child **Additional Materials** • Wide-mouth container • Small cloth	
DAY 2	**Warm Up 1** **Picture Land:** Blastoff!*	Children associate counting down with a decrease in set size.	No additional materials needed	**Building Blocks** Pizza Pizzazz 4
	Engage **Object Land:** Mouse in the Cookie Jar Variation*	Children strengthen their intuitive understanding of subtraction.	**Program Materials** Counters, 5 per team **Additional Materials** • Wide-mouth container • Small cloth	
DAY 3	**Warm Up** **Picture Land:** Count Up*	Children reinforce skills that lay a foundation for mental addition.	No additional materials needed	**Building Blocks** Pizza Pizzazz 4
	Engage **Object Land:** Mouse in the Cookie Jar*	Children recognize that when 1 object is taken away, the size of the set is reduced by 1.	**Program Materials** Counters, 5 per child **Additional Materials** • Wide-mouth container • Small cloth	**e MathTools** Set Tool
DAY 4	**Warm Up** **Picture Land:** Blastoff!*	Children review how to count down from 10 to 1.	No additional materials needed	**Building Blocks** Pizza Pizzazz 4
	Engage **Object Land:** Mouse in the Cookie Jar*	Children count a set of objects and compare amounts.	**Program Materials** Counters, 5 per child **Additional Materials** • Wide-mouth container • Small cloth	
DAY 5	**Warm Up** **Picture Land:** Blastoff!*	Children reinforce the reverse counting sequence.	No additional materials needed	**Building Blocks** Review previous activities
	Review Free-Choice activity	Children review and reinforce skills learned this week and in previous weeks.	Materials will be selected from those used in previous weeks.	

* Includes Challenge Variations

Week 22

Lesson 1

Objective

Children continue to build counting and comparison skills.

Program Materials

Engage
Counters, 5 per child

Additional Materials

Engage
- Wide-mouth container
- Small cloth

Access Vocabulary

cookie jar A container that cookies are kept in
missing Lost, not where it should be

Creating Context

In this lesson, children pretend to get ready for a party. Before introducing the activity, ask English Learners to draw and describe a picture of a party. Keep in mind that not all children have the same kind of cultural experience with children's parties.

BEARY NICE

---

1 Warm Up 5

Skill Building COMPUTING

Count Up

Before beginning **Mouse in the Cookie Jar,** use the **Count Up** activity with the whole group.

Purpose **Count Up** helps children review the counting sequence.

Warm-Up Card 3

Monitoring Student Progress

If . . . children are ready for an additional challenge,

Then . . . have them take turns individually counting up to 10 with appropriate finger displays.

---

2 Engage 30

Concept Building UNDERSTANDING

Mouse in the Cookie Jar

"Today we will pretend that we are getting ready for a party."

Follow the instructions on the Activity Card to play **Mouse in the Cookie Jar.** As children play, ask questions about what is happening in the activity.

Activity Card 12

Purpose **Mouse in the Cookie Jar** teaches children to develop proficiency in counting and comparing quantities.

Monitoring Student Progress

If . . . children have trouble counting a group of 5 objects reliably,

Then . . . use 3 Counters instead of 5 until children gain proficiency with this quantity.

 **Teacher's Note** Encourage children to use this week's vocabulary words as they engage in the activities, discuss math concepts, and make predictions.

Building Blocks For additional practice adding and subtracting hidden objects, children should complete **Building Blocks** Pizza Pizzazz 4.

3 Reflect 10

Extended Response REASONING

Ask the following questions:

- **How did you know who was missing a cookie?** Possible answer: The person who had 4 Counters instead of 5 was missing a cookie.
- **How did you figure that out?** Possible answer: We all counted the Counters in our colors.

 ## 4 Assess

Informal Assessment

Use the Student Assessment Record, **Assessment,** page 100, to record informal observations.

COMPUTING	UNDERSTANDING
Count Up Did the child ❏ respond accurately? ❏ respond quickly? ❏ respond with confidence? ❏ self-correct?	**Mouse in the Cookie Jar** Did the child ❏ make important observations? ❏ extend or generalize learning? ❏ provide insightful answers? ❏ pose insightful questions?

Lesson 2

Objective

Children continue to build a foundation for simple subtraction.

Program Materials

Engage
Counters, 5 per team

Additional Materials

Engage
- Wide-mouth container
- Small cloth

Access Vocabulary

steal To take without permission something that doesn't belong to you
container Something you put things into

Creating Context

English Learners may need a sentence frame to help them talk about the missing Counters. Provide them with one such as the following:

_____ had _____ Counters. One is missing. Now he/she has _____ .

Stand-up JOB

---

1 Warm Up 5

Concept Building COMPUTING

Blastoff!
Before beginning **Mouse in the Cookie Jar,** use the **Blastoff!** activity with the whole group.

Purpose **Blastoff!** reminds children of the association between counting down and a decrease in set size.

Warm-Up Card 4

Monitoring Student Progress

If . . . children have trouble counting down from 10,

Then . . . have them count down from 5 and gradually work up to counting down from 10.

2 Engage 30

Concept Building ENGAGING

Mouse in the Cookie Jar
"Today we will see what happens when the mouse comes and gets into the cookie jar."

Follow the instructions on the Activity Card to play **Mouse in the Cookie Jar** Variation: **Whole-Class Variation.** As the children play, ask questions about what is happening in the activity.

Activity Card 12

Purpose **Mouse in the Cookie Jar** whole-class variation helps children strengthen their intuitive understanding of subtraction.

Monitoring Student Progress

If . . . no one suggests emptying the container and counting the Counters to find out who is missing one,

Then . . . suggest this possibility to children.

 Building Blocks For additional practice adding and subtracting hidden objects, children should complete **Building Blocks** Pizza Pizzazz 4.

3 Reflect   10 ◐

Extended Response **REASONING**

Ask the following questions:

■ **What is this activity about?** Possible answer: trying to figure out whose Counter was stolen.

■ **What facts do we know that help us figure out whose Counter was taken?** Possible answers: the color of Counters each person has; the number of Counters that were put in the jar; the number of Counters that were in the jar after the mouse came.

4 Assess

Informal Assessment

Use the Student Assessment Record, **Assessment,** page 100, to record informal observations.

COMPUTING	ENGAGING
Blastoff!	**Mouse in the Cookie Jar**
Did the child	Did the child
❏ respond accurately?	❏ pay attention to the contributions of others?
❏ respond quickly?	
❏ respond with confidence?	❏ contribute information and ideas?
❏ self-correct?	
	❏ improve on a strategy?
	❏ reflect on and check accuracy of work?

Doggone Good

More Subtraction • Lesson 2   283

Lesson 3

Objective

Children associate removing objects from a set with decreases in set size.

Program Materials

Engage
Counters, 5 per child

Additional Materials

Engage
- Wide-mouth container
- Small cloth

Access Vocabulary

sneak To do something very quietly so that no one notices

donor A person who gives up something because he or she wants to

Creating Context

English Learners may need help when brainstorming the ways a cookie was taken away by the mouse. Ask the class to think of ways to describe what the mouse did. Give them examples such as *takes away, steals, sneaks out,* or *grabs.* Have children act out the mouse's action.

 Warm Up 5

Skill Building COMPUTING

Count Up

Before beginning **Mouse in the Cookie Jar,** use the **Count Up** activity with the whole group.

Purpose Count Up reinforces skills that lay a foundation for mental addition.

Warm-Up Card 3

Monitoring Student Progress

If . . . a child cannot predict the next number up, **Then . . .** ask a classmate to help prompt him or her.

 Engage 30

Concept Building REASONING

Mouse in the Cookie Jar

"Today we will continue to play **Mouse in the Cookie Jar.**"

Follow the instructions on the Activity Card to play **Mouse in the Cookie Jar.** As children play, ask questions about what is happening in the activity.

Purpose Mouse in the Cookie Jar helps children recognize that when 1 object is taken away, the size of the set is reduced by 1.

Activity Card 12

MathTools Use the Set Tool to demonstrate and explore subtraction.

Monitoring Student Progress

If . . . children have trouble predicting whose cookie was taken,	**Then . . .** remind them of how many cookies they put into the jar.

 Building Blocks For additional practice adding and subtracting hidden objects, children should complete **Building Blocks** Pizza Pizzazz 4.

3 Reflect 10

Extended Response REASONING

Ask the following questions:

- **What does this activity remind you of?** Accept all reasonable answers.
- **What do you like about this activity?** Accept all reasonable answers.

4 Assess

Informal Assessment

Use the Student Assessment Record, **Assessment,** page 100, to record informal observations.

COMPUTING	REASONING
Count Up	**Mouse in the Cookie Jar**
Did the child	Did the child
❏ respond accurately?	❏ provide a clear explanation?
❏ respond quickly?	❏ communicate reasons and strategies?
❏ respond with confidence?	❏ choose appropriate strategies?
❏ self-correct?	❏ argue logically?

Lesson 4

Objective

Children count and compare to solve problems.

Program Materials

Engage
Counters, 5 per child

Additional Materials

Engage
- Wide-mouth container
- Small cloth

Access Vocabulary

pay attention To watch and listen carefully; don't get distracted

pretend to sleep To act like you are sleeping when you are not

wake up To stop sleeping

Creating Context

English Learners are often exposed to academic vocabulary in school. However they are less frequently exposed to the English vocabulary of things they do at home, especially if no one in the home speaks English. For this reason, it may be helpful to talk through the sleep routine for this game. Role-play sleeping and waking up to help make these actions more comprehensible.

---

1 Warm Up 5

Skill Building COMPUTING

Blastoff!

Before beginning **Mouse in the Cookie Jar,** use the **Blastoff!** activity with the whole group.

Purpose Blastoff! provides children with an opportunity to review counting down from 10 to 1.

Warm-Up Card 4

Monitoring Student Progress

If . . . children have trouble showing the countdown with their fingers,

Then . . . remind them to use their thumbs to help hold down their fingers. Model finger displays throughout the day.

2 Engage 30

Concept Building ENGAGING

Mouse in the Cookie Jar

"Today the mouse might take more than one cookie, so pay attention."

Follow the instructions on the Activity Card to play **Mouse in the Cookie Jar** Challenge variation. As children play, ask questions about what is happening in the activity.

Activity Card 12

Purpose Mouse in the Cookie Jar teaches children to count a set of objects and compare amounts.

Monitoring Student Progress

If . . . children have trouble recognizing when two Counters are missing,

Then . . . play the original version.

Building **B**locks For additional practice adding and subtracting hidden objects, children should complete **Building Blocks** Pizza Pizzazz 4.

3 Reflect   10

Extended Response REASONING

Ask the following questions:

■ **How would you explain Mouse in the Cookie Jar to someone who has never done it before?** Possible answer: The Counters are cookies. We count them and put them into a jar. Then we pretend to sleep, and a mouse takes a cookie. When we wake up, we have to figure out whose cookie is gone.

■ **What do you notice when you play?** Accept all reasonable answers.

4 Assess

Informal Assessment

Use the Student Assessment Record, **Assessment,** page 100, to record informal observations.

COMPUTING	ENGAGING
Blastoff!	**Mouse in the Cookie Jar**
Did the child	Did the child
❑ respond accurately?	❑ pay attention to the contributions of others?
❑ respond quickly?	
❑ respond with confidence?	
❑ self-correct?	❑ contribute information and ideas?
	❑ improve on a strategy?
	❑ reflect on and check accuracy of work?

Lesson 5
Review

Objective
Children review and reinforce skills learned in this week and in previous weeks.

Program Materials
Engage
See Activity Cards for materials.

Additional Materials
Engage
See Activity Cards for materials.

Creating Context
Discuss instances when English Learners might want to count down or count up. Have them draw pictures and name examples of when counting backward and forward is useful.

A WHALE OF A GOOD JOB!

---

1 Warm Up 5

Skill Practice COMPUTING
Blastoff!
Before beginning the **Free-Choice** activity, use the **Blastoff!** activity with the whole group.

Purpose **Blastoff!** helps children reinforce their knowledge of the reverse counting sequence.

Warm-Up Card 4

Monitoring Student Progress

If . . . children have trouble counting down, **Then . . .** have them repeat short number sequences.

2 Engage 20

Concept Building APPLYING
Free-Choice Activity
For the last day of the week, allow children to choose an activity from the previous weeks. Some activities they may choose include the following:
- Line Land: **Step-by-Step Number Line**
- Sky Land: **Going Down**
- Object Land: **Minus Mouse**

Make a note of the activities children select. Do they prefer easy or challenging activities? If you believe your children would benefit from extra practice on specific skills, choose an activity for them.

3 Reflect 10

Extended Response REASONING

Ask the following questions:

■ **What did you like about playing** Mouse in the Cookie Jar?

■ **Was there anything about playing this game you didn't like?**

■ **Did this game help you do something you couldn't do before? What did it help you do?**

■ **What was easy when you were playing** Mouse in the Cookie Jar?

■ **What was hard when you were playing** Mouse in the Cookie Jar?

■ **What do you remember most about this game?**

■ **What was your favorite part?**

4 Assess 10

A Gather Evidence
Formal Assessment

Have children complete the weekly test, *Assessment,* page 67. Record formal assessment scores on the Student Assessment Record, *Assessment,* page 100.

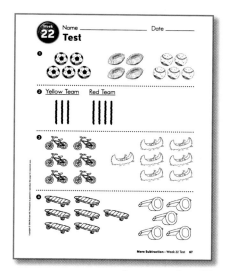

Assessment, p. 67

B Summarize Findings

Review the Student Assessment Records. Determine whether children have Minimal, Basic, or Secure understanding of the concepts presented in Week 22.

C Differentiate Instruction

Based on your observations, use these teaching strategies next week to follow up.

Minimal Understanding

• Repeat the Warm-Up and Engage activities to develop concepts of counting objects.

• Use **Building Blocks** computer activities beginning with Pizza Pizzazz 2 to develop and reinforce numeration concepts.

Basic Understanding

• Repeat Engage activities in subsequent weeks to reinforce basic counting concepts.

• Use **Building Blocks** computer activities beginning with Pizza Pizzazz 4 to reinforce this week's concepts.

Secure Understanding

• Use Challenge variations of **Mouse in the Cookie Jar.**

• Use computer activities to extend children's understanding of the Week 22 concepts.

Depicting Numbers

Week at a Glance

This week, children begin **Number Worlds,** Week 23, and continue to explore Picture Land.

Background

In Picture Land, numbers are represented as a set of dots. When a child is asked, "What number do you have?" he or she might count each dot to determine set size or identify the amount by recognizing the pattern that marks that amount. Thinking about numbers as a product of counting and as a set that has a particular value will become intertwined in the child's sense of number.

How Children Learn

As they begin this week's activities, children should have developed a vocabulary that includes numbers and terms for comparison.

By the end of the week, children should be recognizing numerals more readily.

Skills Focus

- Identify numerals
- Compute quantity
- Use numbers in conversation
- Verify equivalence by counting

Teaching for Understanding

As children engage in these activities, they gain additional experience with abstract representations of numbers and build connections among the different representations.

Observe closely while evaluating the Engage activities assigned for this week.

- Are children making accurate calculations and comparisons?
- Are children identifying quantities and numerals correctly?
- Are children using math vocabulary properly?

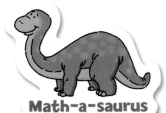

Math at Home

Give one copy of the Letter to Home, page A23, to each child. Complete the activity in class, and then encourage children to share it with their caregivers.

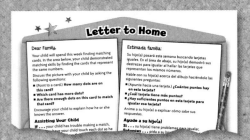

Letter to Home, Teacher Edition, p. A23

Math Vocabulary

equal Having the same amount

match To group together things that are alike

English Learners

SPANISH COGNATES

English	Spanish
numerals	números
cards	cartas
pair	par
equal	igual

ALTERNATE VOCABULARY

numeral The symbol for a number

pair Two cards that show the same number or amount

Week 23 Planner   Depicting Numbers

PACING	LESSON	LEARNING GOALS	MATERIALS	TECHNOLOGY
DAY 1	**Warm Up** **Picture Land:** Name That Numeral*	Children are introduced to unfamiliar numerals.	**Program Materials** • Large Number Cards (1–5) • Large Number Cards (1–12) for the Challenge	**B**uilding **B**locks Space Race
	Engage **Picture Land:** Go Fish*	Children compute quantity.	**Program Materials** • Small Dot Set Cards (1–5), 2 sets per pair • Small Dot Set Cards (1–10), 2 sets per pair for the Challenge • Dot Cube	**ⓔ MathTools** Money and Coins
DAY 2	**Warm Up** **Picture Land:** Name That Numeral*	Children build numeral recognition.	**Program Materials** • Large Number Cards (1–5) • Large Number Cards (1–12) for the Challenge	**B**uilding **B**locks Space Race
	Engage **Picture Land:** Go Fish*	Children use numbers in conversation.	**Program Materials** • Small Dot Set Cards (1–5), 2 sets per pair • Small Dot Set Cards (1–10), 2 sets per pair for the Challenge • Dot Cube	
DAY 3	**Warm Up** **Picture Land:** Name That Numeral Variation*	Children work in teams to identify numbers.	**Program Materials** • Large Number Cards (1–5) • Large Number Cards (1–12) for the Challenge	**B**uilding **B**locks Space Race
	Engage **Picture Land:** Go Fish Variation*	Children gain additional experience with number representations.	**Program Materials** • Small Dot Set Cards (1–5), 2 sets per pair • Small Dot Set Cards (1–10), 2 sets per pair for the Challenge • Dot Cube	
DAY 4	**Warm Up** **Picture Land:** Name That Numeral Variation*	Children identify numerals.	**Program Materials** • Large Number Cards (1–5) • Large Number Cards (1–12) for the Challenge	**B**uilding **B**locks Space Race
	Engage **Picture Land:** Go Fish Variation*	Children build connections between different number representations.	**Program Materials** • Small Dot Set Cards (1–5), 2 sets per pair • Small Dot Set Cards (1–10), 2 sets per pair for the Challenge • Dot Cube	
DAY 5	**Warm Up** **Picture Land:** Name That Numeral*	Children learn to recognize numerals.	**Program Materials** • Large Number Cards (1–5) • Large Number Cards (1–12) for the Challenge	**B**uilding **B**locks Review previous activities
	Review Free-Choice activity	Children review and reinforce skills and concepts learned in this and previous weeks.	Materials will be selected from those used in previous weeks.	

* Includes Challenge Variations

Lesson 1

Objective
Children identify and compute quantity.

Program Materials

Warm Up
- Large Number Cards (1–5)
- Large Number Cards (1–12) for the Challenge

Engage
- Small Dot Set Cards (1–5), 2 sets
- Dot Cube
- Small Dot Set Cards (1–10), 2 sets for the Challenge

Additional Materials
No additional materials needed

Access Vocabulary
go fish To try to catch fish; in this game, to try to get the card you need from the pile
match To pair up two or more objects or numbers that are the same or go together

Creating Context
Help English Learners distinguish between *numeral* and *number*. Explain that although they are related, the word *numeral* describes the number symbol, and the word *number* refers to the quantity of a set.

Pretty Good!

1 Warm Up 5

Skill Building COMPUTING
Name That Numeral
Before beginning **Go Fish**, use the **Name That Numeral** activity with the whole group.

Purpose **Name That Numeral** introduces children to unfamiliar numerals.

Warm-Up Card 5

Monitoring Student Progress
If . . . children do not recognize a numeral,
Then . . . call on an individual child to help identify the number.

2 Engage 30

Concept Building UNDERSTANDING
Go Fish
"Today we are going to do a matching activity."
Follow the instructions on the Activity Card to play **Go Fish**. As children play, ask questions about what is happening in the activity.

Purpose **Go Fish** gives children practice identifying or computing quantity on a Dot Set Card.

Activity Card 18

e MathTools Use the Money and Coins Tool to demonstrate and explore counting.

Monitoring Student Progress

If . . . children have trouble making matches,

Then . . . remind them to count the dots on both cards and to look at the patterns.

Teacher's Note Encourage children to use this week's vocabulary words as they engage in the activities, discuss math concepts, and make predictions.

Building Blocks For additional practice with numbers, children should complete **Building Blocks** Space Race.

3 Reflect 10

Extended Response REASONING

Ask the following questions:

- **How do you know when you have a match?** Accept all reasonable answers.
- **Are there other ways to tell? Explain your idea.** Accept all reasonable answers.

4 Assess

Informal Assessment

Use the Student Assessment Record, **Assessment,** page 100, to record informal observations.

COMPUTING	UNDERSTANDING
Name That Numeral	**Go Fish**
Did the child	Did the child
❑ respond accurately?	❑ make important observations?
❑ respond quickly?	❑ extend or generalize learning?
❑ respond with confidence?	❑ provide insightful answers?
❑ self-correct?	❑ pose insightful questions?

Lesson 2

Objective

Children use math vocabulary.

Program Materials

Warm Up
- Large Number Cards (1–5)
- Large Number Cards (1–12) for the Challenge

Engage
- Small Dot Set Cards (1–5), 2 sets
- Dot Cube
- Small Dot Set Cards (1–10), 2 sets for the Challenge

Additional Materials

No additional materials needed

Access Vocabulary

take turns To wait for all the other players to go before you go again
pile A stack of cards; objects placed on top of one another

Creating Context

Ask English Learners if any of them have ever gone fishing. Ask them to role-play what you have to do to catch a fish. As they dramatize, describe the steps, repeating key words and phrases. Make the connection to the game **Go Fish.** If children need additional practice with basic vocabulary, ask children to help create a variation of **Go Fish,** perhaps with colors or patterns.

1 Warm Up 5

Concept Building COMPUTING
Name That Numeral
Before beginning **Go Fish,** use the **Name That Numeral** activity with the whole group.

Purpose Name That Numeral helps children build numeral recognition.

Warm-Up Card 5

Monitoring Student Progress

| If . . . children have trouble identifying numerals, | Then . . . discuss the shape of the numerals to help children find ways to identify numerals. |

2 Engage 30

Concept Building ENGAGING
Go Fish
"Today we will continue to play **Go Fish.**"

Follow the instructions on the Activity Card to play **Go Fish.** As children play, ask questions about what is happening in the activity.

Purpose Go Fish gives children the opportunity to use numbers in conversation so that they have the information needed to solve problems.

Activity Card 18

Monitoring Student Progress

If . . . children are making mistakes during the activity,

Then . . . play a few rounds in which they answer your questions aloud so that you can identify any mistakes in their reasoning.

 Building Blocks For additional practice with numbers, children should complete **Building Blocks** Space Race.

3 Reflect   10

Extended Response REASONING

Ask the following questions:

- **How would you describe** Go Fish **to someone who has never played?** Possible answer: You take turns asking each other for cards that match your cards. If the person you ask doesn't have the card you ask for, you "go fish" in the pile. The person with the most pairs when all the cards are matched wins.

- **What does this remind you of?** Accept all reasonable answers.

4 Assess

Informal Assessment

Use the Student Assessment Record, **Assessment,** page 100, to record informal observations.

COMPUTING	ENGAGING
Name That Numeral	**Go Fish**
Did the child	Did the child
❑ respond accurately?	❑ pay attention to the contributions of others?
❑ respond quickly?	
❑ respond with confidence?	❑ contribute information and ideas?
❑ self-correct?	
	❑ improve on a strategy?
	❑ reflect on and check accuracy of work?

Week 23

Lesson 3

Objective

Children develop recognition of number representations such as numerals.

Program Materials

Warm Up
- Large Number Cards (1–5)
- Large Number Cards (1–12) for the Challenge

Engage
- Small Dot Set Cards (1–5), 2 sets
- Dot Cube
- Small Dot Set Cards (1–10), 2 sets for the Challenge

Additional Materials

No additional materials needed

Access Vocabulary

facedown Cards on a table with the dots or numerals down (not showing)

answer in your head To think of the answer without saying it aloud

Creating Context

Sentence frames are useful tools to help English Learners structure a response to a question. In this lesson, children must request cards from one another. Provide the frame, "Do you have a _____?" Then model the answer as a simple yes/no or suggest, "Yes, I do," or, "Yes, I have a _____."

---

1 Warm Up 5

Skill Building COMPUTING

Name That Numeral

Before beginning **Go Fish**, use **Name That Numeral** Variation: **Team Name That Numeral** with the whole group.

Purpose **Team Name That Numeral** has children work in teams to identify numbers.

Warm-Up Card 5

Monitoring Student Progress

If . . . one team is having more trouble than the other team is,

Then . . . mix up the teams for the next round.

2 Engage 30

Concept Building REASONING

Go Fish

"Today we will play **Go Fish** using Number Cards."

Follow the instructions on the Activity Card to play **Go Fish** Variation: **Number Go Fish.** As children play, ask questions about what is happening in the activity.

Purpose **Number Go Fish** gives children an opportunity to gain additional experience with number representations.

Activity Card 18

Monitoring Student Progress

If . . . children are not able to recognize numerals,

Then . . . have them play with Dot Set Cards.

 Building Blocks For additional practice with numbers, children should complete **Building Blocks** Space Race.

3 Reflect 10

Extended Response REASONING

Ask the following questions:

- **Who won the game?** Allow each pair of children to answer.

- **How did you figure out who the winner was?** Possible answers: We counted the cards each player had, and the player with the most cards was the winner.

- **How did the winner get more cards than the other player? What happened when he or she was playing that helped him or her get more cards?** Possible answers: When he was asking for a card, he got it from the other player. She found it when she fished.

4 Assess

Informal Assessment

Use the Student Assessment Record, **Assessment,** page 100, to record informal observations.

COMPUTING	REASONING
Name That Numeral	**Go Fish**
Did the child	Did the child
❏ respond accurately?	❏ provide a clear explanation?
❏ respond quickly?	❏ communicate reasons and strategies?
❏ respond with confidence?	❏ choose appropriate strategies?
❏ self-correct?	❏ argue logically?

Bonus!

Week 23

Lesson 4

Objective

Children build connections between different number representations.

Program Materials

Warm Up
- Large Number Cards (1–5)
- Large Number Cards (1–12) for the Challenge

Engage
- Small Dot Set Cards (1–5), 2 sets
- Dot Cube
- Small Dot Set Cards (1–10), 2 sets for the Challenge

Additional Materials

No additional materials needed

Access Vocabulary

shuffle To mix up cards so they are not in order

set of cards A group of cards that has all the counting numbers

Creating Context

Some children may not have experience with card games like **Go Fish.** It may be helpful to review some of the routines and rules that are common in card games such as taking turns, deciding who goes first, keeping the cards out of the other players' sight, and being a good sport.

1 Warm Up   5

Skill Building COMPUTING

Name That Numeral

Before beginning **Go Fish,** use **Name That Numeral** Variation: **Team Name That Numeral** with the whole group.

Purpose Team Name That Numeral helps children learn to identify numerals.

Warm-Up Card 5

Monitoring Student Progress

If . . . neither team identifies a numeral correctly,

Then . . . tell them what the number is and put the card on the bottom of the pile.

2 Engage   30

Concept Building ENGAGING

Go Fish

"Today we will continue to pair up to find numbers that match."

Follow the instructions on the Activity Card to play **Go Fish.** As children play, ask questions about what is happening in the activity.

Purpose Go Fish helps children build connections between dot-set number representations and numerals.

Activity Card 18

Monitoring Student Progress

If . . . children are ready,

Then . . . have them play with the full deck of cards.

 Building Blocks For additional practice with numbers, children should complete **Building Blocks** Space Race.

3 Reflect 10

Extended Response REASONING

Ask the following questions:

- **What was the highest number you matched?** Probable answer: 5

- **How do you know that that is the highest number?** Possible answer: It is higher up when we count.

4 Assess

Informal Assessment

Use the Student Assessment Record, **Assessment,** page 100, to record informal observations.

COMPUTING	ENGAGING
Name That Numeral	**Go Fish**
Did the child	Did the child
❏ respond accurately?	❏ pay attention to the contributions of others?
❏ respond quickly?	
❏ respond with confidence?	❏ contribute information and ideas?
❏ self-correct?	
	❏ improve on a strategy?
	❏ reflect on and check accuracy of work?

Lesson 5

Review

Objective

Children review and reinforce skills and concepts learned in this and previous weeks.

Program Materials

Warm Up
Large Number Cards (1–12)

Engage
See Activity Cards for materials.

Additional Materials

Engage
See Activity Cards for materials.

Creating Context

English Learners benefit from repetition of new vocabulary and sentence patterns. Help them create their own cards to take home so they can practice **Go Fish** with family and friends. Practice in any language is helpful to reinforce mathematics concepts. Practice in English also reinforces new vocabulary.

1 Warm Up

   5

Skill Practice COMPUTING

Name That Numeral

Before beginning the **Free-Choice** activity, use the **Name That Numeral** Challenge activity with the whole group.

Purpose **Name That Numeral** helps children learn to recognize numerals.

Warm-Up Card 5

Monitoring Student Progress

If . . . children are confused by double-digit numbers,

Then . . . wait and introduce these numbers later.

2 Engage

   20

Concept Building APPLYING

Free-Choice Activity

For the last day of the week, allow children to choose an activity from the previous weeks. Some activities they may choose are the following:

- Object Land: **Matching Colors and Shapes**
- Picture Land: **Dog and Bone**
- Picture Land: **Concentration**

Make a note of the activities children select. Do they prefer easy or challenging activities? If you believe your children would benefit from extra practice on specific skills, choose an activity for them.

3 Reflect 10

Extended Response REASONING

Ask the following questions:

- **What did you like about playing** Go Fish?
- **Was there anything about playing this game you didn't like?**
- **Did this game help you do something you couldn't do before? What did it help you do?**
- **What was easy when you were playing** Go Fish?
- **What was hard when you were playing** Go Fish?
- **What do you remember most about this game?**
- **What was your favorite part?**

4 Assess 10

A Gather Evidence

Formal Assessment

Have children complete the weekly test, *Assessment,* page 69. Record formal assessment scores on the Student Assessment Record, *Assessment,* page 100.

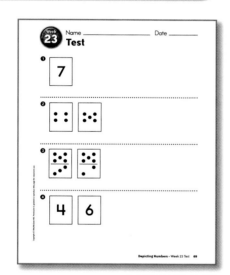

Assessment, p. 69

B Summarize Findings

Review the Student Assessment Records. Determine whether children have Minimal, Basic, or Secure understanding of the concepts presented in Week 23.

C Differentiate Instruction

Based on your observations, use these teaching strategies next week to follow up.

Minimal Understanding

- Repeat the Warm-Up and Engage activities to develop concepts of counting objects.
- Use **Building Blocks** computer activities beginning with Space Race to develop and reinforce numeration concepts.

Basic Understanding

- Repeat Engage activities in subsequent weeks to reinforce basic counting concepts.
- Use **Building Blocks** computer activities beginning with Space Race to reinforce this week's concepts.

Secure Understanding

- Use Challenge variations of **Go Fish.**
- Use computer activities to extend children's understanding of the Week 23 concepts.

Week 24 Comparing Quantities

Week at a Glance

This week, children begin **Number Worlds,** Week 24, and continue to explore Picture Land.

Background

In Picture Land, numbers are represented as a set of dots. When a child is asked, "What number do you have?" he or she might count each dot to determine the size of the set, or identify the amount by recognizing the pattern that marks that amount. Eventually, these ways of thinking about numbers—as a product of counting and as a set that has a particular value—will become intertwined in the child's sense of number.

How Children Learn

As they begin this week's activities, children should understand what dot sets represent and should be somewhat familiar with numerals.

By the end of the week, children should have increased their comparison skills and their numeral recognition.

Skills Focus

- Identify numerals
- Compute and compare quantities
- Check answers

Teaching for Understanding

As children engage in these activities, they continue to build connections between different representations of numbers. The comparisons made in these activities help reinforce children's understanding that numbers that come later in the counting sequence are associated with larger quantities, more complex patterns, and higher numerals than numbers that come earlier in the sequence.

Observe closely while evaluating the Engage activities assigned for this week.

- Are children correctly identifying which card represents a higher quantity?
- Are children able to verify that their comparisons are correct?
- Are children identifying numerals and associating them with the correct quantities?

Math at Home

Give one copy of the Letter to Home, page A24, to each child. Complete the activity in class, and then encourage children to share it with their caregivers.

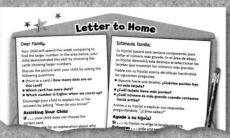

Letter to Home, Teacher Edition, p. A24

Math Vocabulary

faceup Cards on a table with the dots or numerals up (showing)

facedown Cards on a table with the dots or numerals down (not showing)

larger Bigger

English Learners

SPANISH COGNATES

English	Spanish
Bravo!	¡Bravo!
practice	practicar
difference	diferencia
number	número

ALTERNATE VOCABULARY

take turns To wait for all the other players to go before you go again

team A group of people who work together

PACING	LESSON	LEARNING GOALS	MATERIALS	TECHNOLOGY
DAY 1	**Warm Up** **Picture Land:** Name That Numeral*	Children practice identifying numerals.	**Program Materials** Large Number Cards	**Building Blocks** Number Compare 1
	Engage **Picture Land:** Bravo!	Children identify or compute quantities and then compare the quantities.	**Program Materials** Small Dot Set Cards, 2 sets per pair of children	
DAY 2	**Warm Up** **Picture Land:** Name That Numeral Variation*	Children compete in teams as they identify numerals.	**Program Materials** Large Number Cards	**Building Blocks** Number Compare 1
	Engage **Picture Land:** Bravo!	Children compare quantities and check their comparisons.	**Program Materials** Small Dot Set Cards, 2 sets per pair of children	**MathTools** Set Tool
DAY 3	**Warm Up** **Picture Land:** Name That Numeral *	Children gain experience recognizing numerals.	**Program Materials** Large Number Cards	**Building Blocks** Number Compare 1
	Engage **Picture Land:** Bravo! Variation 1	Children compare numerals to determine which one is greater.	**Program Materials** Small Number Cards, 2 sets per pair of children	
DAY 4	**Warm Up** **Picture Land:** Name That Numeral Variation*	Children determine which team can name more numerals.	**Program Materials** Large Number Cards	**Building Blocks** Number Compare 1
	Engage **Picture Land:** Bravo! Variation 2	Children compare quantities to determine who has more dots.	**Program Materials** • Dot Cubes, 1 per child • Counters, 10 per child **Additional Materials** Small containers, 1 per child	
DAY 5	**Warm Up** **Picture Land:** Name That Numeral *	Children build number recognition.	**Program Materials** Large Number Cards	**Building Blocks** Review previous activities
	Review Free-Choice activity	Children review and reinforce concepts and skills from this and other weeks.	Materials will be selected from those used in previous weeks.	

* Includes Challenge Variations

Comparing Quantities

Lesson 1

Objective

Children identify or compute quantities and then compare the quantities.

Program Materials

Warm Up
Large Number Cards

Engage
Small Dot Set Cards, 2 sets per pair of children

Additional Materials

No additional materials needed

Access Vocabulary

faceup Cards on a table with the dots or numerals up (showing)

pile A stack of cards; objects placed on top of one another

Creating Context

It may be helpful to teach comparative and superlative adjectives to English Learners so they can make comparisons. Use groups of objects to show *big, bigger, biggest* and *small, smaller, smallest,* as well as other comparative/superlative adjectives.

Easy as . . .
1 - 2 - 3

1 Warm Up   5

Skill Building   COMPUTING

Name That Numeral

Before beginning **Bravo!** use the **Name That Numeral** activity with the whole group.

Purpose **Name That Numeral** gives children practice identifying numerals.

Warm-Up Card 5

Monitoring Student Progress

If . . . children have trouble identifying a numeral,

Then . . . discuss the lines and shapes that help them recognize the numeral.

2 Engage   30

Concept Building   UNDERSTANDING

Bravo!

"Today you will pair up and play a card game."

Follow the instructions on the Activity Card to play **Bravo!** As children play, ask questions about what is happening in the activity.

Purpose **Bravo!** teaches children to identify or compute quantities and then compare the quantities.

Activity Card 19

Monitoring Student Progress

If . . . children are having trouble determining which card has a higher amount of dots,

Then . . . encourage them to count the dots, touching each dot as they count.

Teacher's Note Encourage children to use this week's vocabulary words as they engage in the activities, discuss math concepts, and make predictions.

Building Blocks For additional practice choosing the larger number, children should complete **Building Blocks** Number Compare 1.

3 Reflect   10

Extended Response   REASONING

Ask the following questions:

- **How did you know who got to keep the cards when you played?** Possible answer: The person who turned over the highest card kept both cards.

- **How can you tell which card has more dots?** Possible answer: We can count the dots.

4 Assess

Informal Assessment

Use the Student Assessment Record, **Assessment,** page 100, to record informal observations.

COMPUTING	UNDERSTANDING
Name That Numeral	**Bravo!**
Did the child	Did the child
❏ respond accurately?	❏ make important observations?
❏ respond quickly?	❏ extend or generalize learning?
❏ respond with confidence?	❏ provide insightful answers?
❏ self-correct?	❏ pose insightful questions?

Math-a-saurus

Lesson 2

Objective

Children compare quantities and check their comparisons for accuracy.

Program Materials

Warm Up
Large Number Cards

Engage
Small Dot Set Cards, 2 sets per pair of children

Additional Materials
No additional materials needed

Access Vocabulary

compare To look at two things to see how they are similar and how they are different
Bravo! Hooray! Yippee!—a cheer of excitement

Creating Context

Help English Learners expand their vocabulary by discussing the ways we can describe an amount that is larger. Make a list of words such as *bigger, greater,* and *more than.* Refer to the list when making comparisons.

---

1 Warm Up

5

Concept Building COMPUTING

Name That Numeral

Before beginning **Bravo!** use **Name That Numeral** Variation: **Team Name That Numeral** with the whole group.

Purpose **Team Name That Numeral** has children compete in teams as they identify numerals.

Warm-Up Card 5

Monitoring Student Progress

| If . . . children shout out the answers instead of listening quietly, | Then . . . remind them to let everyone have a turn and send them to the back of the line. |

---

2 Engage

30

Concept Building ENGAGING

Bravo!

"Today we will see who can get more cards while playing **Bravo!**"

Follow the instructions on the Activity Card to play **Bravo!** As children play, ask questions about what is happening in the activity.

Purpose **Bravo!** helps children learn to compare quantities and check their comparisons.

Activity Card 19

MathTools Use the Set Tool to demonstrate and explore comparing amounts.

Monitoring Student Progress

| If . . . one child is making all the assessments, | Then . . . require that the children take turns deciding which amount is larger. |

Building Blocks For additional practice choosing the larger number, children should complete *Building Blocks* Number Compare 1.

3 Reflect 10 ◐

Extended Response REASONING

Ask the following questions:

- **How can you tell when you need to say "Bravo"?** Possible answer: when the cards have the same amount

- **Could you do it another way? Explain how.** Possible answer: Look to see if the patterns are the same.

4 Assess

Informal Assessment

Use the Student Assessment Record, **Assessment,** page 100, to record informal observations.

COMPUTING	ENGAGING
Name That Numeral	**Bravo!**
Did the child	Did the child
❏ respond accurately?	❏ pay attention to the contributions of others?
❏ respond quickly?	
❏ respond with confidence?	❏ contribute information and ideas?
❏ self-correct?	
	❏ improve on a strategy?
	❏ reflect on and check accuracy of work?

Lesson 3

Objective

Children compare numerals to determine which is greater.

Program Materials

Warm Up
Large Number Cards

Engage
Small Number Cards, 2 sets per pair of children

Additional Materials

No additional materials needed

Access Vocabulary

flip To turn over quickly

Creating Context

English Learners often benefit from discussing concepts in their primary language. By checking with peers, children can deepen conceptual understanding. Encourage all children to discuss what they are doing in class, allowing discussion in English Learners' primary languages.

you hit the SPOT!

---

1 Warm Up   5

Skill Building   COMPUTING

Name That Numeral
Before beginning **Bravo!** use the **Name That Numeral** activity with the whole group.

Purpose   Name That Numeral allows children to gain experience recognizing numerals.

Warm-Up Card 5

Monitoring Student Progress

If . . . children are not ready to work with higher numerals,

Then . . . wait and introduce them slowly as children become ready.

---

2 Engage   30

Concept Building   REASONING

Bravo!
"Today we will play **Bravo!** using Number Cards."

Follow the instructions on the Activity Card to play **Bravo! Variation 1: Number Bravo!** As children play, ask questions about what is happening in the activity.

Purpose   Number Bravo! helps children compare numerals to determine which is greater.

Activity Card 19

Monitoring Student Progress

If . . . children have trouble naming the numerals or understanding the magnitude of each number,

Then . . . review numeral recognition skills with them throughout the day. Make sure they have plenty of opportunities to associate each numeral with the quantity and number name it stands for.

 Building Blocks For additional practice choosing the larger number, children should complete **Building Blocks** Number Compare 1.

3 Reflect 10

Extended Response REASONING

Ask the following questions:

- **How is this version of** Bravo! **different from the other version you played?** Possible answer: In this version, we play with Number Cards instead of Dot Cards.

- **Is the version with Number Cards harder than the version with Dot Cards? Why?** Accept all reasonable answers.

- **How do you know which number is bigger?** Possible answer: The bigger number comes after the smaller number when counting up.

4 Assess

Informal Assessment

Use the Student Assessment Record, **Assessment,** page 100, to record informal observations.

COMPUTING	REASONING
Name That Numeral	**Bravo!**
Did the child	Did the child
❏ respond accurately?	❏ provide a clear explanation?
❏ respond quickly?	
❏ respond with confidence?	❏ communicate reasons and strategies?
❏ self-correct?	❏ choose appropriate strategies?
	❏ argue logically?

Lesson 4

Objective

Children compare quantities to determine who has more dots.

Program Materials

Warm Up

Large Number Cards

Engage

- Dot Cubes, 1 per child
- Counters, 10 per child

Additional Materials

Engage

Small cups or containers, 1 per child

Access Vocabulary

remind To make you think of

pair Two cards that show the same number

Creating Context

Show two different cards from **Bravo!** Ask children to describe how these numbers are different and similar. Make sure they include descriptions of the size of the numbers as well as the differences in the shape of the numerals. Before you begin, review the meaning of the words *different* and *similar*.

---

1 Warm Up 5

Skill Building COMPUTING

Name That Numeral

Before beginning **Bravo!** use **Name That Numeral** Variation: **Team Name That Numeral** with the whole group.

Purpose **Team Name That Numeral** challenges children to see which team can name more numerals.

Warm-Up Card 5

Monitoring Student Progress

If . . . some children are struggling to identify numbers,

Then . . . have them work with a classmate to find ways to differentiate the numerals.

---

2 Engage 30

Concept Building ENGAGING

Bravo!

"Today we will play **Bravo!** using Dot Cubes."

Follow the instructions on the Activity Card to play **Bravo!** Variation 2: **Dot Cube Bravo!** As children play, ask questions about what is happening in the activity.

Purpose **Dot Cube Bravo!** helps children compare quantities to determine who has more dots.

Activity Card 19

Monitoring Student Progress

If . . . children disagree about which Cube has a higher number,

Then . . . encourage them to count the dots.

 Building Blocks For additional practice choosing the larger number, children should complete **Building Blocks** Number Compare 1.

3 Reflect 10

Extended Response REASONING

Ask the following questions:

- **How can you check to see who gets to put a Counter into their cup?** Possible answer: Count the dots.

- **What does Bravo! remind you of?** Accept all reasonable answers.

4 Assess

Informal Assessment

Use the Student Assessment Record, **Assessment,** page 100, to record informal observations.

COMPUTING	ENGAGING
Name That Numeral	**Bravo!**
Did the child	Did the child
❏ respond accurately?	❏ pay attention to the contributions of others?
❏ respond quickly?	
❏ respond with confidence?	❏ contribute information and ideas?
❏ self-correct?	
	❏ improve on a strategy?
	❏ reflect on and check accuracy of work?

Lesson 5
Review

Objective

Children review and reinforce concepts and skills from this and other weeks.

Program Materials

Warm Up
Large Number Cards

Engage
See Activity Cards for materials.

Additional Materials

Engage
See Activity Cards for materials.

Creating Context

English Learners need practice using the new vocabulary. Pair children and ask them to draw two sets of objects that show the *same, larger,* and *smaller* amounts.

✓ Cumulative Assessment

After Lesson 5 is finished, children should complete the Cumulative Assessment, **Assessment,** pages 94–95. Using the key on **Assessment,** page 93, identify incorrect responses. Reteach and review the suggested activities to reinforce concept understanding.

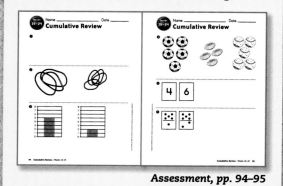

Assessment, pp. 94–95

---

1 Warm Up   5

Skill Practice   COMPUTING

Name That Numeral

Before beginning the **Free-Choice** activity, use the **Name That Numeral** activity with the whole group.

Purpose   Name That Numeral helps children build number recognition.

Warm-Up Card 5

Monitoring Student Progress

If . . . a child is unsure about a numeral,   **Then . . .** ask another child to name it and ask the group to describe how the number looks

2 Engage   20

Concept Building   APPLYING

Free-Choice Activity

For the last day of the week, allow children to choose an activity from the previous weeks. Some activities they may choose are the following:

- Line Land: **The Number Line Game**
- Circle Land: **Skating Party**
- Object Land: **Plus Pup**

Make a note of the activities children select. Do they prefer easy or challenging activities? If you believe your children would benefit from extra practice on specific skills, choose an activity for them.

Reflect   10

Extended Response   REASONING

Ask the following questions:

- **What did you like about playing** Bravo?
- **Was there anything about playing this game you didn't like?**
- **Did this game help you do something you couldn't do before? What did it help you do?**
- **What was easy when you were playing** Bravo?
- **What was hard when you were playing** Bravo?
- **What do you remember most about this game?**
- **What was your favorite part?**

Assess   10

A Gather Evidence

Formal Assessment

Have children complete the weekly test, **Assessment,** page 71. Record formal assessment scores on the Student Assessment Record, **Assessment,** page 100.

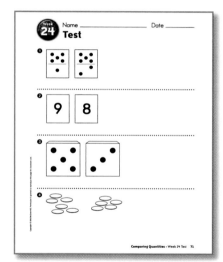

Assessment, p. 71

B Summarize Findings

Review the Student Assessment Records. Determine whether children have Minimal, Basic, or Secure understanding of the concepts presented in Week 24.

C Differentiate Instruction

Based on your observations, use these teaching strategies next week to follow up.

Minimal Understanding

- Repeat the Warm-Up and Engage activities to develop concepts of counting objects.
- Use **Building Blocks** computer activities beginning with Number Compare 1 to develop and reinforce numeration concepts.

Basic Understanding

- Repeat Engage activities in subsequent weeks to reinforce basic counting concepts.
- Use **Building Blocks** computer activities beginning with Number Compare 1 to reinforce this week's concepts.

Secure Understanding

- Use Challenge variations of this week's activities.
- Use computer activities to extend children's understanding of the Week 24 concepts.

Making Comparisons

Week at a Glance

This week, children begin *Number Worlds,* Week 25, and continue to explore Object Land and Line Land.

Background

In Object Land, numbers are represented as groups of objects. This is the first way numbers were represented historically, and this is the first way children naturally learn about numbers. In Object Land, children work with real, tangible objects and with pictures of objects. In Line Land, children explore number lines.

How Children Learn

As they begin this week's activities, children should be able to count collections to compare the collections.

By the end of the week, children should be developing their ability to compare sets accurately by counting, even when the objects that make up the sets are different.

Skills Focus

- Count and identify set size
- Identify primary colors
- Identify shapes
- Compare sets

Teaching for Understanding

As children engage in these activities, they build on their intuitive understanding of relative magnitude and develop strategies to determine how many more objects one set has compared to another.

Observe closely while evaluating the Engage activities assigned for this week.

- Are children identifying shapes and colors correctly?
- Are children counting and accurately determining set size?
- Are children able to assess how many more objects are in one set than in the other?

Math at Home

Give one copy of the Letter to Home, page A25, to each child. Complete the activity in class, and then encourage children to share it with their caregivers.

Letter to Home, Teacher Edition, p. A25

Math Vocabulary

hexagon A shape with six sides

extra More than what is needed

English Learners

SPANISH COGNATES

English	Spanish
comparisons	comparaciones
different	diferente
strategy	estrategia
problem	problema
triangle	triángulo

ALTERNATE VOCABULARY

figure out To think about how to find an answer

set A group of objects or numbers that go together

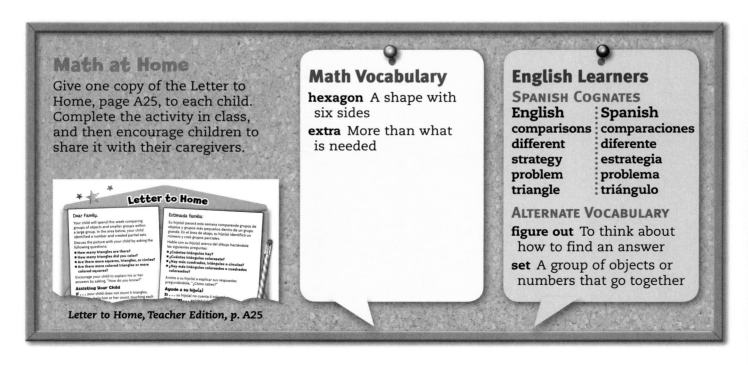

Week 25 Planner | Making Comparisons

PACING	LESSON	LEARNING GOALS	MATERIALS	TECHNOLOGY
DAY 1	**Warm Up** **Line Land:** Count Up*	Children review counting forward.	**Program Materials** Step-by-Step Number Line	**Building Blocks** Number Compare 2
	Engage **Object Land:** Count and Compare	Children identify the number of objects in a set.	**Program Materials** 10 Counters **Prepare Ahead** Tape two 18-inch geometric shapes to the floor.	**MathTools** Set Tool
DAY 2	**Warm Up** **Line Land:** Blastoff!*	Children review the reverse counting sequence.	**Program Materials** Step-by-Step Number Line	**Building Blocks** Number Compare 2
	Engage **Object Land:** Count and Compare	Children discover how many more objects there are in one set than in another set.	**Program Materials** 10 Counters **Prepare Ahead** Tape two 18-in. geometric shapes to the floor.	
DAY 3	**Warm Up** **Line Land:** Count Up*	Children associate counting forward with moving up the number line.	**Program Materials** Step-by-Step Number Line	**Building Blocks** Number Compare 2
	Engage **Object Land:** 2-D Count and Compare	Children compare partial and complete sets of objects.	**Program Materials** Object Land Activity Sheet 1, p. B1, 1 per child **Additional Materials** Red, blue, and green crayons, 1 of each per child	
DAY 4	**Warm Up** **Line Land:** Blastoff!*	Children build a foundation for mental subtraction.	**Program Materials** Step-by-Step Number Line	**Building Blocks** Number Compare 2
	Engage **Object Land:** 2-D Count and Compare Variation*	Children compare sets to identify which have more, less, or the same amount.	**Program Materials** • Object Land Activity Sheet 2, p. B2, 1 per child • Object Land Activity Sheet 3, p. B3, 1 per child for the Challenge	
DAY 5	**Warm Up** **Line Land:** Blastoff!*	Children associate counting back with moving backward on the number line.	**Program Materials** Step-by-Step Number Line	**Building Blocks** Review previous activities
	Review Free-Choice activity	Children review and reinforce concepts and skills learned in this and other weeks.	Materials will be selected from those used in previous weeks.	

* Includes Challenge Variations

Week 25

Making Comparisons

Lesson 1

Objective

Children count to identify the number of objects in a set.

Program Materials

Warm Up
Step-by-Step Number Line

Engage
10 Counters

Prepare Ahead

Engage
Tape two 18-in. geometric shapes to the floor.

Access Vocabulary

collection A group of the same or similar objects

compare To look at things to see how they are similar and how they are different

Creating Context

One way to help English Learners successfully participate in these lessons is to review or introduce vocabulary that describes the attributes of objects and sets children will compare. Select a set and ask children to brainstorm some descriptions of the objects. Use gestures and repeat the words suggested by the children in order to aid comprehension.

EXCELLENT

---

1 Warm Up 5

Skill Building COMPUTING

Count Up

Before beginning **Count and Compare,** use the **Count Up** activity with the whole group.

Purpose **Count Up** gives children the opportunity to review counting forward.

Warm-Up Card 6

Monitoring Student Progress

If . . . a child's counting does not correspond to the steps taken on the Number Line,

Then . . . remind children that they say the next number up only when they step onto the next space on the Number Line.

---

2 Engage 30

Concept Building UNDERSTANDING

Count and Compare

"Today we will use shapes to help us compare two groups of Counters."

Follow the instructions on the Activity Card to play **Count and Compare.** As children play, ask questions about what is happening in the activity.

Activity Card 11

Purpose **Count and Compare** challenges children to identify the number of objects in a set.

MathTools Use the Set Tool to demonstrate and explore counting and comparing.

Monitoring Student Progress

If . . . there are large differences in the skill levels of children in the class,

Then . . . vary the quantity of Counters to be compared each time a new child has a turn so the challenge is appropriate to the child's skill level.

> **Teacher's Note** Encourage children to use this week's vocabulary words as they engage in the activities, discuss math concepts, and make predictions.

Building Blocks For additional practice with comparisons, children should complete **Building Blocks** Number Compare 2.

3 Reflect   10

Extended Response   REASONING

Ask the following questions:

- **How did you figure out which shape had the most (or fewest) Counters?** Possible answer: I counted the Counters in each shape and then I knew that the triangle had more because 3 is bigger than 2.
- **Did anyone figure it out a different way? Please explain.** Answers will vary.

4 Assess

Informal Assessment
Use the Student Assessment Record, **Assessment,** page 100, to record informal observations.

COMPUTING	UNDERSTANDING
Count Up	**Count and Compare**
Did the child	Did the child
❏ respond accurately?	❏ make important observations?
❏ respond quickly?	❏ extend or generalize learning?
❏ respond with confidence?	
❏ self-correct?	❏ provide insightful answers?
	❏ pose insightful questions?

Lesson 2

Objective

Children discover how many more objects are in one set than in another set.

Program Materials

Warm Up
Step-by-Step Number Line

Engage
10 Counters

Prepare Ahead

Engage
Tape two 18-in. geometric shapes to the floor.

Access Vocabulary

stays the same Does not change
whole group Everyone

Creating Context

Providing patterns or structures to English Learners can help them describe their observations when comparing sets of objects. Practice phrases such as *this set has more, this set has less,* and *these sets have the same amount.*

1 Warm Up 5

Concept Building COMPUTING

Blastoff!
Before beginning **Count and Compare,** use the **Blastoff!** activity with the whole group.

Purpose Blastoff! helps children reinforce the reverse counting sequence.

Warm-Up Card 7

Monitoring Student Progress

If . . . a child does not remember the next number down,

Then . . . call on another child to say that number. Allow the child who is counting to take an extra turn to determine whether he or she can say all the numbers now.

2 Engage 30

Concept Building ENGAGING
Count and Compare

"Today we will practice figuring out how many more Counters are in one group than in the other group."

Follow the instructions on the Activity Card to play **Count and Compare.** As children play, ask questions about what is happening in the activity.

Purpose Count and Compare helps children discover how to determine how many more objects are in one set than in another set.

Activity Card 11

Monitoring Student Progress

If . . . children respond to "how many more" by giving the *total* number of Counters instead of the extra number of Counters,

Then . . . ask, "How many Counters would we need to take away to make the number of Counters equal?" Allow several children to answer and test their predictions until you get the correct answer.

 Building Blocks For additional practice with comparisons, children should complete **Building Blocks** Number Compare 2.

3 Reflect 10

Extended Response REASONING

Ask the following questions:

- **What would happen if the shapes were different?** The numbers would stay the same.
- **What does this activity remind you of?** Answers will vary.

4 Assess

Informal Assessment

Use the Student Assessment Record, **Assessment,** page 100, to record informal observations.

COMPUTING	ENGAGING
Blastoff!	**Count and Compare**
Did the child	Did the child
❏ respond accurately?	❏ pay attention to the contributions of others?
❏ respond quickly?	
❏ respond with confidence?	❏ contribute information and ideas?
❏ self-correct?	
	❏ improve on a strategy?
	❏ reflect on and check accuracy of work?

Lesson 3

Objective

Children compare partial and complete sets of objects.

Program Materials

Warm Up
Step-by-Step Number Line

Engage
Object Land Activity Sheet 1, p. B1,
1 per child

Additional Materials

Engage
Red, green, and blue crayons, 1 of each
per child

Access Vocabulary

circle A round shape
triangle A shape with three sides
square A shape with four sides that
are the same size

Creating Context

Many English Learners find it helpful to
review and repeat vocabulary that has been
taught before using it in an activity. Before
playing **2-D Count and Compare,** review
the names of the basic shapes by showing
them to children, saying their names,
and counting the number of sides each
shape has.

---

1 Warm Up   5

Skill Building COMPUTING

Count Up
Before beginning **2-D
Count and Compare,**
use the **Count Up** activity
with the whole group.

Purpose **Count Up**
teaches children to
associate counting
forward with moving
up the number line.

Warm-Up Card 6

Monitoring Student Progress

If . . . a child
can count
easily from
1 to 10,

Then . . . make the activity more
challenging. Have the child stop at
either 7, 8, or 9 and ask him or her,
"How many more steps do you need
to take to get to number 10?"

2 Engage   30

Concept Building REASONING

2-D Count and Compare
"Today we will compare
shapes on a worksheet."

Follow the instructions on
the Activity Card to play
2-D Count and Compare.
As the children play, ask
questions about what is
happening in the activity.

Purpose **2-D Count and
Compare** allows children
to compare partial and
complete sets of objects.

Activity Card 13

Monitoring Student Progress

If . . . children are relying on visible cues to make comparisons,

Then . . . ask questions that encourage them to use numbers when they compare.

 Building Blocks For additional practice with comparisons, children should complete **Building Blocks** Number Compare 2.

3 Reflect   10

Extended Response REASONING

Ask the following questions:

- **What do you notice about this worksheet?** Answers will vary.

- **Is it confusing to count when only some of the shapes are colored? Why or why not?** Accept all reasonable answers.

- **When I asked you which of the colored shapes had more, how did you figure it out?** Answers will vary. Have several children respond.

4 Assess

Informal Assessment

Use the Student Assessment Record, **Assessment,** page 100, to record informal observations.

COMPUTING	REASONING
Count Up Did the child ❑ respond accurately? ❑ respond quickly? ❑ respond with confidence? ❑ self-correct?	**2-D Count and Compare** Did the child ❑ provide a clear explanation? ❑ communicate reasons and strategies? ❑ choose appropriate strategies? ❑ argue logically?

Your Work Hogs the Spotlight

SMILE

Lesson 4

Objective

Children compare sets to identify which have more, less, or the same amount.

Program Materials

Warm Up
Step-by-Step Number Line

Engage
- Object Land Activity Sheet 2, p. B2, 1 per child
- Object Land Activity Sheet 3, p. B3, 1 per child for the Challenge

Additional Materials

No additional materials needed

Access Vocabulary

puppy A baby dog
pet store A shop where you can buy animals and things to take care of animals

Creating Context

Graphic organizers are often beneficial when explaining concepts to English Learners. This lesson relies a great deal on comparisons and comparative language. Draw a three-column chart and review *big,* *bigger,* and *biggest* by drawing different sizes of the same simple shape. Select other comparative adjectives and repeat this process.

---

1 Warm Up 5

Skill Building COMPUTING

Blastoff!
Before beginning **2-D Count and Compare,** use the **Blastoff!** activity with the whole group.

Purpose **Blastoff!** builds a foundation for mental subtraction.

Warm-Up Card 7

Monitoring Student Progress

If . . . children can count down easily from 10 to 1,

Then . . . make the activity more challenging by asking them to stop when they reach a certain number. Have them predict the next number they will step on if they take an additional step backward.

---

2 Engage 30

Concept Building ENGAGING

2-D Count and Compare
"Today we will try to figure out which groups of animals are bigger."

Follow the instructions on the Activity Card to play **2-D Count and Compare** Variation: **Pet Store.** As children play, ask questions about what is happening in the activity.

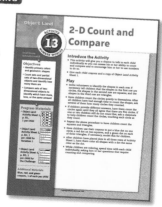

Activity Card 13

Purpose **Pet Store** helps children understand how to make comparisons between sets of different sizes.

Monitoring Student Progress

If . . . children do not agree on the size of a set,

Then . . . have them count again.

 Building Blocks For additional practice with comparisons, children should complete **Building Blocks** Number Compare 2.

3 Reflect 10 ▶

Extended Response REASONING

Ask the following questions:

- **Which group of animals had the most?** puppies
- **How do you know?** Possible answer: Five is bigger than all the other numbers.
- **Which group of animals had the smallest amount?** fish
- **How did you figure that out?** Possible answer: because 1 comes first when you count up

High Flyer!

4 Assess

Informal Assessment

Use the Student Assessment Record, **Assessment,** page 100, to record informal observations.

COMPUTING	ENGAGING
Blastoff!	**2-D Count and Compare**
Did the child	Did the child
❏ respond accurately?	❏ pay attention to the contributions of others?
❏ respond quickly?	
❏ respond with confidence?	❏ contribute information and ideas?
❏ self-correct?	❏ improve on a strategy?
	❏ reflect on and check accuracy of work?

Whooo did a great job? YOU!

Lesson 5

Review

Objective

Children review and reinforce concepts and skills learned in this and other weeks.

Program Materials

Warm Up
Step-by-Step Number Line

Engage
See Activity Cards for materials.

Additional Materials

Engage
See Activity Cards for materials.

Creating Context

English Learners may need practice asking questions in English. For children who speak a tonal language such as Cantonese or Vietnamese, using intonation in English may be especially challenging. Model questions for the children to ask, paying attention to the question form and intonation.

1 **Warm Up** 5

Skill Practice COMPUTING

Blastoff!
Before beginning the **Free-Choice** activity, use the **Blastoff!** activity with the whole group.

Purpose **Blastoff!** demonstrates the association between counting back and moving backward on the number line.

Warm-Up Card 7

Monitoring Student Progress

| **If . . .** children have trouble remembering the reverse-counting sequence, | **Then . . .** practice this with them throughout the day. |

2 **Engage** 20

Concept Building APPLYING

Free-Choice Activity
For the last day of the week, allow children to choose an activity from the previous weeks. Some activities they may choose are the following:

• Picture Land: **Dog and Bone**
• Object Land: **Let's Compare**
• Sky Land: **Beanbag Toss**

Make a note of the activities children select. Do they prefer easy or challenging activities? If you believe your children would benefit from extra practice on specific skills, choose an activity for them.

3 Reflect 10

Extended Response REASONING

Ask the following questions:

- **What did you like about playing** Count and Compare?
- **Was there anything about playing this game you didn't like?**
- **Did this game help you do something you couldn't do before? What did it help you do?**
- **What was easy when you were playing** 2-D Count and Compare?
- **What was hard when you were playing** 2-D Count and Compare?
- **What do you remember most about this game?**
- **What was your favorite part?**

4 Assess 10

A Gather Evidence

Formal Assessment

Have children complete the weekly test, **Assessment,** page 73. Record formal assessment scores on the Student Assessment Record, **Assessment,** page 100.

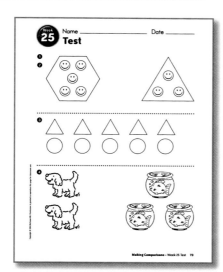

Assessment, p. 73

B Summarize Findings

Review the Student Assessment Records. Determine whether children have Minimal, Basic, or Secure understanding of the concepts presented in Week 25.

C Differentiate Instruction

Based on your observations, use these teaching strategies next week to follow up.

Minimal Understanding

- Repeat the Warm-Up and Engage activities to develop concepts of counting objects.
- Use **Building Blocks** computer activities beginning with Number Compare 1 to develop and reinforce numeration concepts.

Basic Understanding

- Repeat Engage activities in subsequent weeks to reinforce basic counting concepts.
- Use **Building Blocks** computer activities beginning with Number Compare 2 to reinforce this week's concepts.

Secure Understanding

- Use Challenge variations of this week's activities.
- Use computer activities to extend children's understanding of the Week 25 concepts.

A WHALE OF A GOOD JOB!

Adding and Subtracting

Week at a Glance

This week, children begin **Number Worlds,** Week 26, and continue to explore Line Land.

Background

In Line Land, numbers are represented on a horizontal line using numerals. Children transition from the world of small, countable objects and patterns to a world of abstract numbers. Using number lines builds an understanding that you can move forward and backward through the number sequence. This will become the basis for later addition and subtraction.

How Children Learn

As they begin this week's activities, children should be able to identify which number comes next when counting up or back.

By the end of the week, children should be building addition and subtraction skills as they identify set size when it is increased or decreased by 1. They will also be more familiar with the symbols used in formal mathematics.

Skills Focus

- Count from 1 to 10 and from 10 to 1
- Identify set size and associate it with position on a number line
- Associate changes in quantity with movement along a number line
- Compare positions on a number line
- Solve problems

Teaching for Understanding

As children engage in these activities, they build associations between movement on the number line and the plus and minus symbols used in more formal mathematics. This lays a strong foundation for addition and subtraction.

Observe closely while evaluating the Engage activities assigned for this week.

- Are children correctly identifying the changes in set size?
- Are children recognizing which sets have more, less, or the same amount?
- Are children associating numbers with the number line?

Math at Home

Give one copy of the Letter to Home, page A26, to each child. Complete the activity in class, and then encourage children to share it with their caregivers.

Letter to Home, Teacher Edition, p. A26

Math Vocabulary

symbol A letter, number, or picture that has a special meaning or stands for something else

plus sign A symbol that means "add"

minus sign A symbol that means "take away"

English Learners

SPANISH COGNATES

English	Spanish
positions	posiciones
activity	actividad
comparisons	comparaciones
problems	problemas

ALTERNATE VOCABULARY

compare To talk about what is similar and what is different

Week 26 Planner | Adding and Subtracting

PACING	LESSON	LEARNING GOALS	MATERIALS	TECHNOLOGY
DAY 1	**Warm Up** **Line Land:** Count Up*	Children say the numbers in sequence.	**Program Materials** Step-by-Step Number Line	**Building Blocks** Numeral Train Game
	Engage **Line Land:** Plus-Minus Game*	Children associate set size with position on a number line.	**Program Materials** • Number Line Game Board • Pawns, 1 per child • Counters, 10 per child • Line Land +1\−1 Cards • Dot Cube	**MathTools** Number Stairs
DAY 2	**Warm Up** **Line Land:** Blastoff!*	Children build a foundation for mental subtraction.	**Program Materials** Step-by-Step Number Line	**Building Blocks** Numeral Train Game
	Engage **Line Land:** Plus-Minus Game*	Children compare positions on a number line to identify which are higher, lower, or the same.	**Program Materials** • Number Line Game Board • Pawns, 1 per child • Counters, 10 per child Line Land • Line Land +1\−1 Cards • Dot Cube	
DAY 3	**Warm Up** **Line Land:** Line Up*	Children practice sequencing sets.	**Program Materials** • Dot Set Cards (1–5), 1 per child • Dot Set Cards (1–10), 1 per child for the Challenge	**Building Blocks** Numeral Train Game
	Engage **Line Land:** Plus-Minus Game*	Children identify how many there will be if a set is increased or decreased by 1.	**Program Materials** • Number Line Game Board • Pawns, 1 per child • Counters, 10 per child • Line Land +1\−1 Cards • Dot Cube	
DAY 4	**Warm Up** **Line Land:** Count Up*	Children associate counting up with moving forward on the number line.	**Program Materials** Step-by-Step Number Line	**Building Blocks** Numeral Train Game
	Engage **Line Land:** Plus-Minus Game*	Children use comparisons to solve problems.	**Program Materials** • Number Line Game Board • Pawns, 1 per child • Counters, 10 per child • Line Land +1\−1 Cards • Dot Cube	
DAY 5	**Warm Up** **Line Land:** Line Up Variation*	Children identify or compute set size.	**Program Materials** • Number Cards (1–5), 1 per child • Number Cards (1–10), 1 per child for the Challenge	**Building Blocks** Review previous activities
	Review Free-Choice activity	Children review and reinforce skills and concepts learned this week and in previous weeks.	Materials will be selected from those used in previous weeks.	

* Includes Challenge Variations

Lesson 1

Objective

Children associate set size with position on a number line.

Program Materials

Warm Up
Step-by-Step Number Line

Engage
- Number Line Game Board
- Pawns, 1 per child
- Counters, 10 per child
- Line Land +1\−1 Cards
- Dot Cube

Additional Materials

No additional materials needed

Access Vocabulary

sequence The order things go in
number line A line with the number sequence on it that is used to solve math problems

Creating Context

To help English Learners understand the concept of sequence, review counting from 1 to 10. Demonstrate other examples of sequence such as the way you take attendance each day or the way the days of the week follow a sequence on the calendar.

---

1 Warm Up 5

Skill Building COMPUTING

Count Up
Before beginning **Plus-Minus Game,** use the **Count Up** activity with the whole group.

Purpose **Count Up** challenges children to say the numbers in sequence.

Warm-Up Card 6

Monitoring Student Progress

If . . . children have trouble remembering the sequence of numbers,

Then . . . make sure they have plenty of opportunities to practice throughout the day.

2 Engage 30

Concept Building UNDERSTANDING

Plus-Minus Game
"Today we will play a game that is like **The Number Line Game.**"

Follow the instructions on the Activity Card to play **Plus-Minus Game.** As children play, ask questions about what is happening in the activity.

Activity Card 27

Purpose **Plus-Minus Game** helps children associate increasing a quantity with moving forward on a number line and decreasing a quantity with moving backward on a number line.

 **MathTools** Use the Number Stairs to demonstrate and explore adding and subtracting by 1.

Monitoring Student Progress

If . . . a child makes a counting mistake,

Then . . . ask the child to count again.

 **Teacher's Note** Encourage children to use this week's vocabulary words as they engage in the activities, discuss math concepts, and make predictions.

Building Blocks For additional practice with numerals and counting, children should complete *Building Blocks* Numeral Train Game.

3 Reflect 10

Extended Response REASONING

Ask the following questions:

- **In this game, you pick a card after you move your pawn. What do the cards tell you to do?** Possible answer: Put another Counter on my line or take a Counter off my line.

- **How do you know whether to add a Counter or to take one off? How do the cards tell you what to do?** Possible answer: If it is a +1 Card, it means add one; if it is a −1 Card, it means take one away.

4 Assess

Informal Assessment

Use the Student Assessment Record, **Assessment,** page 100, to record informal observations.

COMPUTING	UNDERSTANDING
Count Up	**Plus-Minus Game**
Did the child	Did the child
❑ respond accurately?	❑ make important observations?
❑ respond quickly?	❑ extend or generalize learning?
❑ respond with confidence?	❑ provide insightful answers?
❑ self-correct?	❑ pose insightful questions?

Lesson 2

 Warm Up 5

Objective

Children compare positions on a number line to identify which are higher, lower, or the same.

Program Materials

Warm Up
Step-by-Step Number Line

Engage
- Number Line Game Board
- Pawns, 1 per child
- Counters, 10 per child
- Line Land +1\−1 Cards
- Dot Cube

Additional Materials
No additional materials needed

Access Vocabulary

position Where something is
equal Having the same amount

Creating Context

Some English Learners have difficulty pronouncing the final /s/ in *plus* and *minus*. When English Learners practice using the operation signs have them repeat their pronunciation to make sure that they are correct.

Concept Building COMPUTING
Blastoff!
Before beginning **Plus-Minus Game,** use the **Blastoff!** activity with the whole group.

Purpose **Blastoff!** builds children's foundations for mental subtraction.

Warm-Up Card 7

Monitoring Student Progress

If . . . children need additional practice counting back,

Then . . . have them work on an activity such as **Going Down.**

2 Engage 30

Concept Building ENGAGING
Plus-Minus Game
"Today we will use plus and minus symbols."

Follow the instructions on the Activity Card to play **Plus-Minus Game.** As children play, ask questions about what is happening in the activity.

Purpose **Plus-Minus Game** teaches children to solve problems by comparing positions on a number line to identify which are higher, lower, or the same.

Activity Card 27

Monitoring Student Progress

| **If . . .** children have trouble remembering the difference between *plus* and *minus*, | **Then . . .** remind them of Plus Pup and Minus Mouse. |

 Building Blocks For additional practice with numerals and counting, childen should complete **Building Blocks** Numeral Train Game.

 3 **Reflect** 10

Extended Response REASONING

Ask the following questions:

- **How can you tell which person is winning?** Possible answer: The person closest to the top is winning.

- **Is there a way you could figure out who is winning without looking at the board?** Possible answer: You can use the numbers.

 4 **Assess**

Informal Assessment

Use the Student Assessment Record, **Assessment,** page 100, to record informal observations.

COMPUTING	**ENGAGING**
Blastoff!	**Plus-Minus Game**
Did the child	Did the child
❏ respond accurately?	❏ pay attention to the contributions of others?
❏ respond quickly?	
❏ respond with confidence?	❏ contribute information and ideas?
❏ self-correct?	
	❏ improve on a strategy?
	❏ reflect on and check accuracy of work?

Lesson 3

Objective

Children identify how many there will be if a set is increased or decreased by 1.

Program Materials

Warm Up
- Dot Set Cards (1–5), 1 per child
- Dot Set Cards (1–10), 1 per child for the Challenge

Engage
- Number Line Game Board
- Pawns, 1 per child
- Counters, 10 per child
- Line Land +1\−1 Cards
- Dot Cube

Additional Materials

No additional materials needed

Access Vocabulary

"How do you know?" **"What clues do you have that tell you this is the answer? What helps you to figure out this answer?"**

stay in the same spot Don't move; stay in exactly the same place

Creating Context

Using their primary language to discuss ideas and strategies can be helpful to English Learners. Encourage children to talk with their partners about when to move up, move back, or stay put in the **Plus-Minus Game.**

---

1 Warm Up   5

Skill Building COMPUTING

Line Up

Before beginning **Plus-Minus Game,** use the **Line Up** activity with the whole group.

Purpose **Line Up** gives children practice sequencing sets.

Warm-Up Card 8

Monitoring Student Progress

If . . . a child incorrectly identifies the number on his or her card,

Then . . . have him or her recount the dots.

---

2 Engage   30

Concept Building REASONING

Plus-Minus Game

"Today we will see who can get to the end of the Number Line Game Board first."

Follow the instructions on the Activity Card to play **Plus-Minus Game.** As children play, ask questions about what is happening in the activity.

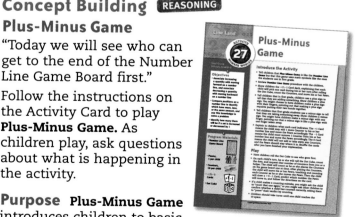

Activity Card 27

Purpose **Plus-Minus Game** introduces children to basic addition and subtraction as they identify how many there will be if a set is increased or decreased by 1.

Monitoring Student Progress

If . . . a child makes a mistake predicting what number his or her Pawn will move to,

Then . . . assess the origin of the mistake by asking the child, "How do you know?"

 Building Blocks For additional practice with numerals and counting, children should complete **Building Blocks** Numeral Train Game.

3 Reflect   10

Extended Response REASONING

Ask the following questions:

- **How would you explain this activity to someone who is not here?** Possible answer: On your turn, you roll the Dot Cube and figure out how many Counters you get. Put the Counters on the board and move your Pawn to the last one. Then take a card and decide if you need to move up, move back, or stay in the same spot. The first person to get to 10 wins.

- **What patterns do you see? How would you describe these patterns?** Accept all reasonable answers.

4 Assess

Informal Assessment

Use the Student Assessment Record, **Assessment,** page 100, to record informal observations.

COMPUTING	REASONING
Line Up	**Plus-Minus Game**
Did the child	Did the child
❏ respond accurately?	❏ provide a clear explanation?
❏ respond quickly?	❏ communicate reasons and strategies?
❏ respond with confidence?	❏ choose appropriate strategies?
❏ self-correct?	❏ argue logically?

Lesson 4

Objective

Children use comparisons to solve problems.

Program Materials

Warm Up

Step-by-Step Number Line

Engage

- Number Line Game Board
- Pawns, 1 per child
- Counters, 10 per child
- Line Land +1\−1 Cards
- Dot Cube

Additional Materials

No additional materials needed

Access Vocabulary

plus sign The symbol that means "add"
minus sign The symbol that means "take away"

Creating Context

English Learners can become more proficient with academic language through role-play and dramatization. Work with the plus and minus signs by having English Learners hold up a small card with the appropriate sign on it while others play the game. When children don't agree, ask them to explain why they recommend the sign they are showing.

1 Warm Up 5

Skill Building COMPUTING

Count Up

Before beginning **Plus-Minus Game,** use the **Count Up** activity with the whole group.

Purpose Count Up demonstrates the association between counting up and moving forward on the number line.

Warm-Up Card 6

Monitoring Student Progress

If . . . children are ready for a challenge,

Then . . . have them stop at numbers that are close to 10 and predict how many more steps they need to take to reach 10. Ask the whole class to make their own silent predictions. After a few children answer aloud, ask the child on the Number Line to test one of the predictions by taking that many steps to see if he or she lands on 10.

2 Engage 30

Concept Building ENGAGING

Plus-Minus Game

"Today we will continue to play **Plus-Minus Game.**"

Follow the instructions on the Activity Card to play **Plus-Minus Game.** As the children play, ask questions about what is happening in the activity.

Activity Card 27

Purpose **Plus-Minus Game** helps children associate a plus sign with moving forward on the number line and a minus sign with moving backward.

Monitoring Student Progress

If . . . children focus on the position of a Pawn when answering questions,

Then . . . ask them questions that require numeric answers.

 For additional practice with numerals and counting, children should complete **Building Blocks** Numeral Train Game.

3 Reflect 10

Extended Response REASONING

Ask the following questions:

■ **What do you notice when you play** Plus-Minus Game? Accept all reasonable answers.

■ **What does it remind you of?** Accept all reasonable answers.

4 Assess

Informal Assessment

Use the Student Assessment Record, **Assessment,** page 100, to record informal observations.

COMPUTING	ENGAGING
Count Up	**Plus-Minus Game**
Did the child	Did the child
❏ respond accurately?	❏ pay attention to the contributions of others?
❏ respond quickly?	
❏ respond with confidence?	❏ contribute information and ideas?
❏ self-correct?	
	❏ improve on a strategy?
	❏ reflect on and check accuracy of work?

Dino-might!

No "lion"- Math is fun!

Week 26

Adding and Subtracting

Lesson 5

Review

Objective

Children will review and reinforce skills and concepts learned in this week and in previous weeks.

Program Materials

Warm Up
- Number Cards 1–5, 1 per child
- Number Cards 1–10, 1 per child for the Challenge

Engage
See Activity Cards for materials.

Additional Materials

Engage
See Activity Cards for materials.

Creating Context

Review the comparative terms *more, less,* and *same* with English Learners. Use the number line to demonstrate and explain the definitions. Have children use objects to create sets that show *more, less,* and *the same* amount.

Bonus!

---

1 Warm Up

5

Skill Building COMPUTING

Line Up

Before beginning the **Free-Choice** activity, use **Line Up** Variation: **Number Card Line Up** with the whole group.

Purpose **Number Card Line Up** helps children identify or compute set size.

Warm-Up Card 8

Monitoring Student Progress

If . . . children have trouble recognizing the numerals,

Then . . . use the Dot Set Card variation.

---

2 Engage

20

Concept Building APPLYING

Free-Choice Activity

For the last day of the week, allow children to choose an activity from the previous weeks. Some activities they may choose are the following:
- Line Land: **Next Number Up**
- Object Land: **Plus Pup**
- Object Land: **The Mouse in the Cookie Jar**

Make a note of the activities children select. Do they prefer easy or challenging activities? If you believe your children would benefit from extra practice on specific skills, choose an activity for them.

Reflect 10

Extended Response REASONING

Ask the following questions:

- **What did you like about playing** Plus-Minus Game?
- **Was there anything about playing this game you didn't like?**
- **Did this game help you do something you couldn't do before? What did it help you do?**
- **What was easy when you were playing** Plus-Minus Game?
- **What was hard when you were playing** Plus-Minus Game?
- **What do you remember most about this game?**
- **What was your favorite part?**

 Assess 10

A Gather Evidence

Formal Assessment

Have children complete the weekly test, **Assessment,** page 75. Record formal assessment scores on the Student Assessment Record, **Assessment,** page 100.

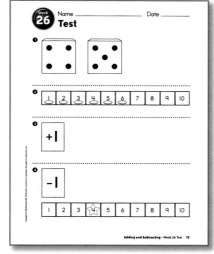

Assessment, p. 75

B Summarize Findings

Review the Student Assessment Records. Determine whether children have Minimal, Basic, or Secure understanding of the concepts presented in Week 26.

C Differentiate Instruction

Based on your observations, use these teaching strategies next week to follow up.

Minimal Understanding

- Repeat the Warm-Up and Engage activities to develop concepts of counting objects.
- Use **Building Blocks** computer activities beginning with Numeral Train Game to develop and reinforce numeration concepts.

Basic Understanding

- Repeat Engage activities in subsequent weeks to reinforce basic counting concepts.
- Use **Building Blocks** computer activities beginning with Numeral Train Game to reinforce this week's concepts.

Secure Understanding

- Use Challenge variations of **Plus-Minus Game.**
- Use computer activities to extend children's understanding of the Week 26 concepts.

Week 27 More Graphs

Week at a Glance

This week, children begin **Number Worlds,** Week 27, and continue to explore Sky Land.

Background

In Sky Land, numbers are represented along a vertical scale such as a thermometer. This gives children an opportunity to develop the language of height as it is used in mathematics. Children learn that this language can be used interchangeably with the language used in Line Land, helping them consolidate the knowledge from both lands.

How Children Learn

As they begin this week's activities, children should be able to determine which of two objects or lines is longer by comparing them.

By the end of the week, children should be more focused on using numbers to help make comparisons.

Skills Focus

- Count to identify set size
- Record quantity as units on a bar graph
- Associate increasing a quantity with increasing the height of a scale measure
- Use a graph to record and compare quantities

Teaching for Understanding

As children engage in these activities, they develop an understanding of the relationship between a set size and scale height. The bar graphs also lay a foundation for understanding charts and graphs when these topics are more formally introduced at higher grades.

Observe closely while evaluating the Engage activities assigned for this week.

- Are children counting to find the correct set size?
- Are children recording quantity on the bar graphs and using the graphs to make comparisons?

Math at Home

Give one copy of the Letter to Home, page A27, to each child. Complete the activity in class, and then encourage children to share it with their caregivers.

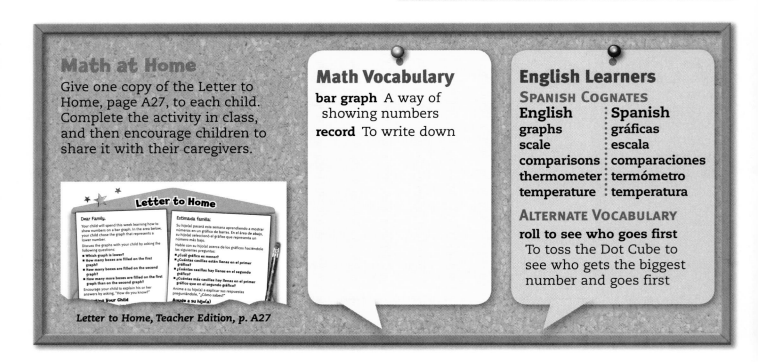

Letter to Home, Teacher Edition, p. A27

Math Vocabulary

bar graph A way of showing numbers

record To write down

English Learners

SPANISH COGNATES

English	Spanish
graphs	gráficas
scale	escala
comparisons	comparaciones
thermometer	termómetro
temperature	temperatura

ALTERNATE VOCABULARY

roll to see who goes first To toss the Dot Cube to see who gets the biggest number and goes first

PACING	LESSON	LEARNING GOALS	MATERIALS	TECHNOLOGY
DAY 1	**Warm Up** **Sky Land:** Count Up Variation*	Children develop the ability to count on from a number other than 1.	**Program Materials** Classroom Thermometer	**Building Blocks** Dino Shop 3
	Engage **Sky Land:** Drop the Counter*	Children count to identify set size.	**Program Materials** • Counters, 5 per group • Sky Land Activity Sheet 1, p. B8, 1 per child • Dot Cubes, 1 per group **Additional Materials** • Paper cups, 1 per group • Crayons, 1 per child	**e MathTools** Graphing Tool
DAY 2	**Warm Up** **Sky Land:** Blastoff! Variation 1*	Children associate counting down with decreasing the height of a scale measure.	**Program Materials** Classroom Thermometer	**Building Blocks** Dino Shop 3
	Engage **Sky Land:** Drop the Counter*	Children record quantity as units on a bar graph.	**Materials** Same as Lesson 1	
DAY 3	**Warm Up** **Sky Land:** Count Up Variation*	Children count and stop at a specified number.	**Program Materials** Classroom Thermometer	**Building Blocks** Dino Shop 3
	Engage **Sky Land:** Drop the Counter*	Children use a graph to record and compare quantities.	**Materials** Same as Lesson 1	
DAY 4	**Warm Up** **Sky Land:** Blastoff! Variation 2*	Children count down from 10 to 1.	**Additional Materials** No additional materials needed	**Building Blocks** Dino Shop 3
	Engage **Sky Land:** Drop the Counter*	Children associate increasing a quantity with increasing the height of a scale measure.	**Materials** Same as Lesson 1	
DAY 5	**Warm Up** **Sky Land:** Blastoff! Variation 2*	Children develop a foundation for subtraction.	**Additional Materials** No additional materials needed	**Building Blocks** Review previous activities
	Review Free-Choice activity	Children review and reinforce skills and concepts learned this week and in previous weeks.	Materials will be selected from those used in previous weeks.	

* Includes Challenge Variations

Lesson 1

Objective

Children identify set size by counting objects.

Program Materials

Warm Up
Classroom Thermometer

Engage
- Counters, 5 per group
- Sky Land Activity Sheet 1, p. B8, 1 per child
- Dot Cubes, 1 per group

Additional Materials

Engage
- Paper cups or other containers, 1 per group
- Crayons, 1 per child

Access Vocabulary

wink To close and open one eye quickly
point To use your finger to show something

Creating Context

Pointing and winking are considered impolite in some cultures. Help all children understand that these gestures are used here as signals for the game. If you would prefer, substitute other signals for pointing and winking.

---

1 Warm Up 5

Skill Building COMPUTING

Count Up

Before beginning **Drop the Counter,** use **Count Up** Variation: **Pointing and Winking** with the whole group.

Purpose **Pointing and Winking** develops children's ability to count on from a number other than 1.

Warm-Up Card 9

Monitoring Student Progress

If . . . children are having trouble counting on from where you stop,

Then . . . remind them to listen carefully and count silently while you count aloud.

---

2 Engage 30

Concept Building UNDERSTANDING

Drop the Counter

"Today we will keep track of how many Counters drop into a container and how many fall on the floor."

Follow the instructions on the Activity Card to play **Drop the Counter.** As children play, ask questions about what is happening in the activity.

Activity Card 31

Purpose **Drop the Counter** reinforces the idea that children should count to identify set size.

MathTools Use the Graphing Tool to demonstrate and explore bar graphs.

Monitoring Student Progress

| **If . . .** children have trouble counting the boxes on the graphs, | **Then . . .** have them alternate colors when they color each box. |

Teacher's Note Encourage children to use this week's vocabulary words as they engage in the activities, discuss math concepts, and make predictions.

Building Blocks For additional practice adding small numbers, children should complete **Building Blocks** Dino Shop 3 (1–5).

3 Reflect 10

Extended Response [REASONING]

Ask the following questions:

- **Did more Counters land in the container or on the floor?** Answers will vary.
- **How do you know?** Possible answer: The one with more Counters has a taller graph.
- **How can you check to see if you are correct?** Possible answer: We can count.

4 Assess

Informal Assessment

Use the Student Assessment Record, **Assessment,** page 100, to record informal observations.

COMPUTING	UNDERSTANDING
Count Up	**Drop the Counter**
Did the child	Did the child
❏ respond accurately?	❏ make important observations?
❏ respond quickly?	❏ extend or generalize learning?
❏ respond with confidence?	❏ provide insightful answers?
❏ self-correct?	❏ pose insightful questions?

FIN-tastic!

Lesson 2

Objective

Children record quantity as units on a bar graph.

Program Materials

Warm Up
Classroom Thermometer

Engage
- Counters, 5 per group
- Sky Land Activity Sheet 1, p. B8, 1 per child
- Dot Cubes, 1 per group

Additional Materials

Engage
- Paper cups or other containers, 1 per group
- Crayons, 1 per child

Access Vocabulary

chilly Very cold
so far Up until now

Creating Context

The words *chilly* and *chili* may be confused by some English Learners. Use a shivering gesture to introduce *chilly*. Show or draw a chili pepper to demonstrate *chili*.

1 Warm Up 5

Concept Building COMPUTING

Blastoff!
Before beginning **Drop the Counter**, use **Blastoff! Variation 1: It's a Chilly Day** with the whole group.

Purpose **It's a Chilly Day** teaches children to associate counting down with decreasing the height of a scale measure.

Warm-Up Card 10

Monitoring Student Progress

If . . . children are not ready to count down from 10,

Then . . . have them count down from 7 or 5.

2 Engage 30

Concept Building ENGAGING

Drop the Counter
"Today we will see if you get more Counters into the container than onto the floor."

Follow the instructions on the Activity Card to play **Drop the Counter.** As children play, ask questions about what is happening in the activity.

Purpose **Drop the Counter** teaches children basic graphing skills as they record quantity on a bar graph.

Activity Card 31

Monitoring Student Progress

If . . . children have trouble comparing their results with one another,

Then . . . remind them to use numbers to decide who dropped more Counters into the container.

 Building Blocks For additional practice adding small numbers, children should complete **Building Blocks** Dino Shop 3 (1–5).

3 Reflect

10

Extended Response REASONING

Ask questions such as the following:

- **Who dropped the most Counters into the container?** Children should name the child who had the highest number.

- **How many Counters did he or she drop into the container?** Possible answer: 5

- **Who dropped the most Counters onto the floor?** Children in each group should name the child who had the highest number. Children can also name the child who had the highest number in the class.

4 Assess

Informal Assessment

Use the Student Assessment Record, **Assessment,** page 100, to record informal observations.

COMPUTING	ENGAGING
Blastoff!	**Drop the Counter**
Did the child	Did the child
❑ respond accurately?	❑ pay attention to the contributions of others?
❑ respond quickly?	
❑ respond with confidence?	❑ contribute information and ideas?
❑ self-correct?	
	❑ improve on a strategy?
	❑ reflect on and check accuracy of work?

EXCELLENT

Dino-might!

Lesson 3

Objective

Children use a graph to compare quantities.

Program Materials

Warm Up
Classroom Thermometer

Engage
- Counters, 5 per group
- Sky Land Activity Sheet 1, p. B8, 1 per child
- Dot Cubes, 1 per group

Additional Materials

Engage
- Paper cups or other containers, 1 per group
- Crayons, 1 per child

Access Vocabulary

count aloud To say the numbers when you count
height How tall or short something is

Creating Context

Young children are still acquiring English and often add *-ed* to all past-tense verbs. While children play **Drop the Counter,** point out that the past-tense form of *fall* is *fell* by using it in sentences, such as "How many counters fell into the container?" and "How many fell outside the container?" If children say *falled*, remind them to use *fell*.

1 Warm Up 5

Skill Building COMPUTING
Count Up
Before beginning **Drop the Counter,** use **Count Up** Variation: **Pointing and Winking** with the whole group.

Purpose Pointing and Winking teaches children to count and stop at a specified number.

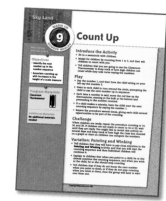

Warm-Up Card 9

Monitoring Student Progress

If . . . children have trouble stopping when you wink at them,

Then . . . try using a more obvious cue such as clapping.

2 Engage 30

Concept Building REASONING
Drop the Counter
"Today we will compare to see who gets the most Counters into the container."

Follow the instructions on the Activity Card to play **Drop the Counter.** As children play, ask questions about what is happening in the activity.

Purpose Drop the Counter teaches children to use a graph to compare quantities.

Activity Card 31

Monitoring Student Progress

If . . . children are making inaccurate comparisons,

Then . . . make sure that they are not skipping rectangles when filling in their graphs.

 Building Blocks For additional practice adding small numbers, children should complete **Building Blocks** Dino Shop 3 (1−5).

3 Reflect
10

Extended Response [REASONING]

Ask questions such as the following:

- **When else have we used graphs to compare numbers?** Possible answer: when we played **Beanbag Toss**

- **Why does it make sense to use graphs now?** Accept all reasonable answers.

- **What do the graphs tell us?** Accept all reasonable answers.

4 Assess

Informal Assessment

Use the Student Assessment Record, **Assessment,** page 100, to record informal observations.

COMPUTING	REASONING
Count Up	**Drop the Counter**
Did the child	Did the child
❏ respond accurately?	❏ provide a clear explanation?
❏ respond quickly?	❏ communicate reasons and strategies?
❏ respond with confidence?	❏ choose appropriate strategies?
❏ self-correct?	❏ argue logically?

Lesson 4

Objective
Children associate increasing a quantity with increasing the height of a scale measure.

Program Materials
Engage
- Counters, 5 per group
- Dot Cubes, 1 per group
- Sky Land Activity Sheet 1, p. B8, 1 per child

Additional Materials
Engage
- Paper cups or other containers, 1 per group
- Crayons, 1 per child

Access Vocabulary
shrinking Getting smaller
color To use a crayon or marker to fill in a drawing

Creating Context
Movement and gestures can help young English Learners clarify the meaning of new concepts and vocabulary. An example of this kind of reinforcement is the **Incredible Shrinking People** variation. The actions emphasize the meaning of the concept, and the physical action creates a strong memory of the new words.

1 Warm Up   5

Skill Building   COMPUTING
Blastoff!
Before beginning **Drop the Counter,** use **Blastoff!** Variation 2: **Incredible Shrinking People** with the whole group.

Purpose **Incredible Shrinking People** helps children count down from 10 to 1.

Warm-Up Card 10

Monitoring Student Progress

| If . . . children are ready to integrate counting down and counting up, | Then . . . try the Challenge variation. |

2 Engage   30

Concept Building   ENGAGING
Drop the Counter
"Today we will see what happens on a graph when numbers go up."

Follow the instructions on the Activity Card to play **Drop the Counter.** As children play, ask questions about what is happening in the activity.

Purpose **Drop the Counter** helps children associate increasing quantity with increasing the height of a scale measure.

Activity Card 31

Monitoring Student Progress

If . . . children are not able to use the Activity Sheet properly,	**Then . . .** encourage them to organize their counters on the graph then color the appropriate rectangles.

 Building Blocks For additional practice adding small numbers, children should complete *Building Blocks* Dino Shop 3 (1–5).

3 Reflect   10

Extended Response   REASONING

Ask questions such as the following:

- **How would you explain this activity to someone who has never done it before?** Possible answer: We drop Counters into a container. Then we count to see how many go into the container and how many fall onto the floor, and color that many boxes on the graph. We compare to see who dropped the most Counters into the container. Allow several children to answer.

- **Did anything about this activity surprise you? If so, why?** Accept all reasonable answers.

 ## 4 Assess

Informal Assessment

Use the Student Assessment Record, **Assessment,** page 100, to record informal observations.

COMPUTING	ENGAGING
Blastoff!	**Drop the Counter**
Did the child	Did the child
❑ respond accurately?	❑ pay attention to the contributions of others?
❑ respond quickly?	
❑ respond with confidence?	❑ contribute information and ideas?
❑ self-correct?	
	❑ improve on a strategy?
	❑ reflect on and check accuracy of work?

Doggone Good

Lesson 5
Review

Objective

Children review and reinforce skills and concepts learned in this and previous weeks.

Program Materials

Engage
See Activity Cards for materials.

Additional Materials

Engage
See Activity Cards for materials.

Creating Context

Ask the child who dropped the most counters into the container to demonstrate and explain how he or she did it. Help beginning level English Learners by narrating the child's procedure for the whole group. Use the past-tense verbs *dropped* and *fell* to reinforce the concepts discussed in Lesson 3.

---

1 Warm Up   5

Skill Practice COMPUTING
Blastoff!

Before beginning the **Free-Choice** activity, use **Blastoff!** Variation 2: **Incredible Shrinking People** with the whole group.

Purpose Incredible Shrinking People helps children develop a foundation for subtraction.

Warm-Up Card 10

Monitoring Student Progress

If . . . children have trouble counting down while shrinking their bodies,

Then . . . determine whether the problem lies in saying the count-down sequence or in matching body movement to this sequence. Provide extra practice to correct the problem you identify.

---

2 Engage   20

Concept Building APPLYING
Free-Choice Activity

For the last day of the week, allow children to choose an activity from the previous weeks. Some activities they may choose include the following:

- Sky Land: **Going Up**
- Sky Land: **Beanbag Toss**
- Line Land: **Plus-Minus Game**

Make a note of the activities children select. Do they prefer easy or challenging activities? If you believe your children would benefit from extra practice on specific skills, choose an activity for them.

Reflect 10

Extended Response REASONING

Ask questions such as the following:

- **What did you like about playing** Drop the Counter?
- **Was there anything about playing this game you didn't like?**
- **Did this game help you do something you couldn't do before? What did it help you do?**
- **What was easy when you were playing** Drop the Counter?
- **What was hard when you were playing** Drop the Counter?
- **What do you remember most about this game?**
- **What was your favorite part?**

 Assess 10

A Gather Evidence

Formal Assessment

Have children complete the weekly test, **Assessment,** page 77. Record formal assessment scores on the Student Assessment Record, **Assessment,** page 100.

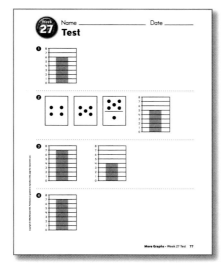

Assessment, p. 77

B Summarize Findings

Review the Student Assessment Records. Determine whether children have Minimal, Basic, or Secure understanding of the concepts presented in Week 27.

C Differentiate Instruction

Based on your observations, use these teaching strategies next week to follow up.

Minimal Understanding

- Repeat the Warm-Up and Engage activities to develop concepts of counting objects.
- Use **Building Blocks** computer activities beginning with Dino Shop 2 to develop and reinforce numeration concepts.

Basic Understanding

- Repeat Engage activities in subsequent weeks to reinforce basic counting concepts.
- Use **Building Blocks** computer activities beginning with Dino Shop 3 to reinforce this week's concepts.

Secure Understanding

- Use Challenge variations of **Drop the Counter.**
- Use computer activities to extend children's understanding of the Week 27 concepts.

Higher and Lower

Week at a Glance

This week, children begin **Number Worlds,** Week 28, and continue to explore Sky Land.

Background

In Sky Land, numbers are represented along a vertical scale. This gives children an opportunity to develop the language of height as it is used in math. Children learn that this language and the language used in Line Land can be used interchangeably, which helps them consolidate concept knowledge.

How Children Learn

As children begin this week's activities, they should be able to calculate sums by counting all objects.

By the end of the week, children should be reinforcing their ability to understand addition as adding two parts to get a whole. For example, when asked, "You have 3 Counters and get 2 more. How many in all?" the child will first count out 3, then 2, and then the child will count all 5.

Skills Focus

- Identify or compute set size and associate it with a point on a scale
- Associate changes in quantity with movement on a scale
- Compare points on a scale
- Solve problems
- Identify how high the scale measure will be if a set is increased or decreased by 1

Teaching for Understanding

As children engage in these activities, they see the effects of changes in quantity on a scale measure, and they develop an understanding of the meaning of +1 and −1.

Observe closely while evaluating the Engage activities assigned for this week.

- Are children counting correctly?
- Are children reflecting the understanding that they move up when they add and move down when they subtract?
- Are children making accurate comparisons and using that information to solve problems?

Math at Home

Give one copy of the Letter to Home, page A28, to each child. Complete the activity in class, and then encourage children to share it with their caregivers.

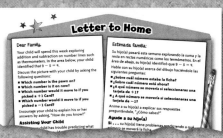

Letter to Home, Teacher Edition, p. A28

Math Vocabulary

direction The way in which you move (up or down)

far A long way away

English Learners

SPANISH COGNATES

English	Spanish
problems	problemas
count	contar
incredible	increíble
subtraction	sustracción

ALTERNATE VOCABULARY

hurry up To go faster

passengers People riding in something such as an elevator or car

PACING	LESSON	LEARNING GOALS	MATERIALS	TECHNOLOGY
DAY 1	**Warm Up** **Sky Land:** Count Up Variation*	Children review counting skills.	**Program Materials** Classroom Thermometer	**Building Blocks** Number Snapshots 4
	Engage **Sky Land:** Hurry Up Passengers, Going Up*	Children associate changes in quantity with movement up and down scale measures.	**Program Materials** • Elevator Game Boards, 1 per child • Pawns, 1 per child • Dot Cube • Up/Down +1\−1 Cards	
DAY 2	**Warm Up** **Sky Land:** Blastoff! Variation 1*	Children reinforce the reverse counting sequence.	**Program Materials** Classroom Thermometer	**Building Blocks** Number Snapshots 4
	Engage **Sky Land:** Hurry Up Passengers, Going Up*	Children measure direction and distance.	**Program Materials** • Elevator Game Boards, 1 per child • Pawns, 1 per child • Dot Cube • Up/Down +1\−1 Cards	**MathTools** Number Stairs
DAY 3	**Warm Up** **Sky Land:** Count Up Variation*	Children build the ability to count on.	**Program Materials** Classroom Thermometer	**Building Blocks** Number Snapshots 4
	Engage **Sky Land:** Hurry Up Passengers, Going Down*	Children explore basic addition and subtraction.	**Program Materials** • Elevator Game Boards, 1 per child • Pawns, 1 per child • Dot Cube • Up/Down +1\−1 Cards	
DAY 4	**Warm Up** **Sky Land:** Blastoff!*	Children learn real-world applications for counting down.	**Program Materials** Classroom Thermometer	**Building Blocks** Number Snapshots 4
	Engage **Sky Land:** Hurry Up Passengers, Going Down*	Children make comparisons to solve problems.	**Program Materials** • Elevator Game Boards, 1 per child • Pawns, 1 per child • Dot Cube • Up/Down +1\−1 Cards	
DAY 5	**Warm Up** **Sky Land:** Blastoff! Variation 2*	Children build a foundation for subtraction.	**Additional Materials** No additional materials needed	**Building Blocks** Review previous activities
	Review Free-Choice activity	Children review and reinforce skills and concepts from this and previous weeks.	Materials will be selected from those used in previous weeks.	

* Includes Challenge Variations

Lesson 1

Objective

Children associate changes in quantity with movement up and down scale measures.

Program Materials

Warm Up
Classroom Thermometer

Engage
- Elevator Game Boards, 1 per child
- Pawns, 1 per child
- Dot Cube
- Up/Down +1\−1 Cards

Additional Materials

No additional materials needed

Access Vocabulary

pawn Game piece
thermometer A tool used to tell how hot or how cold something is

Creating Context

Visual cues such as gestures help English Learners comprehend math concepts and vocabulary. Role-play riding an elevator, gesturing to show going up while counting rhythmically and ringing a bell or a triangle at each floor.

Pretty Good!

1 Warm Up 5

Skill Building COMPUTING

Count Up

Before beginning **Hurry Up Passengers, Going Up,** use **Count Up** Variation: **Pointing and Winking** with the whole group.

Purpose **Pointing and Winking** helps children review counting skills.

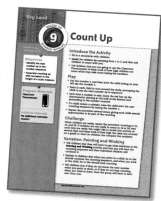

Warm-Up Card 9

Monitoring Student Progress

If . . . a child has trouble thinking of the next number when you point to him or her,

Then . . . make sure you give the child enough time to think of the answer.

2 Engage 30

Concept Building UNDERSTANDING

Hurry Up Passengers, Going Up

"Today we will play a game like **Going Up.**"

Follow the instructions on the Activity Card to play **Hurry Up Passengers, Going Up.** As children play, ask questions about what is happening in the activity.

Purpose **Hurry Up Passengers, Going Up** helps children associate changes in quantity with movement up and down scale measures.

Activity Card 32

Monitoring Student Progress

| If . . . children are confused by the elevator, | Then . . . explain that this activity is like **Plus-Minus Game.** |

 Building Blocks For additional practice immediately recognizing the number of objects in a small group, children should complete **Building Blocks** Number Snapshots 4.

3 Reflect 10

Extended Response REASONING

Ask the following questions:

■ **What is this activity about?** Possible answer: getting an elevator to the top of the board first

■ **What helps you get to the top first?** Possible answers: rolling big numbers; getting +1 Cards

■ **How would you describe the activity to someone who is not here?** Possible answer: It's like **Going Up** except you pick a card after you move and go up, down, or stay in the same place.

4 Assess

Use the Student Assessment Record, **Assessment,** page 100, to record informal observations.

COMPUTING	UNDERSTANDING
Count Up	**Hurry Up Passengers, Going Up**
Did the child	Did the child
❏ respond accurately?	❏ make important observations?
❏ respond quickly?	❏ extend or generalize learning?
❏ respond with confidence?	❏ provide insightful answers?
❏ self-correct?	❏ pose insightful questions?

No "lion"– Math is fun!

Lesson 2

Objective

Children compare positions on a scale to identify which are higher or lower.

Program Materials

Warm Up
Classroom Thermometer

Engage
- Elevator Game Boards, 1 per child
- Pawns, 1 per child
- Dot Cube
- Up/Down +1\−1 Cards

Additional Materials
No additional materials needed

Access Vocabulary
pick To choose
winner The person playing the game who gets to the end first

Creating Context
English Learners may need to review ordinal numbers as they work with positions on a vertical scale. Explain that while first, second, and third have special names, the ordinal numbers after them are counting numbers with -*th* on the end such as fourth, fifth, sixth, and so on.

---

1 Warm Up 5

Concept Building COMPUTING
Blastoff!
Before beginning **Hurry Up Passengers, Going Up,** use **Blastoff!** Variation 1: **It's a Chilly Day** with the whole group.

Purpose **It's a Chilly Day** reinforces children's knowledge of the reverse counting sequence.

Warm-Up Card 10

Monitoring Student Progress

If . . . children have trouble remembering the reverse counting sequence,

Then . . . practice counting down before beginning the activity.

---

2 Engage 30

Concept Building ENGAGING
Hurry Up Passengers, Going Up
"Today we will see who can get to the tenth floor first."

Follow the instructions on the Activity Card to play **Hurry Up Passengers, Going Up.** As children play, ask questions about what is happening in the activity.

Purpose **Hurry Up Passengers, Going Up** teaches children to compare positions on a scale to identify which are higher or lower.

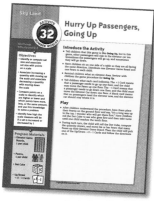

Activity Card 32

 **MathTools** Use the Number Stairs to demonstrate and explore adding and subtracting 1 unit.

Monitoring Student Progress

If . . . children have trouble calculating where to move,

Then . . . have them use Counters as they did when playing **Plus-Minus Game.**

 Teacher's Note Encourage children to use this week's vocabulary words as they engage in the activities, discuss math concepts, and make predictions.

Building Blocks For additional practice immediately recognizing the number of objects in a small group, children should complete **Building Blocks** Number Snapshots 4.

---

 3 Reflect 10

Extended Response REASONING

Ask questions such as the following:

- **How did you know who was winning?** Possible answer: The person who was winning was the one closest to the top.

- **Did anyone figure it out a different way?** Possible answer: The person winning is on the highest number.

 **4 Assess**

Informal Assessment

Use the Student Assessment Record, **Assessment,** page 100, to record informal observations.

COMPUTING	ENGAGING
Blastoff!	**Hurry Up Passengers, Going Up**
Did the child	Did the child
❑ respond accurately?	❑ pay attention to the contributions of others?
❑ respond quickly?	
❑ respond with confidence?	❑ contribute information and ideas?
❑ self-correct?	❑ improve on a strategy?
	❑ reflect on and check accuracy of work?

 your Work Hogs the Spotlight

Lesson 3

Objective

Children explore basic addition and subtraction.

Program Materials

Warm Up
Classroom Thermometer

Engage
- Elevator Game Boards, 1 per child
- Pawns, 1 per child
- Dot Cube
- Up/Down +1\−1 Cards

Additional Materials

No additional materials needed

Access Vocabulary

mistake Wrong answer
ground floor The floor the elevator leaves from

Creating Context

Young English Learners can be confused by the varied uses of *up* and *down* in English phrases. Give examples such as *count up*, *count down*, *hurry up*, and *quiet down*. Ask children if they can think of any frequently used phrases that include *up* or *down*. Point out that, in these phrases, *up* and *down* do not really refer to location.

1 Warm Up 5

Skill Building COMPUTING
Count Up
Before beginning **Hurry Up Passengers, Going Down,** use **Count Up** Variation: **Pointing and Winking** with the whole group.

Purpose **Pointing and Winking** builds children's ability to count on.

Warm-Up Card 9

Monitoring Student Progress

If . . . a counting mistake is made,

Then . . . have the child who made the mistake start over the sequence by counting from 1.

2 Engage 30

Concept Building REASONING
Hurry Up Passengers, Going Down

"Today we will see who is fastest at going from the tenth floor to the first floor."

Follow the instructions on the Activity Card to play **Hurry Up Passengers, Going Down.** As children play, ask questions about what is happening in the activity.

Purpose **Hurry Up Passengers, Going Down** gives children the opportunity to explore basic addition and subtraction.

Activity Card 33

Monitoring Student Progress

If . . . children are having trouble remembering which direction to move when they pick a card,

Then . . . review what each card indicates and what actions must be performed. Monitor children's game play to make sure they understand.

 Teacher's Note The elevator car descending down a shaft provides a vivid illustration of subtraction as a quantity being taken away from its current position. When children are encouraged to talk about the change they have just enacted using the elevator, and to use numbers to do so, they will have an opportunity to use many of the concepts developed in previous activities and to build connections among them.

Building Blocks For additional practice immediately recognizing the number of objects in a small group, children should complete **Building Blocks** Number Snapshots 4.

3 Reflect 10

Extended Response REASONING

Ask questions such as the following:

■ **What differences do you notice between** Hurry Up Passengers, Going Down **and** Hurry Up Passengers, Going Up? Possible answer: The goal is different.

■ **Who got to the bottom first? How did he or she get to the bottom before anyone else?** Possible answers: He got a lot of −1 Cards. She rolled big numbers.

4 Assess

Informal Assessment

Use the Student Assessment Record, **Assessment,** page 100, to record informal observations.

COMPUTING	REASONING
Count Up	**Hurry Up Passengers, Going Down**
Did the child	Did the child
❑ respond accurately?	❑ provide a clear explanation?
❑ respond quickly?	❑ communicate reasons and strategies?
❑ respond with confidence?	❑ choose appropriate strategies?
❑ self-correct?	❑ argue logically?

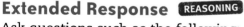

Lesson 4

Objective

Children make comparisons to solve problems.

Program Materials

Warm Up
Classroom Thermometer

Engage
- Elevator Game Boards, 1 per child
- Pawns, 1 per child
- Dot Cube
- Up/Down +1\−1 Cards

Additional Materials

No additional materials needed

Access Vocabulary

hurry up To go faster
passengers People riding in something such as an elevator or car

Creating Context

Young children who are still acquiring English often add -ed to all past-tense verbs. Point out that the past-tense forms of *win* and *go* are *won* and *went* by using them in sentences such as "Who won?" If children say *winned* or *goed,* refer to the correct usage.

KEEP IT UP!

---

1 Warm Up 5

Skill Building COMPUTING
Blastoff!

Before beginning **Hurry Up Passengers, Going Down,** use **Blastoff!** Variation 1: **It's a Chilly Day** with the whole group.

Purpose **It's a Chilly Day** Variation 1 demonstrates real-world applications for counting down.

Warm-Up Card 10

Monitoring Student Progress

If . . . children are ready,

Then . . . use the Challenge variation.

---

2 Engage 30

Concept Building ENGAGING
Hurry Up Passengers, Going Down

"Today we will play **Hurry Up Passengers, Going Down.**"

Follow the instructions on the Activity Card to play **Hurry Up Passengers, Going Down.** As children play, ask questions about what is happening in the activity.

Purpose **Hurry Up Passengers, Going Down** teaches children to make comparisons and to solve problems.

Activity Card 33

Monitoring Student Progress

If . . . children have trouble remembering that the goal is to get to the ground floor first,

Then . . . use a stuffed animal or some other object to help mark the goal.

 Building Blocks For additional practice immediately recognizing the number of objects in a small group, children should complete **Building Blocks** Number Snapshots 4.

3 Reflect 10

Extended Response REASONING

Ask questions such as the following:

- **Who went the farthest distance?** Students should name the child who won.
- **Did anyone go the same distance?** Answers will vary.

4 Assess

Informal Assessment

Use the Student Assessment Record, **Assessment,** page 100, to record informal observations.

COMPUTING	ENGAGING
Blastoff!	**Hurry Up Passengers, Going Down**
Did the child	Did the child
❏ respond accurately?	❏ pay attention to the contributions of others?
❏ respond quickly?	
❏ respond with confidence?	❏ contribute information and ideas?
❏ self-correct?	
	❏ improve on a strategy?
	❏ reflect on and check accuracy of work?

You've got it DOWN

you're doing SWIMMINGLY

Week 28

Lesson 5
Review

Objective

Children review and reinforce skills and concepts learned in this week and in previous weeks.

Program Materials

Engage
See Activity Cards for materials.

Additional Materials

Engage
See Activity Cards for materials.

Creating Context

The words *higher* and *lower* are words of comparison. Review the patterns of regular comparatives and superlatives in English. Introduce the examples *high, higher,* and *highest* and *low, lower,* and *lowest*. Have students work with *big, small, long,* and *short*.

1 Warm Up   5

Skill Practice COMPUTING
Blastoff!
Before beginning the **Free-Choice** activity, use **Blastoff!** Variation 2: **Incredible Shrinking People** with the whole group.

Purpose **Incredible Shrinking People** helps children build a foundation for subtraction.

Warm-Up Card 10

Monitoring Student Progress

If . . . children are ready,

Then . . . try the Challenge variation to help them integrate counting up and counting down.

2 Engage   20

Concept Building APPLYING
Free-Choice Activity
For the last day of the week, allow children to choose an activity from the previous weeks. Some activities they may choose include the following:
- Circle Land: **Skating Party**
- Picture Land: **Bravo!**
- Sky Land: **Drop the Counter**

Make a note of the activities children select. Do they prefer easy or challenging activities? If you believe your children would benefit from extra practice on specific skills, choose an activity for them.

3 Reflect   10

Extended Response REASONING

Ask questions such as the following:

- **What did you like about playing** Hurry Up Passengers, Going Up?
- **Was there anything about playing this game you didn't like?**
- **Did this game help you do something you couldn't do before? What did it help you do?**
- **What was easy when you were playing** Hurry Up Passengers, Going Down?
- **What was hard when you were playing** Hurry Up Passengers, Going Down?
- **What do you remember most about this game?**
- **What was your favorite part?**

4 Assess   10

A Gather Evidence

Formal Assessment

Have children complete the weekly test, *Assessment,* page 79. Record formal assessment scores on the Student Assessment Record, *Assessment,* page 100.

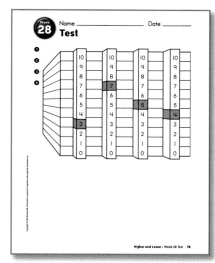

Assessment, p. 79

B Summarize Findings

Review the Student Assessment Records. Determine whether children have Minimal, Basic, or Secure understanding of the concepts presented in Week 28.

C Differentiate Instruction

Based on your observations, use these teaching strategies next week to follow up.

Minimal Understanding
- Repeat the Warm-Up and Engage activities to develop concepts of counting objects.
- Use *Building Blocks* computer activities beginning with Number Snapshots 2 to develop and reinforce numeration concepts.

Basic Understanding
- Repeat Engage activities in subsequent weeks to reinforce basic counting concepts.
- Use *Building Blocks* computer activities beginning with Number Snapshots 4 to reinforce this week's concepts.

Secure Understanding
- Use Challenge variations of this week's activities.
- Use computer activities to extend children's understanding of the Week 28 concepts.

Week 29 More Rotations

Week at a Glance

This week, children begin **Number Worlds,** Week 29, and continue to explore Circle Land.

Background

In Circle Land, children continue to explore numbers by examining dials. Children develop spatial intuitions in Circle Land that become the foundation for understanding many concepts in mathematics that deal with circular motion— such as pie charts, time, and number bases.

How Children Learn

As they begin this week's activities, children should be able to find the number that is just after or just before another number by counting up from 1.

By the end of the week, children should have improved skills that will let them immediately determine the number just before or just after another number by counting back or by counting on.

Skills Focus

- Identify or compute set size
- Associate set size with points on a dial
- Associate changes in quantity with movement on a dial
- Compare points on a dial
- Solve problems
- Identify set size when adding or subtracting 1

Teaching for Understanding

As children engage in these activities, they build basic addition and subtraction skills and recognize that distance traveled around a dial is greater when a set is increased and less when a set is decreased.

Observe closely while evaluating the Engage activities assigned for this week.

- Are children computing numbers correctly?
- Are children making valid comparisons using positions on the dial?
- Are children able to identify set size when a given set is increased by 1 or decreased by 1?

Math at Home

Give one copy of the Letter to Home, page A29, to each child. Complete the activity in class, and then encourage children to share it with their caregivers.

Letter to Home, Teacher Edition, p. A29

Math Vocabulary

stumble Trip; fall

near Close to

dial A circle with numbers and a pointer such as a clock

English Learners

SPANISH COGNATES

English	Spanish
circle	círculo
dial	dial
problem	problema
revolutions	revoluciones

ACCESS VOCABULARY

hand Another word for the pointer on a dial or clock

skating party A party at a skating rink

PACING	LESSON	LEARNING GOALS	MATERIALS	TECHNOLOGY
DAY 1	**Warm Up** **Circle Land:** Count Up*	Children practice counting up.	**Program Materials** Classroom Dial	**Building Blocks** Before and After Math
	Engage **Circle Land:** Ice, Be Nice at Our Skating Party	Children associate set size with points on a dial.	**Program Materials** • Skating Party Game Board • Pawns, 1 per child • Dot Cube • Skating +1\−1 Cards • 20 Award Cards	**e MathTools** Stopwatch Tool
DAY 2	**Warm Up** **Circle Land:** Blastoff!*	Children build a foundation for mental subtraction.	**Program Materials** Classroom Dial	**Building Blocks** Before and After Math
	Engage **Circle Land:** Ice, Be Nice at Our Skating Party	Children compare points on a dial.	**Program Materials** • Skating Party Game Board • Pawns, 1 per child • Dot Cube • Skating +1\−1 Cards • 20 Award Cards	
DAY 3	**Warm Up** **Circle Land:** Blastoff!*	Children associate counting back with moving counter-clockwise on a dial.	**Program Materials** Classroom Dial	**Building Blocks** Before and After Math
	Engage **Circle Land:** Ice, Be Nice at Our Skating Party	Children identify how many there will be if a set is increased or decreased by 1.	**Program Materials** • Skating Party Game Board • Pawns, 1 per child • Dot Cube • Skating +1\−1 Cards • 20 Award Cards	
DAY 4	**Warm Up** **Circle Land:** Count Up*	Children build a foundation for mental addition.	**Program Materials** Classroom Dial	**Building Blocks** Before and After Math
	Engage **Circle Land:** Ice, Be Nice at Our Skating Party	Children solve problems involving set size.	**Program Materials** • Skating Party Game Board • Pawns, 1 per child • Dot Cube • Skating +1\−1 Cards • 20 Award Cards	
DAY 5	**Warm Up** **Circle Land:** Blastoff!*	Children reinforce the reverse counting sequence.	**Program Materials** Classroom Dial	**Building Blocks** Review previous activities
	Review Free-Choice activity	Children will review and reinforce skills and concepts from this week and in previous weeks.	Materials will be selected from those used in previous weeks.	

* Includes Challenge Variations

Lesson 1

Objective

Children identify how many there will be if a set is increased or decreased by 1.

Program Materials

Warm Up
Classroom Dial
Engage
- Skating Party Game Boards, 1 per group
- Pawns, 1 per child
- Dot Cubes, 1 per group
- Skating +1\−1 Cards, 1 set per group
- Award Cards, 20 per group

Additional Materials
No additional materials needed

Access Vocabulary

rink An oval, ice covered area for ice-skaters to skate on
pick Choose

Creating Context

Many young children have neither been ice-skating nor been to an ice-skating rink. Use a picture to show how all skaters go around the rink in the same direction to avoid crashing into each other. Mention that ice-skaters must wear warm clothing. Children who live in warm climates may be unfamiliar with scarves, gloves, and other cold-weather clothing. Point out such clothing using pictures or by bringing in these items so that children can acquire this vocabulary.

---

1 Warm Up 5

Skill Building COMPUTING
Count Up
Before beginning **Ice, Be Nice at Our Skating Party,** use the **Count Up** activity with the whole group.

Purpose **Count Up** gives children practice counting up.

Warm-Up Card 11

Monitoring Student Progress

If . . . children have trouble associating counting up with movement on the dial,

Then . . . as they count, have them rotate their arms like the arrow on the dial.

---

2 Engage 30

Concept Building UNDERSTANDING
Ice, Be Nice at Our Skating Party
"Today we will do an activity similar to **Skating Party.**"
Follow the instructions on the Activity Card to play **Ice, Be Nice at Our Skating Party.** As children play, ask questions about what is happening in the activity.

Purpose **Ice, Be Nice at Our Skating Party** teaches children to associate set size with points on a dial.

Activity Card 35

MathTools Use the Stopwatch Tool to demonstrate and explore using dials.

Monitoring Student Progress

If . . . children are having trouble remembering which direction to move,

Then . . . remind them that when they roll they move up to higher numbers.

Teacher's Note Encourage children to use this week's vocabulary words as they engage in the activities, discuss math concepts, and make predictions.

Building Blocks For additional practice, children should complete **Building Blocks** Before and After.

 3 Reflect 10

Extended Response **REASONING**

Ask the following questions:

- **Who went around the rink the most times?** Children should choose the child who won.

- **How do you know?** Possible answer: He or she had the most Award Cards at the end of the game.

- **How come that child went around the rink more times than the other children? What helped him or her go farther than the rest of you?** Possible answers: He rolled a lot of big numbers. She got a lot of +1 Cards.

Informal Assessment

Use the Student Assessment Record, **Assessment,** page 100, to record informal observations.

COMPUTING	UNDERSTANDING
Count Up	**Ice, Be Nice at Our Skating Party**
Did the child	Did the child
❏ respond accurately?	❏ make important observations?
❏ respond quickly?	
❏ respond with confidence?	❏ extend or generalize learning?
❏ self-correct?	❏ provide insightful answers?
	❏ pose insightful questions?

Good for Ewe!

Lesson 2

Objective

Children make comparisons to solve problems involving set size.

Program Materials

Warm Up
Classroom Dial
Engage
- Skating Party Game Boards, 1 per group
- Pawns, 1 per child
- Dot Cubes, 1 per group
- Skating +1\−1 Cards, 1 set per group
- Award Cards, 20 per group

Additional Materials
No additional materials needed

Access Vocabulary

clockwise The direction the hand on a dial moves so that it moves toward the bigger numbers
counter-clockwise The direction on a dial that moves from larger numbers to smaller numbers

Creating Context

Ice, Be Nice at Our Skating Party uses a rhyme in the title. Not all languages feature words that rhyme. Young English Learners may need extra examples to learn the concept of rhyme. Practice rhyming words to familiarize English Learners with listening for rhyme.

1 Warm Up   5

Concept Building   COMPUTING

Blastoff!
Before beginning **Ice, Be Nice at Our Skating Party**, use the **Blastoff!** activity with the whole group.

Purpose Blastoff! helps children build a foundation for mental subtraction.

Warm-Up Card 12

Monitoring Student Progress

If . . . children have trouble counting individually,   **Then . . .** have them count as a group while volunteers move the hand on the dial.

2 Engage   30

Concept Building   UNDERSTANDING

Ice, Be Nice at Our Skating Party
"Today we will see who can go around the Skating Party Game Board the most times."

Follow the instructions on the Activity Card to play **Ice, Be Nice at Our Skating Party.** As children play, ask questions about what is happening in the activity.

Purpose Ice, Be Nice at Our Skating Party teaches children to compare points on a dial.

Activity Card 35

Monitoring Student Progress

If . . . children are having trouble determining the winner,

Then . . . remind them to count their Award Cards and see who has the most.

 Building Blocks For additional practice, children should complete **Building Blocks** Before and After.

3 Reflect 10

Extended Response REASONING

Ask questions such as the following:

- **What does this activity remind you of?** Accept all reasonable answers.

- **How would you explain how to play** Ice, Be Nice at Our Skating Party **to someone who isn't here?** Allow several children to answer. Possible answer: We take turns moving around the board by rolling the Dot Cube and picking cards; when we pass 0 we pick up an Award Card; when the teacher says stop, we all count our cards to see who has the most.

- **What do you like about this game?** Answers will vary.

4 Assess

Informal Assessment

Use the Student Assessment Record, **Assessment,** page 100, to record informal observations.

COMPUTING	UNDERSTANDING
Blastoff!	**Ice, Be Nice at Our Skating Party**
Did the child	Did the child
❏ respond accurately?	❏ pay attention to the contributions of others?
❏ respond quickly?	
❏ respond with confidence?	❏ contribute information and ideas?
❏ self-correct?	❏ improve on a strategy?
	❏ reflect on and check accuracy of work?

Easy as . . . 1 - 2 - 3

Lesson 3

Objective
Children associate set size with points on a dial.

Program Materials
Warm Up
Classroom Dial
Engage
- Skating Party Game Boards, 1 per group
- Pawns, 1 per child
- Dot Cubes, 1 per group
- Skating +1\−1 Cards, 1 set per group
- Award Cards, 20 per group

Additional Materials
No additional materials needed

Access Vocabulary
mistake Wrong answer
harder Takes more work to do

Creating Context
Young children may not yet know how to tell time on a clock, but there are many other dials that they may be familiar with such as the dial on a radio. Suggest that children look for and draw pictures of dials they have seen at home, in the classroom, or in the school.

---

1 Warm Up 5

Concept Building COMPUTING
Blastoff!
Before beginning **Ice, Be Nice at Our Skating Party,** use the **Blastoff!** activity with the whole group.

Purpose Blastoff! teaches children to associate counting back with moving counterclockwise on a dial.

Warm-Up Card 12

Monitoring Student Progress

If . . . children are ready, **Then . . .** use the Challenge variation.

---

2 Engage 30

Concept Building REASONING
Ice, Be Nice at Our Skating Party
"Today we will compare Award Cards to see who wins."

Follow the instructions on the Activity Card to play **Ice, Be Nice at Our Skating Party.** As children play ask questions about what is happening in the activity.

Purpose Ice, Be Nice at Our Skating Party helps children identify how many there will be if a set is increased or decreased by 1.

Activity Card 35

Monitoring Student Progress

If . . . children are not moving the correct number of spaces,

Then . . . remind them to count aloud and to listen for mistakes when others count. A common mistake is counting the space they are resting on as number 1 for their next move.

 Building Blocks For additional practice, children should complete **Building Blocks** Before and After.

 3 Reflect 10

Extended Response REASONING

Ask questions such as the following:

- **Do the Skating +1\−1 Cards make this activity harder or easier?** Accept all reasonable answers.

- **Why do you say that?** Answers will vary.

- **Do the +1\−1 Cards make it easier or harder to win?** Answers will vary.

- **Why do you think that?** Possible answer: They make it easier if you get a lot of +1 Cards and harder if you get a lot of −1 Cards.

 4 Assess

Informal Assessment

Use the Student Assessment Record, **Assessment,** page 100, to record informal observations.

COMPUTING	REASONING
Blastoff! Did the child	**Ice, Be Nice at Our Skating Party** Did the child
❑ respond accurately?	❑ provide a clear explanation?
❑ respond quickly?	❑ communicate reasons and strategies?
❑ respond with confidence?	❑ choose appropriate strategies?
❑ self-correct?	❑ argue logically?

STEADY JOB!

Lesson 4

Objective

Children build a foundation for mental subtraction.

Program Materials

Warm Up
Classroom Dial
Engage

- Skating Party Game Boards, 1 per group
- Pawns, 1 per child
- Dot Cubes, 1 per group
- Skating +1\−1 Cards, 1 set per group
- Award Cards, 20 per group

Additional Materials

No additional materials needed

Access Vocabulary

pawn An object that shows where you are on a game board
"time is up" Stop playing

Creating Context

Many young children do not have much experience with board games. Introduce English Learners to some of the routines used in board games such as determining who goes first, how to take turns, how to win, and what happens in case of a tie.

---

1 Warm Up 5

Skill Building COMPUTING
Count Up

Before beginning **Ice, Be Nice at Our Skating Party,** use the **Count Up** activity with the whole group.

Purpose **Count Up** builds a foundation for mental addition.

Warm-Up Card 11

Monitoring Student Progress

If . . . children have trouble predicting which number comes next,

Then . . . have them recount from 0.

---

2 Engage 30

Concept Building ENGAGING
Ice, Be Nice at Our Skating Party

"Today we will continue playing **Ice, Be Nice at Our Skating Party.**"

Follow the instructions on the Activity Card to play **Ice, Be Nice at Our Skating Party.** As children play, ask questions about what is happening in the activity.

Purpose **Ice, Be Nice at Our Skating Party** challenges children to solve problems involving set size.

Activity Card 35

Monitoring Student Progress

If . . . children are having trouble figuring out which direction to move when they pick a card,

Then . . . review what each card indicates and the actions that are associated with each symbol and icon.

 For additional practice, children should complete **Building Blocks** Before and After.

3 Reflect 10

Extended Response REASONING

Ask questions such as the following:

■ **How do you know how many spaces to move?** Possible answer: I count the dots on the Dot Cube and do what the Skating Card says.

■ **How can you check your answer?** Possible answer: count again

■ **How do you know what direction to move when you pick a card?** Allow several students to answer. Possible answers: Go forward to a bigger number if the card says +1, and go backward to a smaller number if the card says −1. Go forward if I skated well, and go backward if I fall.

4 Assess

Informal Assessment

Use the Student Assessment Record, **Assessment,** page 100, to record informal observations.

COMPUTING	ENGAGING
Count Up	**Ice, Be Nice at Our Skating Party**
Did the child	Did the child
❏ respond accurately?	❏ pay attention to the contributions of others?
❏ respond quickly?	
❏ respond with confidence?	❏ contribute information and ideas?
❏ self-correct?	
	❏ improve on a strategy?
	❏ reflect on and check accuracy of work?

Lesson 5

Review

Objective

Children review and reinforce skills and concepts learned in this week and in previous weeks.

Program Materials

Warm Up
Classroom Dial

Engage
See Activity Cards for materials.

Additional Materials

Engage
See Activity Cards for materials.

Creating Context

To aid English Learners, use gestures to demonstrate the meanings of some words. For example, *bumpy, fast, smooth,* and *slowly* can be demonstrated through gestures.

1 **Warm Up** 5

Concept Building COMPUTING

Blastoff!
Before beginning the **Free-Choice** activity, use the **Blastoff!** activity with the whole group.

Purpose **Blastoff!** reinforces the reverse counting sequence.

Warm-Up Card 12

Monitoring Student Progress

If . . . children are not counting together, **Then . . .** remind the class not to count until the volunteer has moved the hand on the dial to that number.

2 **Engage** 20

Concept Building APPLYING

Free-Choice Activity
For the last day of the week, allow children to choose an activity from the previous weeks. Some activities they may choose include the following:

- Line Land: **Next Number Up**
- Line Land: **Plus-Minus Game**
- Sky Land: **Hurry Up Passengers, Going Down**

Make a note of the activities children select. Do they prefer easy or challenging activities? If you believe your children would benefit from extra practice on specific skills, choose an activity for them.

3 Reflect 10

Extended Response REASONING

Ask questions such as the following:

- **What did you like about playing** Ice, Be Nice at Our Skating Party?
- **Was there anything about playing this game you didn't like?**
- **Did this game help you do something you couldn't do before? What did it help you do?**
- **What was easy when you were playing** Ice, Be Nice at Our Skating Party?
- **What was hard when you were playing** Ice, Be Nice at Our Skating Party?
- **What do you remember most about this game?**
- **What was your favorite part?**

4 Assess 10

A Gather Evidence

Formal Assessment

Have children complete the weekly test, **Assessment,** page 81. Record formal assessment scores on the Student Assessment Record, **Assessment,** page 100.

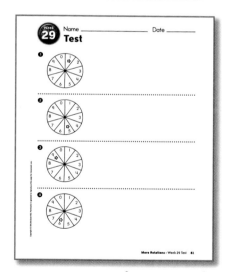

Assessment, p. 81

B Summarize Findings

Review the Student Assessment Records. Determine whether children have Minimal, Basic, or Secure understanding of the concepts presented in Week 29.

C Differentiate Instruction

Based on your observations, use these teaching strategies next week to follow up.

Minimal Understanding

- Repeat the Warm-Up and Engage activities to develop concepts of counting objects.
- Use **Building Blocks** computer activities beginning with Before and After to develop and reinforce numeration concepts.

Basic Understanding

- Repeat Engage activities in subsequent weeks to reinforce basic counting concepts.
- Use **Building Blocks** computer activities beginning with Before and After to reinforce this week's concepts.

Secure Understanding

- Use Challenge variations of this week's activities.
- Use computer activities to extend children's understanding of the Week 29 concepts.

Sequencing Numbers

Week at a Glance

This week, children begin **Number Worlds,** Week 30, and continue to explore Picture Land.

Background

In Picture Land, numbers are represented as sets of dots. When a child is asked, "What number do you have?" he or she might count each dot to determine the set size or identify the amount by recognizing the pattern that marks that amount. Thinking about numbers as a product of counting and as a set that has a particular value will become intertwined in the child's sense of number.

Teaching for Understanding

As children engage in these activities, they strengthen their understanding of the position and magnitude of each numeral.

Observe closely while evaluating the Engage activities assigned for this week.

- Are children identifying numerals correctly?
- Are children putting numerals in order?

How Children Learn

As they begin this week's activities, children should be making accurate comparisons of smaller groups by counting.

By the end of the week, children should be better able to order units and collections. For example, when children are given cards with the numerals 1–6 on them, they will put them in order.

You hit the SPOT!

High Flyer!

Math at Home

Give one copy of the Letter to Home, page A30, to each child. Complete the activity in class, and then encourage children to share it with their caregivers.

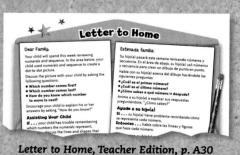

Letter to Home, Teacher Edition, p. A30

Math Vocabulary

connect To draw a line between two things

left The side where writing starts

right The opposite of left

English Learners

SPANISH COGNATES

English	Spanish
sequence	secuencia
numbers	números
position	posición
connect	conectar

ACCESS VOCABULARY

numerals Number symbols

strategy A plan for how to solve a problem

Week 30 Planner Sequencing Numbers

PACING	LESSON	LEARNING GOALS	MATERIALS	TECHNOLOGY
DAY 1	**Warm Up** **Picture Land:** Name That Numeral*	Children demonstrate knowledge of numerals.	**Program Materials** • Large Number Cards (1–5) • Large Number Cards (1–12) for the Challenge	**Building Blocks** Countdown Crazy **e MathTools** Number Line
	Engage **Picture Land:** Sequencing Numbers*	Children sequence numerals.	**Program Materials** • Small Number Cards (1–5), 1 set per child • Small Number Cards (1–10), 1 set per child for the Challenge	
DAY 2	**Warm Up** **Picture Land:** Name That Numeral*	Children identify numerals.	**Program Materials** • Large Number Cards (1–5) • Large Number Cards (1–12) for the Challenge	**Building Blocks** Countdown Crazy
	Engage **Picture Land:** Sequencing Numbers*	Children strengthen their understanding of the value and magnitude of numerals.	**Program Materials** • Small Number Cards (1–5), 1 set per child • Small Number Cards (1–10), 1 set per child for the Challenge	
DAY 3	**Warm Up** **Picture Land:** Name That Numeral Variation*	Children work in teams to recognize numerals.	**Program Materials** • Large Number Cards (1–5) • Large Number Cards (1–12) for the Challenge	**Building Blocks** Countdown Crazy
	Engage **Picture Land:** Connecting the Dots*	Children connect numbers in sequence.	**Program Materials** Picture Land Activity Sheet 1, p. B4, 1 per child **Additional Materials** Pencils or crayons, 1 per child	
DAY 4	**Warm Up** **Picture Land:** Name That Numeral*	Children gain experience recognizing numerals.	**Program Materials** • Large Number Cards (1–5) • Large Number Cards (1–12) for the Challenge	**Building Blocks** Countdown Crazy
	Engage **Picture Land:** Connecting the Dots*	Children build an understanding of the position of each numeral in the number sequence.	**Program Materials** Picture Land Activity Sheet 2, p. B5, 1 per child **Additional Materials** Pencils or crayons, 1 per child	
DAY 5	**Warm Up** **Picture Land:** Name That Numeral Variation*	Children develop proficiency in naming numerals.	**Program Materials** • Large Number Cards (1–5) • Large Number Cards (1–12) for the Challenge	**Building Blocks** Review previous activities
	Review Free-Choice activity	Children will review and reinforce skills and concepts from this week and in previous weeks.	Materials will be selected from those used in previous weeks.	

* Includes Challenge Variations

Lesson 1

Objective

Children strengthen their understanding of the value and magnitude of numerals.

Program Materials

Warm Up
- Large Number Cards (1–5)
- Large Number Cards (1–12) for the Challenge

Engage
- Small Number Cards (1–5), 1 set per child
- Small Number Cards (1–12), 1 set per child for the Challenge

Additional Materials

No additional materials needed

Access Vocabulary

set of cards Cards with all the numbers you need to play a game

"on your own" Without help

Creating Context

When giving instructions for an activity or game, it is important to check with English Learners frequently to make sure they understand. Demonstrate each step and ask children to role-play these steps to be sure they understand what to do.

Way to hop to it!

1 Warm Up 5

Skill Building COMPUTING

Name That Numeral

Before beginning **Sequencing Numbers,** use the **Name That Numeral** activity with the whole group.

Purpose Name That Numeral gives children an opportunity to demonstrate knowledge of numerals.

Warm-Up Card 5

Monitoring Student Progress

If . . . a child is unsure about a numeral,

Then . . . ask another child to name it and discuss the numeral's appearance with the class.

2 Engage 30

Concept Building UNDERSTANDING

Sequencing Numbers

"Today we will put Number Cards in order."

Follow the instructions on the Activity Card to play **Sequencing Numbers.** As children play, ask questions about what is happening in the activity.

Purpose Sequencing Numbers teaches children to put numerals in order.

Activity Card 20

 **MathTools** Use the Number Line to demonstrate and explore the number sequence.

Monitoring Student Progress

| **If . . .** children have trouble sequencing numerals, | **Then . . .** model the process for them. |

 Teacher's Note Encourage children to use this week's vocabulary words as they engage in the activities, discuss math concepts, and make predictions.

Building Blocks For additional practice sequencing, children should complete **Building Blocks** Countdown Crazy.

3 Reflect 10

Extended Response REASONING

Ask the following questions:

- **What was the highest number you had?** 5
- **Where did you put it when you put the numbers in order?** Possible answer: at the end of the line
- **What was the smallest number you had?** 1
- **Where did you put it when you put the numbers in order?** Possible answer: at the beginning of the line

4 Assess

Informal Assessment

Use the Student Assessment Record, **Assessment,** page 100, to record informal observations.

COMPUTING	UNDERSTANDING
Name That Numeral	**Sequencing Numbers**
Did the child	Did the child
❑ respond accurately?	❑ make important observations?
❑ respond quickly?	❑ extend or generalize learning?
❑ respond with confidence?	❑ provide insightful answers?
❑ self-correct?	❑ pose insightful questions?

Lesson 2

Objective
Children identify and sequence numerals.

Program Materials
Warm Up
- Large Number Cards (1–5)
- Large Number Cards (1–12) for the Challenge

Engage
- Small Number Cards (1–5), 1 set per child
- Small Number Cards (1–12), 1 set per child for the Challenge

Additional Materials
No additional materials needed

Access Vocabulary
whole group Everyone
next The one after this one

Creating Context
There are many words in English that have multiple meanings. English Learners are sometimes confused by the use of a familiar word in an unfamiliar context. For example, in this lesson cards with large numbers and small numbers are discussed. Remind children that this description refers to the amount, not the size, of the cards. Encourage children to ask for clarification if they are unsure.

Math-a-saurus

---

1 Warm Up 5

Concept Building `COMPUTING`
Name That Numeral
Before beginning **Sequencing Numbers,** use the **Name That Numeral** activity with the whole group.

Purpose **Name That Numeral** helps children identify numerals.

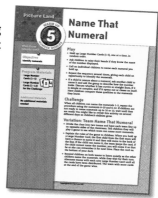
Warm-Up Card 5

Monitoring Student Progress

If . . . children can name numerals 1–5,	Then . . . repeat the procedure using the numerals 6–10 or 6–12.

---

2 Engage 30

Concept Building `ENGAGING`
Sequencing Numbers
"Today we will try to make sure that the numbers are in order."

Follow the instructions on the Activity Card to play **Sequencing Numbers.** As children play, ask questions about what is happening in the activity.

Purpose **Sequencing Numbers** strengthens children's understanding of the value and magnitude of numerals.

Activity Card 20

Monitoring Student Progress

If . . . children cannot remember which number comes next,

Then . . . have them count up from 1 to help them remember.

 Building Blocks For additional practice sequencing, children should complete **Building Blocks** Countdown Crazy.

3 Reflect 10

Extended Response REASONING

Ask questions such as the following:

- **How did you know what order to put the numbers in?** Possible answer: I counted.
- **What does this remind you of?** Accept all reasonable answers.

4 Assess

Informal Assessment

Use the Student Assessment Record, **Assessment,** page 100, to record informal observations.

COMPUTING	ENGAGING
Name That Numeral	**Sequencing Numbers**
Did the child	Did the child
❏ respond accurately?	❏ pay attention to the contributions of others?
❏ respond quickly?	
❏ respond with confidence?	❏ contribute information and ideas?
❏ self-correct?	❏ improve on a strategy?
	❏ reflect on and check accuracy of work?

Bee-u-tiful WORK

No "lion"—Math is fun!

Lesson 3

Objective

Children build an understanding of the position of each numeral in the number sequence.

Program Materials

Warm Up
- Large Number Cards (1–5)
- Large Number Cards (1–12) for the Challenge

Engage
- Picture Land Activity Sheet 1, p. B4, 1 per child
- Picture Land Activity Sheet 3, p. B6, 1 per child for the Challenge

Additional Materials

Engage
Pencils or crayons, 1 per child

Access Vocabulary

numerals Number symbols
connect To draw a line between two things

Creating Context

Ask children if they have ever heard the expression "connect the dots." Explain that it often refers to a puzzle in which you draw a line from one dot to another in counting order to create a picture. Tell them that people sometimes suggest that a mystery can be solved by following the clues so they say, "just connect the dots." Help them understand the relationship between the two phrases.

1 Warm Up 5

Skill Building COMPUTING

Name That Numeral

Before beginning **Connecting the Dots**, use **Name That Numeral** Variation: **Team Name That Numeral** with the whole group.

Purpose Team Name That Numeral has children work in teams to recognize numerals.

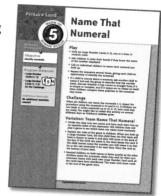

Warm-Up Card 5

Monitoring Student Progress

If . . . children are comfortable with the game, **Then . . .** add a time limit to make it challenging.

2 Engage 30

Concept Building REASONING

Connecting the Dots

"Today we will make a picture by putting numbers in order."

Follow the instructions on the Activity Card to play **Connecting the Dots.** As children play, ask questions about what is happening in the activity.

Purpose Connecting the Dots gives children an incentive to connect numbers in sequence.

Activity Card 21

Monitoring Student Progress

If . . . children have trouble keeping track of the numerals as they draw lines,

Then . . . have them circle the numbers they are connecting.

Building **B**locks For additional practice sequencing, children should complete **Building Blocks** Countdown Crazy.

3 Reflect   10 ⊙

Extended Response **REASONING**

Ask questions such as the following:

- **Could you have done this another way? Why or why not?** Possible answer: No, it wouldn't make the picture then.

- **How do you know when you're done?** Possible answer: All the dots are connected.

- **What did you like about this activity?** Allow several children to answer.

4 Assess

Informal Assessment

Use the Student Assessment Record, **Assessment,** page 100, to record informal observations.

COMPUTING	REASONING
Name That Numeral	**Connecting the Dots**
Did the child	Did the child
❏ respond accurately?	❏ provide a clear explanation?
❏ respond quickly?	❏ communicate reasons and strategies?
❏ respond with confidence?	
❏ self-correct?	❏ choose appropriate strategies?
	❏ argue logically?

Way to hop to it!

Bonus!

Lesson 4

Objective

Children connect numbers in sequence.

Program Materials

Warm Up
- Large Number Cards (1–5)
- Large Number Cards (1–12) for the Challenge

Engage
- Picture Land Activity Sheet 2, p. B5, 1 per child
- Picture Land Activity Sheet 4, p. B7, 1 per child for the Challenge

Additional Materials

Engage
Pencils or crayons, one per child

Access Vocabulary

dots Small round spots that often make a design or pattern
mistake Error, incorrect answer
strategy A plan of action to solve a problem

Creating Context

Help children demonstrate their grasp of the skills used in this lesson by asking, "What do you think would happen if you connected the dots while counting backward? Would you get the same picture?" Consider giving children an additional activity sheet so that they can test their predictions.

---

1 Warm Up 5

Skill Building `COMPUTING`

Name That Numeral

Before beginning **Connecting the Dots,** use the **Name That Numeral** activity with the whole group.

Purpose **Name That Numeral** gives children experience recognizing numerals.

Warm-Up Card 5

Monitoring Student Progress

If . . . children need help with other skills,

Then . . . use a different Warm-Up activity.

---

2 Engage 30

Concept Building `ENGAGING`

Connecting the Dots

"Today we will make another picture by connecting the dots."

Follow the instructions on the Activity Card to play **Connecting the Dots.** As children play, ask questions about what is happening in the activity.

Activity Card 21

Purpose **Connecting the Dots** helps children build an understanding of the position of each numeral in the number sequence.

Monitoring Student Progress

If . . . children have trouble remembering the sequence to use,

Then . . . let them put Number Cards in order and then reference the cards as they do the Activity Sheet.

 Building Blocks For additional practice sequencing, children should complete **Building Blocks** Countdown Crazy.

 Teacher's Note Encourage children to use this week's vocabulary words as they engage in the activities, discuss math concepts, and make predictions.

3 Reflect 10

Extended Response REASONING

Ask questions such as the following:

- **What do you notice about this activity?** Accept all reasonable answers.

- **What does it remind you of?** Answers will vary.

- **Was it hard or easy for you to do this activity?** Allow several children to answer, and use their answers to assess their progress.

4 Assess

Informal Assessment

Use the Student Assessment Record, **Assessment,** page 100, to record informal observations.

COMPUTING	ENGAGING
Name That Numeral	**Connecting the Dots**
Did the child	Did the child
❏ respond accurately?	❏ pay attention to the contributions of others?
❏ respond quickly?	
❏ respond with confidence?	❏ contribute information and ideas?
❏ self-correct?	
	❏ improve on a strategy?
	❏ reflect on and check accuracy of work?

Lesson 5
Review

Objective

Children review and reinforce skills and concepts learned in this and previous weeks.

Program Materials

Warm Up
- Large Number Cards (1–5)
- Large Number Cards (1–12) for the Challenge

Engage
See Activity Cards for materials.

Additional Materials

Engage
See Activity Cards for materials.

Creating Context

In addition to assessing children's understanding of math concepts, complete an observation of their proficiency in math vocabulary.

✔ Cumulative Assessment

After Lesson 5 is complete, children should complete the Cumulative Assessment, **Assessment,** pages 97–98. Using the key on **Assessment,** page 96, identify incorrect responses. Reteach and review the suggested activities to reinforce concept understanding.

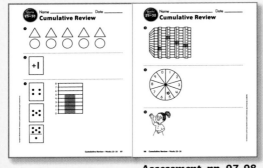

Assessment, pp. 97–98

1 Warm Up   5

Skill Practice COMPUTING
Name That Numeral

Before beginning the **Free-Choice** activity, use **Name That Numeral** Variation: **Team Name That Numeral** with the whole group.

Purpose **Team Name That Numeral** helps children develop proficiency in naming numerals.

Warm-Up Card 5

Monitoring Student Progress

If . . . a child cannot name a number,   **Then . . .** name it for the class so they can remember it for next time.

2 Engage   20

Concept Building APPLYING
Free-Choice Activity

For the last day of the week, allow the children to choose an activity from the previous weeks. Some activities they may choose include the following:

- Object Land: **Count and Compare**
- Line Land: **Plus-Minus Game**
- Circle Land: **Ice, Be Nice at Our Skating Party**

Make a note of the activities children select. Do they prefer easy or challenging activities? If you believe your children would benefit from extra practice on specific skills, choose an activity for them.

3 Reflect

10

Extended Response REASONING

Ask questions such as the following:

■ **What did you like about playing** Sequencing Numbers?

■ **Was there anything about playing this game you didn't like?**

■ **Did this game help you do something you couldn't do before? What did it help you do?**

■ **What was easy when you were playing** Connecting the Dots?

■ **What was hard when you were playing** Connecting the Dots?

■ **What do you remember most about this game?**

■ **What was your favorite part?**

4 Assess

10

A Gather Evidence

Formal Assessment

Have children complete the weekly test, *Assessment,* page 83. Record formal assessment scores on the Student Assessment Record, *Assessment,* page 100.

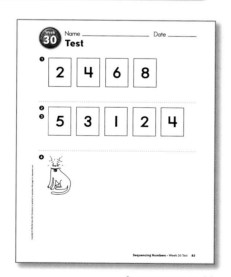

Assessment, p. 83

B Summarize Findings

Review the Student Assessment Records. Determine whether children have Minimal, Basic, or Secure understanding of the concepts presented in Week 30.

C Differentiate Instruction

Based on your observations, use these teaching strategies next week to follow up.

Minimal Understanding

• Repeat the Warm-Up and Engage activities to develop concepts of counting objects.

• Use *Building Blocks* computer activities beginning with Countdown Crazy to develop and reinforce numeration concepts.

Basic Understanding

• Repeat Engage activities in subsequent weeks to reinforce basic counting concepts.

• Use *Building Blocks* computer activities beginning with Countdown Crazy to reinforce this week's concepts.

Secure Understanding

• Use Challenge variations of this week's activities.

• Use computer activities to extend children's understanding of the Week 30 concepts.

Appendix

Letter to Home

Dear Family,

Your child will spend this week learning and reviewing counting. In the area below, your child demonstrated counting skills by drawing a circle for each number counted.

Discuss the picture with your child by asking the following questions:

- **How many circles did you draw?**
- **How do you know?**
- **Will you count to check?**

Help your child further develop counting skills by working together to count objects around your home.

Assisting Your Child

If . . . your child has difficulty counting,
Then . . . have the child repeat short sequences (1, 2, 3) after you.

This week, your child will continue using the *Number Worlds* program to master these concepts while counting objects and sounds.

Estimada familia:

Su hijo(a) pasará esta semana aprendiendo y revisando la forma de contar. En el área de abajo, su hijo(a) demostró sus destrezas para contar al dibujar un círculo por cada número que contó.

Hable con su hijo(a) acerca del dibujo haciéndole las siguientes preguntas:

- **¿Cuántos círculos dibujaste?**
- **¿Cómo sabes?**
- **¿Vas a contar para comprobar?**

Ayude a su hijo(a) a desarrollar destrezas para contar trabajando juntos para contar objetos en la casa.

Ayude a su hijo(a)

Si . . . su hijo(a) tiene dificultad para contar,
Entonces . . . pida a su hijo(a) que repita secuencias cortas (1, 2, 3) después de usted.

Esta semana su hijo(a) continuará usando el programa *El mundo de los números* para dominar estos conceptos mientras cuenta objetos y sonidos.

Draw a circle for each number you can count up to.

Letter to Home

Dear Family,

Your child will spend this week learning and reviewing colors and counting. In the area below, your child demonstrated knowledge of colors and counting skills by coloring the number of shapes requested.

Discuss the picture with your child by asking the following questions:

- **How many red circles are there?**
- **How many yellow circles? How many blue circles?**
- **How do you know?**
- **Will you count to check?**

Encourage your child to count and to identify colors when answering your questions.

Assisting Your Child

If . . . your child has trouble identifying colors, **Then . . .** review the colors using objects in the room.

This week, your child will continue using the *Number Worlds* program to master these concepts while sorting and counting.

Estimada familia:

Su hijo(a) pasará esta semana aprendiendo y revisando los colores y contando. En el área de abajo, su hijo(a) demostró conocimiento de los colores y destrezas para contar al colorear el número de figuras solicitado.

Hable con su hijo(a) acerca del dibujo haciéndole las siguientes preguntas:

- **¿Cuántos círculos rojos hay?**
- **¿Cuántos círculos amarillos? ¿Cuántos círculos azules?**
- **¿Cómo sabes?**
- **¿Vas a contar para comprobar?**

Anime a su hijo(a) a contar e identificar los colores cuando contesta sus preguntas.

Ayude a su hijo(a)

Si . . . su hijo(a) tiene problemas para identificar los colores, **Entonces . . .** revise los colores usando objetos en el cuarto.

Esta semana su hijo(a) continuará usando el programa *El mundo de los números* para dominar estos conceptos mientras clasifica y cuenta.

Color 1 circle red, 2 circles yellow, 3 circles blue, and 4 circles green.

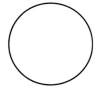

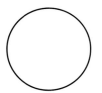

Letter to Home

Dear Family,

Your child will spend this week sorting groups of objects and comparing them to see which have *more* or *less*. In the area below, your child demonstrated knowledge of colors and counting skills by coloring the number of shapes requested.

Discuss the picture with your child by asking the following questions:

- **How many red squares are there?**
- **How many blue squares are there?**
- **Are there more red squares or more blue squares?**
- **How do you know?**

Encourage your child to count and to use the terms *more* and *less* when answering your questions.

Assisting Your Child

If . . . your child has trouble counting the squares,

Then . . . have your child count objects instead of pictures.

This week, your child will continue using the **Number Worlds** program to master these concepts while sorting by color and counting.

Estimada familia:

Su hijo(a) pasará esta semana clasificando grupos de objetos y comparándolos para ver cuál tiene *más* o *menos*. En el área de abajo, su hijo(a) demostró conocimiento de los colores y destrezas para contar al colorear el número de figuras solicitado.

Hable con su hijo(a) acerca del dibujo haciéndole las siguientes preguntas:

- **¿Cuántos cuadros rojos hay?**
- **¿Cuántos cuadros azules hay?**
- **¿Hay más cuadros rojos o más cuadros azules?**
- **¿Cómo sabes?**

Anime a su hijo(a) a contar y usar los términos *más* y *menos* cuando contesta sus preguntas.

Ayude a su hijo(a)

Si . . . su hijo(a) tiene problemas para contar los cuadros,

Entonces . . . pida a su hijo(a) que cuente objetos en vez de dibujos.

Esta semana su hijo(a) continuará usando el programa **El mundo de los números** para dominar estos conceptos mientras clasifica por color y cuenta.

Color 1 square red and 3 squares blue.

   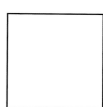

Letter to Home

Dear Family,

Your child will spend this week sorting and matching objects with cards that show the same number of dots. In the area below, your child demonstrated matching skills by ringing the requested card and drawing a matching number of circles.

Discuss the picture with your child by asking the following questions:

- **How many dots are on the square you ringed?**
- **How many circles did you draw?**
- **Does the number of circles match the number of dots?**
- **How do you know?**

Encourage your child to count and to use the terms *more* and *less* when answering your questions.

Assisting Your Child

If . . . your child has trouble identifying the correct card,

Then . . . have your child count the dots on each card.

This week, your child will continue using the *Number Worlds* program to master these concepts while matching objects.

Estimada familia:

Su hijo(a) pasará esta semana clasificando e igualando objetos con tarjetas que muestran el mismo número de puntos. En el área de abajo, su hijo(a) demostró destrezas para igualar al tomar la carta indicada y dibujar un número igual de círculos.

Hable con su hijo(a) acerca del dibujo haciéndole las siguientes preguntas:

- **¿Cuántos puntos hay en el cuadro que tomaste?**
- **¿Cuántos círculos dibujaste?**
- **¿El número de círculos es igual al número de puntos?**
- **¿Cómo sabes?**

Anime a su hijo(a) a contar y usar los términos *más* y *menos* cuando contesta sus preguntas.

Ayude a su hijo(a)

Si . . . su hijo(a) tiene problemas para identificar la tarjeta correcta,

Entonces . . . pida a su hijo(a) que cuente los puntos en cada tarjeta.

Esta semana su hijo(a) continuará usando el programa *El mundo de los números* para dominar estos conceptos mientras iguala objetos.

Ring the card that shows 2 dots. Then draw a matching number of circles.

 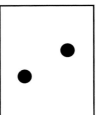 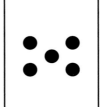

Letter to Home

Dear Family,

Your child will spend this week identifying shapes. In the area below, your child showed his or her knowledge of shapes by coloring the shapes specific colors.

Discuss the picture with your child by asking the following questions:

- **Which one is a circle?**
- **What makes it look like a circle?**
- **Which shape is a rectangle? A triangle? A square?**
- **How do you know?**

Encourage your child to identify shapes around your home.

Assisting Your Child

If . . . your child has trouble identifying the correct shape,
Then . . . discuss ways to recognize shapes.

This week, your child will continue using the **Number Worlds** program to master these concepts while discussing shapes.

Estimada familia:

Su hijo(a) pasará esta semana identificando figuras. En el área de abajo, su hijo(a) demostró su conocimiento de las figuras coloreándolas de colores específicos.

Hable con su hijo(a) acerca del dibujo haciéndole las siguientes preguntas:

- **¿Cuál es el círculo?**
- **¿Qué lo hace ver como círculo?**
- **¿Cuál figura es un rectángulo? ¿Un triángulo? ¿Un cuadrado?**
- **¿Cómo sabes?**

Anime a su hijo(a) a identificar las figuras que hay en su casa.

Ayude a su hijo(a)

Si . . . su hijo(a) tiene problemas para identificar la figura correcta,
Entonces . . . hable sobre las formas de reconocer las figuras.

Esta semana su hijo(a) continuará usando el programa **El mundo de los números** para dominar estos conceptos mientras habla de las figuras.

Color the square red, the circle blue, the triangle green, and the rectangle yellow.

   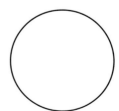

Letter to Home

Dear Family,

Your child will spend this week matching colors and shapes. In the area below, your child demonstrated knowledge of shapes by matching the shapes.

Discuss the picture with your child by asking the following questions:

- (Point to a shape.) **Can you tell me what shape this is?**
- **Which shape is a triangle?**
- **Which circle is bigger?**
- **Which rectangle is smaller?**

Encourage your child to explain how he or she knows the answer.

Assisting Your Child

If . . . your child has trouble identifying a specific shape,

Then . . . have your child try to make the shape with his or her body.

This week, your child will continue using the *Number Worlds* program to master these concepts while matching shapes and colors.

Estimada familia:

Su hijo(a) pasará esta semana igualando colores y figuras. En el área de abajo, su hijo(a) demostró conocimiento de las figuras al igualarlas.

Hable con su hijo(a) acerca del dibujo haciéndole las siguientes preguntas:

- (Apunte a una figura) **¿Puedes decirme qué figura es?**
- **¿Cuál figura es un triángulo?**
- **¿Cuál círculo es más grande?**
- **¿Cuál círculo es más chico?**

Anime a su hijo(a) a explicar cómo sabe la respuesta.

Ayude a su hijo(a)

Si . . . su hijo(a) tiene problemas identificando una figura específica,

Entonces . . . pida a su hijo(a) que trate de hacer la figura con su cuerpo.

Esta semana su hijo(a) continuará usando el programa *El mundo de los números* para dominar estos conceptos mientras iguala figuras y colores.

Draw lines to connect the matching shapes.

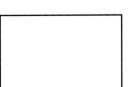

Letter to Home

Dear Family,

Your child will spend this week learning that the time required to perform an action increases with the number of actions. In the area below, your child applied this concept by drawing a specific number of shapes.

Discuss the picture with your child by asking the following questions:

- **Are there more circles or more triangles?**
- **How do you know?**
- **Will you count to make sure?**
- **Can you clap four times?**

Encourage your child to use numbers and comparisons when answering your questions.

Assisting Your Child

If . . . your child does not stop at 4 when clapping,
Then . . . remind him or her to count up to 4 and then stop.

This week, your child will continue using the *Number Worlds* program to master these concepts by following directions on cards that tell him or her how many times to perform an action.

Estimada familia:

Su hijo(a) pasará esta semana aprendiendo que el tiempo requerido para desempeñar una acción aumenta más con el número de acciones. En el área de abajo, su hijo(a) aplicó este concepto al dibujar un número específico de figuras.

Hable con su hijo(a) acerca del dibujo haciéndole las siguientes preguntas:

- **¿Hay más círculos o más triángulos?**
- **¿Cómo sabes?**
- **¿Vas a contar para estar seguro(a)?**
- **¿Puedes aplaudir cuatro veces?**

Anime a su hijo(a) a usar los números y comparaciones cuando contesta sus preguntas.

Ayude a su hijo(a)

Si . . . su hijo(a) no se detiene en 4 cuando aplaude,
Entonces . . . recuérdele que debe contar hasta 4 y detenerse.

Esta semana su hijo(a) continuará usando el programa *El mundo de los números* para dominar estos conceptos siguiendo instrucciones en tarjetas que le dicen cuántas veces debe realizar una acción.

Draw one circle then draw four triangles.

Letter to Home

Dear Family,

Your child will spend this week counting out objects and making sure that everyone has one. In the area below, your child applied this ability by giving each family member an apple.

Discuss the picture with your child by asking the following questions:

- **How many apples are there?**
- **How do you know?**
- **Is there one for each person?**
- **Will you count to make sure?**

Encourage your child to use numbers and comparisons when answering your questions.

Assisting Your Child

If . . . your child needs more help,
Then . . . have him or her help with tasks such as setting the table so that everyone has a plate.

This week, your child will continue using the **Number Worlds** program to master these concepts while counting out a given number of objects.

Estimada familia:

Su hijo(a) irá contando esta semana objetos y se asegurará de que cada quién tenga uno. En el área de abajo, su hijo(a) aplicó esta destreza al darle a cada miembro de la familia una manzana.

Hable con su hijo(a) acerca del dibujo haciéndole las siguientes preguntas:

- **¿Cuántas manzanas hay?**
- **¿Cómo sabes?**
- **¿Hay una para cada persona?**
- **¿Vas a contar para estar seguro?**

Anime a su hijo(a) a usar números y comparaciones cuando contesta sus preguntas.

Ayude a su hijo(a)

Si . . . su hijo(a) necesita más ayuda,
Entonces . . . pida a su hijo(a) que ayude con tareas, como poner la mesa, para que cada quien tenga un plato.

Esta semana su hijo(a) continuará usando el programa **El mundo de los números** para dominar estos conceptos mientras va contando un número dado de objetos.

Draw one apple for each person in your family.

Letter to Home

Dear Family,

Your child will spend this week identifying counting errors. In the area below, your child applied this concept by finding a mistake and identifying which number should have been counted.

Discuss the picture with your child by asking the following questions:

- **How many circles did you draw?**
- **How do you know?**
- **Will you count to make sure?**

Encourage your child to identify counting mistakes by repeating the activity, and choose a different number to skip each time.

Assisting Your Child

If . . . your child makes a mistake when counting,
Then . . . have your child start counting from 1 again.

This week, your child will continue using the **Number Worlds** program to master these concepts by detecting errors when the teacher counts.

Estimada familia:

Su hijo(a) pasará esta semana identificando errores al contar. En el área de abajo, su hijo(a) aplicó esto al encontrar un error e identificar cuál número debería haberse contado.

Hable con su hijo(a) acerca del dibujo haciéndole las siguientes preguntas:

- **¿Cuántos círculos dibujaste?**
- **¿Cómo sabes?**
- **¿Vas a contar para comprobar?**

Anime a su hijo(a) a identificar los errores al contar repitiendo la actividad, seleccionando un número diferente para omitir cada vez.

Ayude a su hijo(a)

Si . . . su hijo(a) comete un error cuando cuenta,
Entonces . . . pida a su hijo(a) que empiece a contar desde el 1 otra vez.

Esta semana su hijo(a) continuará usando el programa **El mundo de los números** para dominar estos conceptos al detectar errores cuando el maestro cuenta.

I will count to five. *1, 2, 3, 5.* What number did I skip? Draw that many circles.

Letter to Home

Dear Family,

Your child will spend this week matching cards that show the same number of objects. In the area below, your child applied this by drawing a ball for each dog.

Discuss the picture with your child by asking the following questions:

■ **How many dogs are there?**
■ **Will you count to make sure?**
■ **How many balls did you draw?**
■ **Are there enough balls for your dogs?**

Encourage your child to explain his or her answers by asking, "How do you know?"

Assisting Your Child

If . . . your child makes a mistake when counting,
Then . . . have your child touch each picture as he or she counts.

This week, your child will continue using the *Number Worlds* program to master these concepts by matching cards to make sure that each dog has a bone.

Estimada familia:

Su hijo(a) pasará esta semana igualando tarjetas que muestren el mismo número de objetos. En el área de abajo, su hijo(a) aplicó esto al dibujar una pelota para cada perro.

Hable con su hijo(a) acerca del dibujo haciéndole las siguientes preguntas:

■ **¿Cuántos perros hay?**
■ **¿Vas a contar para verificar?**
■ **¿Cuántas pelotas dibujaste?**
■ **¿Hay suficientes pelotas para tus perros?**

Anime a su hijo(a) a explicar sus repuestas al preguntar, "¿Cómo sabes?"

Ayude a su hijo(a)

Si . . . su hijo(a) comete un error al contar,
Entonces . . . pida a su hijo(a) que toque cada dibujo mientras va contando.

Esta semana su hijo(a) continuará usando el programa *El mundo de los números* para dominar estos conceptos igualando tarjetas para asegurarse de que cada perro tenga un hueso.

Draw 1 ball for each dog.

Letter to Home

Dear Family,

Your child will spend this week finding matching cards. In the area below, your child demonstrated matching skills by finding the cards that show the same numbers of dots.

Discuss the picture with your child by asking the following questions:

- (Point to a card.) **How many dots are on this card?**
- **Which card has more dots?**
- **Are there enough dots on this card to match that card?**

Encourage your child to explain how he or she knows the answer.

Assisting Your Child

If . . . your child has trouble making a match, **Then . . .** have your child touch each dot as he or she counts and show the number with his or her fingers so that your child remembers what number the card represents.

This week, your child will continue using the *Number Worlds* program to master these concepts.

Estimada familia:

Su hijo(a) pasará esta semana buscando tarjetas que sean iguales. En el área de abajo, su hijo(a) demostró sus destrezas para igualar al hallar las tarjetas que muestran el mismo número de puntos.

Hable con su hijo(a) acerca del dibujo haciéndole las siguientes preguntas:

- (Apunte hacia una tarjeta.) **¿Cuántos puntos hay en esta tarjeta?**
- **¿Cuál tarjeta tiene más puntos?**
- **¿Hay suficientes puntos en esta tarjeta para que sea igual a esa tarjeta?**

Anime a su hijo(a) a explicar cómo sabe la respuesta.

Ayude a su hijo(a)

Si . . . su hijo(a) tiene problemas para igualar, **Entonces . . .** pida a su hijo(a) que toque cada punto mientras cuenta y muestre el número con sus dedos para que su hijo(a) recuerde el número que representa la tarjeta.

Esta semana, su hijo(a) continuará usando el programa *El mundo de los números* para dominar estos conceptos.

Draw lines to connect the matching Dot Set Cards.

Letter to Home

Dear Family,

Your child will spend this week exploring the number line. In the area below, your child found a number on the number line and identified the next number up.

Discuss the picture with your child by asking the following questions:

- **Which number is blue?**
- **Which number is yellow?**
- **Is 3 bigger than 4?**
- **Which number comes before the blue box?**

Encourage your child to explain his or her answers by asking, "How do you know?"

Assisting Your Child

If . . . your child needs help identifying the next number up,
Then . . . count together to discover it.

This week, your child will continue using the *Number Worlds* program to master these concepts by moving along a child-sized number line.

Estimada familia:

Su hijo(a) pasará esta semana explorando la recta numérica. En el área de abajo, su hijo(a) encontró un número sobre la recta numérica e identificó el siguiente número hacia arriba.

Hable con su hijo(a) acerca del dibujo haciéndole las siguientes preguntas:

- **¿Cuál número es azul?**
- **¿Cuál número es amarillo?**
- **¿Es 3 mayor que 4?**
- **¿Cuál número está antes de la casilla azul?**

Anime a su hijo(a) a explicar sus respuestas al preguntar, "¿Cómo sabes?"

Ayude a su hijo(a)

Si . . . su hijo(a) necesita ayuda para identificar el siguiente número hacia arriba,
Entonces . . . cuenten juntos para descubrirlo.

Esta semana, su hijo(a) continuará usando el programa *El mundo de los números* para dominar estos conceptos moviéndose sobre una recta numérica para niños.

Color the 3 square blue. Color the next number yellow.

| 1 | 2 | 3 | 4 | 5 | 6 | 7 | 8 | 9 | 10 |

Letter to Home

Dear Family,

Your child will spend this week exploring the positions of numbers on the number line. In the area below, your child applied this concept by drawing a circle to show the position of 4 on the number line.

Discuss the picture with your child by asking the following questions:

- **How many circles are on the number line?**
- **Which number is the last circle on?**
- **How do you know?**

Encourage your child to continue to explore this concept by having him or her draw another circle and answer the questions again.

Assisting Your Child

If . . . your child is not identifying the correct number,

Then . . . have your child count each circle.

This week, your child will continue using the **Number Worlds** program to master these concepts by lining up with classmates on a child-sized number line.

Estimada familia:

Su hijo(a) pasará esta semana explorando las posiciones de los números en la recta numérica. En el área de abajo, su hijo(a) aplicó este concepto al dibujar un círculo para mostrar la posición del 4 en la recta numérica.

Hable con su hijo(a) acerca del dibujo haciéndole las siguientes preguntas:

- **¿Cuántos círculos hay en la recta numérica?**
- **¿Sobre cuál número está el último círculo?**
- **¿Cómo sabes?**

Anime a su hijo(a) a continuar explorando este concepto haciendo que dibuje otro círculo y conteste las preguntas otra vez.

Ayude a su hijo(a)

Si . . . su hijo(a) no está identificando el número correcto,

Entonces . . . pida a su hijo(a) que cuente cada círculo.

Esta semana, su hijo(a) continuará usando el programa **El mundo de los números** para dominar estos conceptos alineándose con sus compañeros sobre una recta numérica para niños.

Draw a circle in the 4 square.

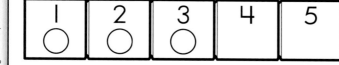

Letter to Home

Dear Family,

Your child will spend this week learning that counting up a number is the same as adding 1 to the set. In the area below, your child applied this concept by putting cards in order based on the numbers of dots on the cards.

Discuss the picture with your child by asking the following questions:

■ **Which number comes first?**
■ **Which number is the next number up?**
■ **Now which number is the next number up?**
■ **Is this number bigger than the last number?**

Encourage your child to explain his or her answers by asking, "How do you know?"

Assisting Your Child

If . . . your child needs extra practice,
Then . . . ask him or her to count sets of objects and to put them in order.

This week, your child will continue using the **Number Worlds** program to master these concepts.

Estimada familia:

Su hijo(a) pasará esta semana aprendiendo que contar un número hacia arriba es lo mismo 1 al grupo. En el área de abajo, su hijo(a) aplicó este concepto poniendo tarjetas en orden basándose en el número de puntos en las tarjetas.

Hable con su hijo(a) acerca del dibujo haciéndole las siguientes preguntas:

■ **¿Cuál número va primero?**
■ **¿Cuál es el siguiente número hacia arriba?**
■ **¿Ahora cuál número es el siguiente hacia arriba?**
■ **¿Es este número mayor que el último número?**

Anime a su hijo(a) a explicar sus respuestas al preguntar, "¿Cómo sabes?"

Ayude a su hijo(a)

Si . . . su hijo(a) necesita más práctica,
Entonces . . . pídale que cuente grupos de objetos y los ordene.

Esta semana, su hijo(a) continuará usando el programa **El mundo de los números** para dominar estos conceptos.

Ring the card that goes first. Draw a line under the card that comes next. Draw two lines under the next number up.

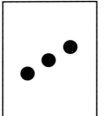 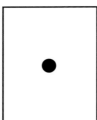 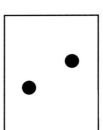

Letter to Home

Dear Family,

Your child will spend this week using the number line to make comparisons. In the area below, your child compared the positions of three stars and identified the stars on the smallest and largest numbers.

Discuss the picture with your child by asking the following questions:

- **Which number is the red star on?**
- **Which number is the green star on?**
- **Is the green star on a number that is bigger or a number that is smaller than the star you didn't color?**

Encourage your child to explain his or her answers by asking, "How do you know?"

Assisting Your Child

If . . . your child needs help making comparisons,

Then . . . have your child count up the number line, touching each box.

This week, your child will continue using the *Number Worlds* program to master these concepts by moving along a child-sized number line.

Estimada familia:

Su hijo(a) pasará esta semana usando la recta numérica para hacer comparaciones. En el área de abajo, su hijo(a) comparó las posiciones de las tres estrellas e identificó las estrellas en el número más pequeño y el más grande.

Hable con su hijo(a) acerca del dibujo haciéndole las siguientes preguntas:

- **¿Sobre cuál número está la estrella roja?**
- **¿Sobre cuál número está la estrella verde?**
- **¿La estrella verde está sobre un número que es mayor o un número que es menor que la estrella que no coloreaste?**

Anime a su hijo(a) a explicar su respuesta al preguntarle, "¿Cómo sabes?"

Ayude a su hijo(a)

Si . . . su hijo(a) necesita ayuda haciendo comparaciones,

Entonces . . . pida a su hijo(a) que cuente hacia arriba en la recta numérica, tocando cada casilla.

Esta semana, su hijo(a) continuará usando el programa *El mundo de los números* para dominar estos conceptos moviéndose sobre una recta numérica para niños.

Color the star in the box with the biggest number red. Color the star in the box with the smallest number green.

Letter to Home

Dear Family,

Your child will spend this week exploring how numbers go up and down number lines. In the area below, your child added 3 to 3.

Discuss the picture with your child by asking the following questions:

- **Which number was the thermometer on?**
- **Which number is it on now?**
- **How many spaces would you need to move to go to the top?**

Encourage your child to explain his or her answers by asking, "How do you know?"

Assisting Your Child

If . . . your child has trouble coloring up to the correct box,

Then . . . count together, and have your child mark each box as you count and then go back to color the boxes.

This week, your child will continue using the *Number Worlds* program to master these concepts by competing to be the first to reach the goal.

Estimada familia:

Su hijo(a) pasará esta semana explorando cómo los números van hacia arriba y hacia abajo en las rectas numéricas. En el área de abajo, su hijo(a) le sumó 3 a 3.

Hable con su hijo(a) acerca del dibujo haciéndole las siguientes preguntas:

- **¿En cuál número estaba el termómetro?**
- **¿En cuál número está ahora?**
- **¿Cuántos espacios necesitas mover para llegar hasta arriba?**

Anime a su hijo(a) a explicar sus respuestas preguntándole, "¿Cómo sabes?"

Ayude a su hijo(a)

Si . . . su hijo(a) tiene problemas coloreando para llegar a la casilla correcta,

Entonces . . . cuenten juntos y pida a su hijo(a) que marque cada casilla mientras usted cuenta y después regrese y coloree las casillas.

Esta semana, su hijo(a) continuará usando el programa *El mundo de los números* para dominar estos conceptos al competir para ser el primero en alcanzar la meta.

Color the thermometer to show it going up by 3 more.

Letter to Home

Dear Family,

Your child will spend this week exploring how numbers move around dials. In the area below, your child identified the position of the number 4.

Discuss the picture with your child by asking the following questions:

- **Which number is the arrow pointing to?**
- **Which number would it be on if you moved up 3?**
- **Which number would it be on if you moved back 2?**

Encourage your child to explain his or her answers by asking, "How do you know?"

Assisting Your Child

If . . . your child has trouble finding the correct number,

Then . . . encourage your child to count up to 4 while pointing at each number.

This week, your child will continue using the *Number Worlds* program to master these concepts by competing to see who can make the most rotations in a set time period.

Estimada familia:

Su hijo(a) pasará esta semana explorando cómo se mueven los números en los diales. En el área de abajo, su hijo(a) identificó la posición del número 4.

Hable con su hijo(a) acerca del dibujo haciéndole las siguientes preguntas:

- **¿Hacia cuál número está apuntando la flecha?**
- **¿En qué número estaría si la mueves 3 hacia adelante?**
- **¿En qué número estaría si la mueves 2 hacia atrás?**

Anime a su hijo(a) a explicar sus respuestas preguntándole, "¿Cómo sabes?"

Ayude a su hijo(a)

Si . . . su hijo(a) tiene problemas para hallar el número correcto,

Entonces . . . anime a su hijo(a) a contar hacia arriba hasta 4 mientras apunta a cada número.

Esta semana, su hijo(a) continuará usando el programa *El mundo de los números* para dominar estos conceptos al competir para ver quien puede hacer más rotaciones en un período de tiempo establecido.

Draw an arrow on the dial to point to 4.

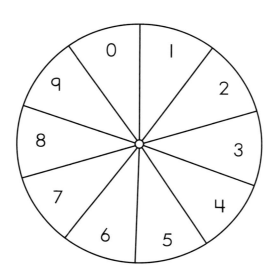

Letter to Home

Dear Family,

Your child will spend this week exploring addition. In the area below, your child drew pictures that can be used to build addition skills when you ask the questions below.

Discuss the picture with your child by asking the following questions:

- **How many red squares are there?**
- **When you pretend to be Plus Pup and add a square, how many squares do you think you will have?**
- **How many squares do you have?**

Encourage your child to explain his or her answers by asking, "How do you know?"

Assisting Your Child

If . . . your child has trouble adding 1, **Then . . .** encourage your child to count the squares he or she drew.

This week, your child will continue using the **Number Worlds** program to master these concepts by predicting what will happen when one object is added to a group of objects.

Estimada familia:

Su hijo(a) pasará esta semana explorando la suma. En el área de abajo, su hijo(a) hizo dibujos que se pueden usar para mejorar las destrezas de suma cuando usted pregunta las siguientes preguntas.

Hable con su hijo(a) acerca del dibujo haciéndole las siguientes preguntas:

- **¿Cuántos cuadrados rojos hay?**
- **Imagínate que eres Tío Más y sumas un cuadro, ¿cuántos cuadros piensas que tendrás?**
- **¿Cuántos cuadros tienes?**
- **¿Cuántos cuadros tienes?**

Anime a su hijo(a) a explicar sus respuestas preguntado, "¿Cómo sabes?"

Ayude a su hijo(a)

Si . . . su hijo(a) tiene problemas para sumar 1, **Entonces . . .** anime a su hijo(a) a contar los cuadros que dibujó.

Esta semana, su hijo(a) continuará usando el programa **El mundo de los números** para dominar estos conceptos al predecir lo que sucederá cuando se suma un objeto a un grupo de objetos.

Draw 3 red squares. Pretend you are Plus Pup, and put another square in the box using a blue crayon.

Letter to Home

Dear Family,

Your child will spend this week exploring subtraction. In the area below, your child drew pictures that can be used to build subtraction skills.

Discuss the picture with your child by asking the following questions:

- **How many triangles are there?**
- **How many triangles do you think you will have if Minus Mouse takes 1 away?**
- (Cross out a triangle.) **How many triangles do you have now?**

Encourage your child to explain his or her answers by asking, "How do you know?"

Assisting Your Child

If . . . your child has trouble subtracting, **Then . . .** work with him or her using real objects before using drawings.

This week, your child will continue using the *Number Worlds* program to master these concepts by predicting what will happen when one object is subtracted from a group of objects.

Estimada familia:

Su hijo(a) pasará esta semana explorando la resta. En el área de abajo, su hijo(a) hizo dibujos que se pueden usar para mejorar las destrezas de resta.

Hable con su hijo(a) acerca del dibujo haciéndole las siguientes preguntas:

- **¿Cuántos triángulos hay?**
- **¿Cuántos triángulos crees que habrá si Tío Menos quita 1?**
- (Tache un triángulo.) **¿Cuántos triángulos tienes ahora?**

Anime a su hijo(a) a explicar sus respuestas preguntado, "¿Cómo sabes?"

Ayude a su hijo(a)

Si . . . su hijo(a) tiene problemas para restar, **Entonces . . .** trabaje con él o ella usando objetos reales antes de usar dibujos.

Esta semana, su hijo(a) continuará usando el programa *El mundo de los números* para dominar estos conceptos al predecir lo que sucederá cuando se resta un objeto de un grupo de objetos.

Draw 5 triangles. If Minus Mouse takes 1 away, how many do you have?

Letter to Home

Dear Family,

Your child will spend this week learning that using numbers is the best way to compare sets of objects. In the area below, your child chose the grouping with more circles by counting, rather than using size to decide.

Discuss the picture with your child by asking the following questions:

■ **Which picture has more circles?**
■ **How many circles are in the first picture?**
■ **How many circles are in the second picture?**
■ **Is 2 bigger than 3?**

Encourage your child to explain his or her answers by asking, "How do you know?"

Assisting Your Child

If . . . your child has trouble determining which group is larger,
Then . . . encourage him or her to count each circle.

This week, your child will continue using the **Number Worlds** program to master these concepts by counting and comparing objects of different sizes.

Estimada familia:

Su hijo(a) pasará esta semana aprendiendo que usar números es la mejor forma de comparar grupos de objetos. En el área de abajo, su hijo(a) seleccionó el agrupamiento con más círculos contándolos en vez de decidir por el tamaño.

Hable con su hijo(a) acerca del dibujo haciéndole las siguientes preguntas:

■ **¿Cuál dibujo tiene más círculos?**
■ **¿Cuántos círculos hay en el primer dibujo?**
■ **¿Cuántos círculos hay en el segundo dibujo?**
■ **¿Es 2 más grande que 3?**

Anime a su hijo(a) a explicar sus respuestas preguntándole, "¿Cómo sabes?"

Ayude a su hijo(a)

Si . . . su hijo(a) tiene problemas para determinar cuál grupo es más grande,
Entonces . . . anímelo(a) a contar cada círculo.

Esta semana, su hijo(a) continuará usando el programa **El mundo de los números** para dominar estos conceptos contando y comparando objetos de diferentes tamaños.

Ring the group with more circles.

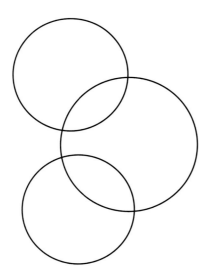

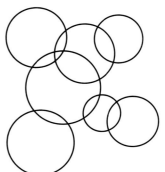

Letter to Home

Dear Family,

Your child will spend this week learning about how numbers move up and down as you count up and down. In the area below, your child chose the graph that represents a higher number.

Discuss the graphs with your child by asking the following questions:

- **How many boxes are filled on the first graph?**
- **How many boxes are filled on the second graph?**
- **How many more boxes are filled on the first graph than on the second graph?**

Encourage your child to explain his or her answers by asking, "How do you know?"

Assisting Your Child

If . . . your child has trouble determining "how many more,"
Then . . . ask, "How many *extra* boxes are on the first graph?"

This week, your child will continue using the **Number Worlds** program to master these concepts by using graphs to keep track of beanbags.

Estimada familia:

Su hijo(a) pasará esta semana aprendiendo que los números van hacia arriba y hacia abajo conforme se cuenta hacia arriba y hacia abajo. En el área de abajo, su hijo(a) seleccionó la gráfica que representa un número más alto.

Hable con su hijo(a) acerca de las gráficas haciéndole las siguientes preguntas:

- **¿Cuántas casillas están llenas en el primer gráfico?**
- **¿Cuántas casillas están llenas en el segundo gráfico?**
- **¿Cuántas más casillas llenas hay en el primer gráfico que en el segundo?**

Anime a su hijo(a) a explicar sus respuestas preguntándole, "¿Cómo sabes?"

Ayude a su hijo(a)

Si . . . su hijo(a) tiene problemas para determinar "cuántas más",
Entonces . . . pregunte "¿Cuántas casillas extra hay en el primer gráfico?"

Esta semana, su hijo(a) continuará usando el programa **El mundo de los números** para dominar estos conceptos usando gráficos para seguir la pista de cojines.

Ring the graph that shows more than the other graph.

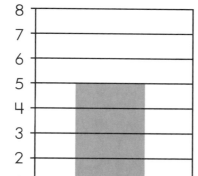

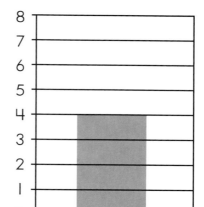

Letter to Home

Dear Family,

Your child will spend this week exploring subtraction. In the area below, your child drew pictures that can be used to build subtraction skills.

Discuss the picture with your child by asking the following questions:

■ **How many circles were there before Minus Mouse came?**

■ **How many circles do you think you have now that Minus Mouse took one away?**

Encourage your child to explain his or her answers by asking, "How do you know?"

Assisting Your Child

If . . . your child has trouble subtracting, **Then . . .** work with him or her using real objects before using drawings.

This week, your child will continue using the **Number Worlds** program to master these concepts by predicting what will happen when one object is subtracted from a group of objects.

Estimada familia:

Su hijo(a) pasará esta semana explorando la resta. En el área de abajo, su hijo(a) hizo dibujos que pueden usarse para mejorar las destrezas de resta.

Hable con su hijo(a) acerca del dibujo haciéndole las siguientes preguntas:

■ **¿Cuántos círculos había antes de que viniera Tío Menos?**

■ **¿Cuántos círculos crees que tienes ahora que Tío menos se llevó uno?**

Anime a su hijo(a) a explicar sus respuestas preguntándole, "¿Cómo sabes?"

Ayude a su hijo(a)

Si . . . su hijo(a) tiene problemas para restar, **Entonces . . .** trabaje con él o ella usando objetos reales antes de usar dibujos.

Esta semana, su hijo(a) continuará usando el programa **El mundo de los números** para dominar estos conceptos prediciendo lo que sucederá cuando se resta un objeto de un grupo de objetos.

Draw 4 circles. Pretend you are Minus Mouse, and erase 1. How many circles do you have now?

Letter to Home

Dear Family,

Your child will spend this week finding matching cards. In the area below, your child demonstrated matching skills by finding the cards that represent the same numbers.

Discuss the picture with your child by asking the following questions:

- (Point to a card.) **How many dots are on this card?**
- **Which card has more dots?**
- **Are there enough dots on this card to match that card?**

Encourage your child to explain how he or she knows the answer.

Assisting Your Child

If . . . your child has trouble making a match, **Then . . .** have your child touch each dot as he or she counts and show the number with his or her fingers so that he or she remembers what number the card represents.

This week, your child will continue using the **Number Worlds** program to master these concepts while trying to match the most cards.

Estimada familia:

Su hijo(a) pasará esta semana buscando tarjetas iguales. En el área de abajo, su hijo(a) demostró sus destrezas para igualar al hallar las tarjetas que representan los mismos números.

Hable con su hijo(a) acerca del dibujo haciéndole las siguientes preguntas:

- (Apunte hacia una tarjeta.) **¿Cuántos puntos hay en esta tarjeta?**
- **¿Cuál tarjeta tiene más puntos?**
- **¿Hay suficientes puntos en esta tarjeta para igualar esa tarjeta?**

Anime a su hijo(a) a explicar cómo sabe sus respuestas.

Ayude a su hijo(a)

Si . . . su hijo(a) tiene problemas para igualar, **Entonces . . .** pida a su hijo(a) que toque cada punto mientras cuenta y muestre el número con los dedos, así su hijo(a) recordará qué numero representa la tarjeta.

Esta semana, su hijo(a) continuará usando el programa **El mundo de los números** para dominar estos conceptos mientras trata de igualar las más tarjetas posibles.

Draw lines to connect the matching Dot Set Cards.

Letter to Home

Dear Family,

Your child will spend this week comparing to find the larger number. In the area below, your child demonstrated this skill by choosing the cards showing larger numbers.

Discuss the picture with your child by asking the following questions:

- (Point to a card.) **How many dots are on this card?**
- **Which card has more dots?**
- **Which number is higher when we count up?**

Encourage your child to explain his or her answers by asking, "How do you know?"

Assisting Your Child

If . . . your child does not choose the correct card,

Then . . . ask for an explanation about why your child chose the card, and help him or her find an explanation that is correct.

This week, your child will continue using the **Number Worlds** program to master these concepts while trying to collect the most cards.

Estimada familia:

Su hijo(a) pasará esta semana comparando para hallar el número más grande. En el área de abajo, su hijo(a) demostró esta destreza al seleccionar las tarjetas que muestran los números más grandes.

Hable con su hijo(a) acerca del dibujo haciéndole las siguientes preguntas:

- (Apunte hacia una tarjeta). **¿Cuántos puntos hay en esta tarjeta?**
- **¿Cuál tarjeta tiene más puntos?**
- **¿Cuál número es más grande cuando contamos hacia arriba?**

Anime a su hijo(a) a explicar sus respuestas preguntándole, "¿Cómo sabes?"

Ayude a su hijo(a)

Si . . . su hijo(a) no elige la tarjeta correcta,

Entonces . . . pida una explicación de por qué seleccionó esa tarjeta y ayúdelo(a) a hallar una explicación que sea correcta.

Esta semana, su hijo(a) continuará usando el programa **El mundo de los números** para dominar estos conceptos mientras trata de recoger las más tarjetas posibles.

For each pair, ring the card showing the bigger number.

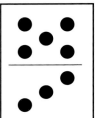

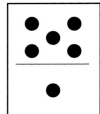

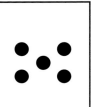

Letter to Home

Dear Family,

Your child will spend this week comparing groups of objects and smaller groups within a large group. In the area below, your child identified a number and created partial sets.

Discuss the picture with your child by asking the following questions:

- **How many triangles are there?**
- **How many triangles did you color?**
- **Are there more squares, triangles, or circles?**
- **Are there more colored triangles or more colored squares?**

Encourage your child to explain his or her answers by asking, "How do you know?"

Assisting Your Child

If . . . your child does not count 5 triangles, **Then . . .** help him or her count, touching each triangle as you count.

This week, your child will continue using the **Number Worlds** program to master these concepts through the use of drawings and real objects.

Estimada familia:

Su hijo(a) pasará esta semana comparando grupos de objetos y grupos más pequeños dentro de un grupo grande. En el área de abajo, su hijo(a) identificó un número y creó grupos parciales.

Hable con su hijo(a) acerca del dibujo haciéndole las siguientes preguntas:

- **¿Cuántos triángulos hay?**
- **¿Cuántos triángulos coloreaste?**
- **¿Hay más cuadrados, triángulos o círculos?**
- **¿Hay más triángulos coloreados o cuadrados coloreados?**

Anime a su hijo(a) a explicar sus respuestas preguntándole, "¿Cómo sabes?"

Ayude a su hijo(a)

Si . . . su hijo(a) no cuenta 5 triángulos, **Entonces . . .** ayúdelo(a) a contar, tocando cada triángulo mientras cuenta.

Esta semana, su hijo(a) continuará usando el programa **El mundo de los números** para dominar estos conceptos a través del uso de dibujos y objetos reales.

Color 1 circle blue, 2 squares red, and 3 triangles green.

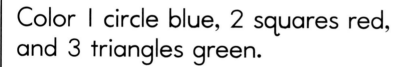

Letter to Home

Dear Family,

Your child will spend this week using basic addition and subtraction in the same activity. In the area below, your child used the information he or she was given to identify where to move.

Discuss the picture with your child by asking the following questions:

- **Which number does the Dot Cube show?**
- **Which number do you move to?**
- **If you went +1 from 5, which number would you be on?**
- **If you moved −1 from 5, which number would you be on?**

Encourage your child to explain his or her answers by asking, "How do you know?"

Assisting Your Child

If . . . your child does not know +1 and −1, **Then . . .** try writing +1 and −1 so your child has a visual cue.

This week, your child will continue using the **Number Worlds** program to master these concepts.

Estimada familia:

Su hijo(a) pasará esta semana empezando a usar la suma y la resta básicas en la misma actividad. En el área de abajo, su hijo(a) usó la información que se le dio para identificar adónde moverse.

Hable con su hijo(a) acerca del dibujo haciéndole las siguientes preguntas:

- **¿Cuál número muestra el dado?**
- **¿A cuál número te mueves?**
- **Si fueras +1 desde el 5, ¿en qué número estarías?**
- **Si te movieras −1 desde el 5, ¿en qué número estarías?**

Anime a su hijo(a) a explicar sus respuestas preguntándole, "¿Cómo sabes?"

Ayude a su hijo(a)

Si . . . su hijo(a) no sabe +1 y −1, **Entonces . . .** trate escribiendo +1 y −1 para que su hijo(a) tenga una pista visual.

Esta semana, su hijo(a) continuará usando el programa **El mundo de los números** para dominar estos conceptos.

Start at 3. Color the number line to show where you move when your Dot Cube looks like this:

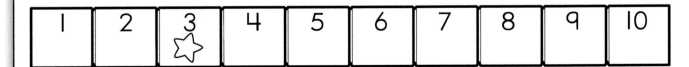

Letter to Home

Dear Family,

Your child will spend this week learning how to show numbers on a bar graph. In the area below, your child chose the graph that represents a lower number.

Discuss the graphs with your child by asking the following questions:

- **Which graph is lower?**
- **How many boxes are filled on the first graph?**
- **How many boxes are filled on the second graph?**
- **How many more boxes are filled on the first graph than on the second graph?**

Encourage your child to explain his or her answers by asking, "How do you know?"

Assisting Your Child

If . . . your child has trouble determining "how many more,"

Then . . . ask, "How many *extra* boxes are on the first graph?"

This week, your child will continue using the **Number Worlds** program to master these concepts by comparing graphs.

Estimada familia:

Su hijo(a) pasará esta semana aprendiendo a mostrar números en un gráfico de barras. En el área de abajo, su hijo(a) seleccionó el gráfico que representa un número más bajo.

Hable con su hijo(a) acerca de los gráficos haciéndole las siguientes preguntas:

- **¿Cuál gráfico es menor?**
- **¿Cuántas casillas están llenas en el primer gráfico?**
- **¿Cuántas casillas hay llenas en el segundo gráfico?**
- **¿Cuántas más casillas hay llenas en el primer gráfico que en el segundo gráfico?**

Anime a su hijo(a) a explicar sus respuestas preguntándole, "¿Cómo sabes?"

Ayude a su hijo(a)

Si . . . su hijo(a) tiene problemas determinando "cuántas más",

Entonces . . . pregunte, "¿Cuántas casillas extra hay en el primer gráfico?"

Esta semana, su hijo(a) continuará usando el programa **El mundo de los números** para dominar estos conceptos comparando su gráfico con el de sus compañeros.

Ring the graph that shows less than the other graph.

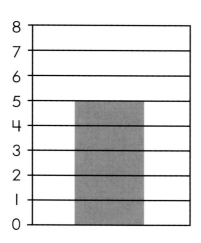

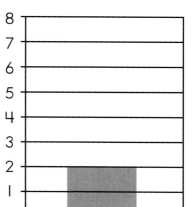

Letter to Home

Dear Family,

Your child will spend this week exploring addition and subtraction on number lines such as thermometers. In the area below, your child identified that $9 - 5 = 4$.

Discuss the picture with your child by asking the following questions:

- **Which number is the pawn on?**
- **Which number is it on now?**
- **Which number would it move to if you picked a +1 Card?**
- **Which number would it move to if you picked a −1 Card?**

Encourage your child to explain his or her answers by asking, "How do you know?"

Assisting Your Child

If . . . your child has trouble predicting what number the pawn would move to,

Then . . . have your child point to the boxes as he or she counts.

This week, your child will continue using the **Number Worlds** program to master these concepts.

Estimada familia:

Su hijo(a) pasará esta semana explorando la suma y la resta en rectas numéricas como los termómetros. En el área de abajo, su hijo(a) identificó que $9 - 5 = 4$.

Hable con su hijo(a) acerca del dibujo haciéndole las siguientes preguntas:

- **¿Sobre cuál número estaba la ficha?**
- **¿Sobre cuál número está ahora?**
- **¿A qué número se movería si seleccionaras una tarjeta de +1?**
- **¿A qué número se movería si seleccionaras una tarjeta de −1?**

Anime a su hijo(a) a explicar sus respuestas preguntándole, "¿Cómo sabes?"

Ayude a su hijo(a)

Si . . . su hijo(a) tiene problemas prediciendo a qué número se moverá la ficha,

Entonces . . . pida a su hijo(a) que apunte a las casillas mientras va contando.

Esta semana, su hijo(a) continuará usando el programa **El mundo de los números** para dominar estos conceptos.

Your pawn is on 9. Color the thermometer to show where it would be if you rolled a 5 and went down.

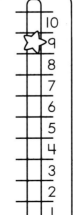

Letter to Home

Dear Family,

Your child will spend this week exploring addition and subtraction on dials. In the area below, your child identified the change in position that happens when numbers are added.

Discuss the picture with your child by asking the following questions:

- **Which number would you move to?**
- **Is 6 a bigger number than 9?**
- **What happens when you move −1?**

Encourage your child to explain his or her answers by asking, "How do you know?"

Assisting Your Child

If . . . your child has trouble finding the correct number,
Then . . . encourage your child to count up while pointing at each number.

This week, your child will continue using the **Number Worlds** program to master these concepts by competing to see who can make the most rotations in a set time period.

Estimada familia:

Su hijo(a) pasará esta semana explorando la suma y la resta en diales. En el área de abajo, su hijo(a) identificó el cambio en posición que sucede cuando se suman números.

Hable con su hijo(a) acerca del dibujo haciéndole las siguientes preguntas:

- **¿Hacia cuál número te moverías?**
- **¿Es 6 un número más grande que 9?**
- **¿Qué sucede cuando mueves −1?**

Anime a su hijo(a) a explicar sus respuestas preguntándole, "¿Cómo sabes?"

Ayude a su hijo(a)

Si . . . su hijo(a) tiene problemas hallando el número correcto,
Entonces . . . anime a su hijo(a) a contar hacia arriba mientras apunta a cada número.

Esta semana, su hijo(a) continuará usando el programa **El mundo de los números** para dominar estos conceptos compitiendo para ver quién hace más rotaciones en un período de tiempo establecido.

Color the number that shows where you would be if you are on 6 and you rolled a 3.

Letter to Home

Dear Family,

Your child will spend this week reviewing numerals and sequence. In the area below, your child used numerals and sequence to create a dot-to-dot picture.

Discuss the picture with your child by asking the following questions:

■ **Which number comes first?**
■ **Which number comes last?**
■ **How do you know which number to move to next?**

Encourage your child to explain his or her answers by asking, "How do you know?"

Assisting Your Child

If . . . your child has trouble remembering which numbers the numerals represent,
Then . . . discuss the lines and shapes that make each number.

This week, your child will continue using the **Number Worlds** program to master these concepts by completing sequencing activities.

Estimada familia:

Su hijo(a) pasará esta semana revisando números y secuencia. En el área de abajo, su hijo(a) usó números y secuencia para crear un dibujo de punto en punto.

Hable con su hijo(a) acerca del dibujo haciéndole las siguientes preguntas:

■ **¿Cuál es el primer número?**
■ **¿Cuál es el último número?**
■ **¿Cómo sabes a qué número ir después?**

Anime a su hijo(a) a explicar sus respuestas preguntándole, "¿Cómo sabes?"

Ayude a su hijo(a)

Si . . . su hijo(a) tiene problema recordando cómo se representa cada número,
Entonces . . . hable sobre las líneas y figuras que hace cada número.

Esta semana, su hijo(a) continuará usando el programa **El mundo de los números** para dominar estos conceptos al completar actividades de secuencia.

Connect the dots.

Name _____

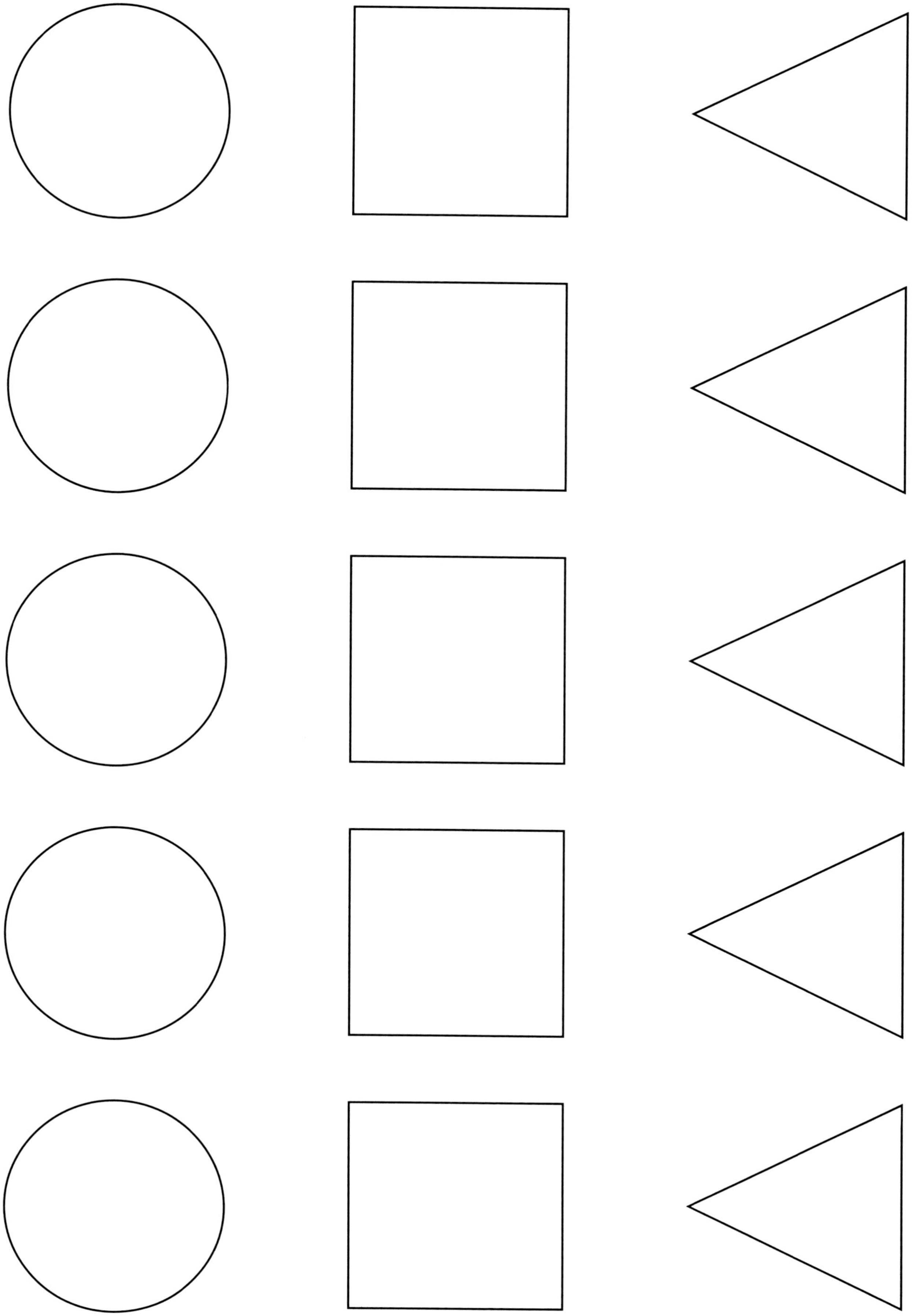

Shapes

Pet Store

Name _____

Animal Center

Circus Tent

Name _____

B4 Picture Land • Activity Sheet 1

Fish

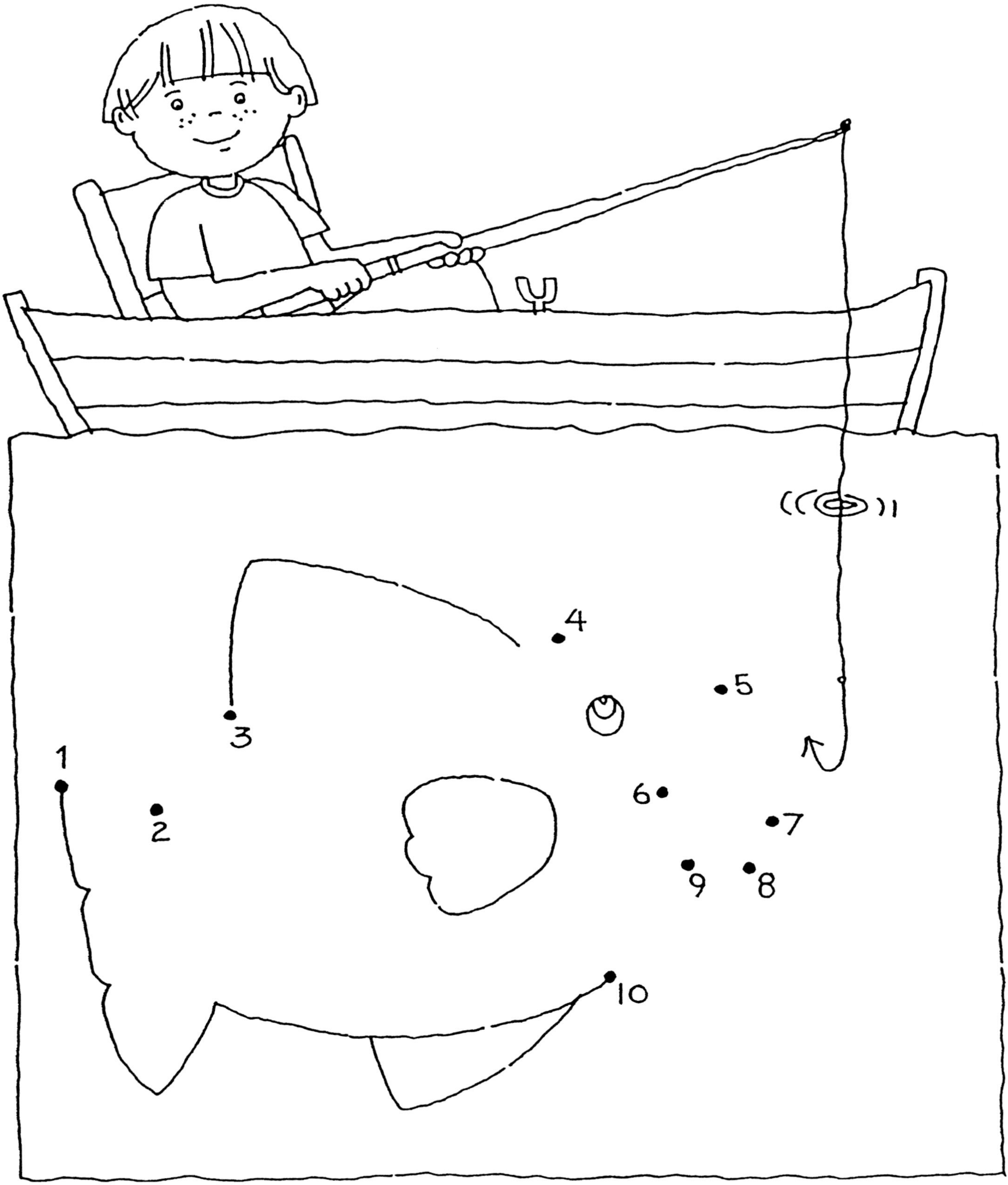

1

4

5

3

1

2

6

7

9   8

10

Doghouse

Name _____

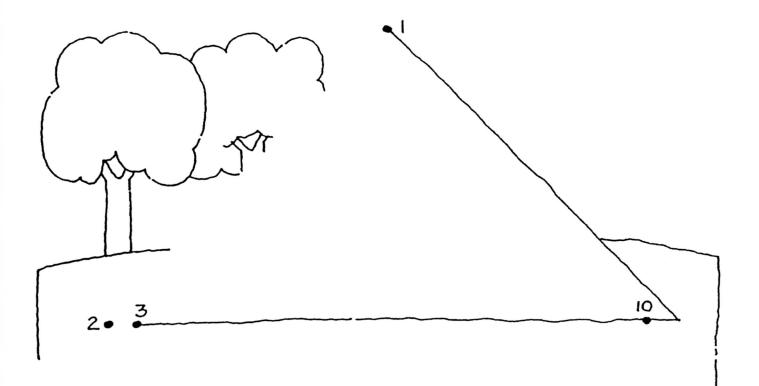

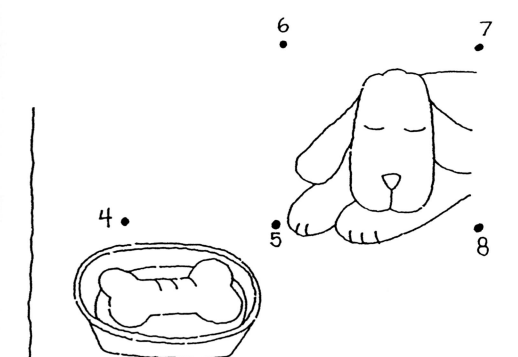

Rocket

Name _____

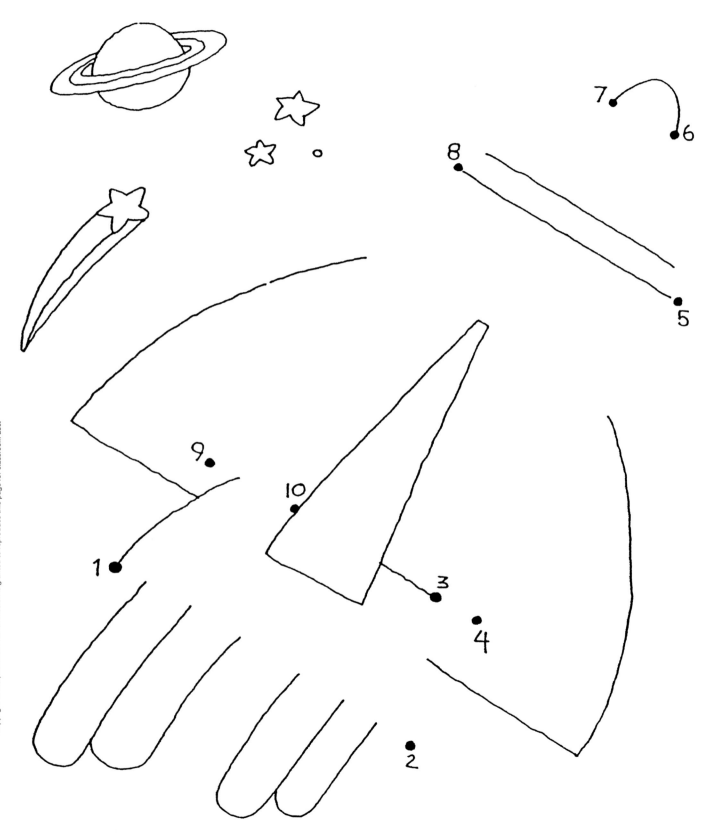

Drop the Counter
Record Form

Name _____

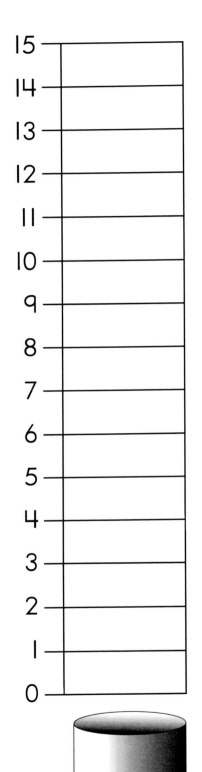

15 —
14 —
13 —
12 —
11 —
10 —
9 —
8 —
7 —
6 —
5 —
4 —
3 —
2 —
1 —
0 —

15 —
14 —
13 —
12 —
11 —
10 —
9 —
8 —
7 —
6 —
5 —
4 —
3 —
2 —
1 —
0 —

About Mathematics Intervention

What is an intervention?

An intervention is any instructional or practice activity designed to help students who are not making adequate progress.

- For struggling students, this requires an acceleration of development over a sufficient period of time.
- If the problem is small, the intervention might be brief.
- If the problem or lag in development is large, the intervention may last for weeks, months, or all year.

How is *Number Worlds* mathematics intervention different?

Number Worlds instruction is:
- More explicit
- More systematic
- More intense
- More supportive

What is the three-tier approach?

Layers of intervention are organized in tiers to enable schools to respond to student needs.

Tier 1
Core Instruction: high quality, comprehensive instruction for all students. It reduces the number of students who later become at-risk for academic problems.

Tier 2
Supplemental Intervention: Addresses essential content for students who are not making adequate progress in the core program. It reduces the need for more intensive intervention.

Tier 3
Intensive Intervention: Increases in intensity and duration for students with low-level skills and a sustained lack of adequate progress within Tiers 1 and 2.

Number Worlds is designed for both Tier 2 and Tier 3 students. Tier 2 students may spend a brief time in the program learning a key concept and then be quickly reintegrated back into the core instructional program. Levels A–C are also appropriate for Tier 1 students.

Students in Tier 3 will most likely need to complete the entire **Number Worlds** curriculum at their learning levels. Tier 3 requires intensive intervention for students with low skills and a sustained lack of adequate progress within Tiers 1 and 2. Teaching at this level is more intensive and includes more explicit instruction that is designed to meet the individual needs of struggling students. Group size is smaller, and the duration of daily instruction is longer.

Building Number Sense with *Number Worlds:*

A Mathematics Program for Young Children

Sharon Griffin

What is number sense? We all know number sense when we see it but, if asked to define what it is and what it consists of, most of us, including the teachers among us, would have a much more difficult time. Yet this is precisely what we need to know to teach number sense effectively. Consider the answers three kindergarten children provide when asked the following question from the Number Knowledge Test (Griffin & Case, 1997): Which is bigger: seven or nine?"

> Brie responds quickly, saying "Nine." When asked how she figured it out, she says, "Well, you go, 'seven' (pause) 'eight', 'nine' (putting up two fingers while saying the last two numbers). That means nine has two more than seven. So it's bigger."
>
> Leah says, hesitantly, "Nine?" When asked how she figured it out, she says, "Because nine's a big number."
>
> Caitlin looks genuinely perplexed, as if the question was not a sensible thing to ask, and says, "I don't know."

Kindergarten teachers will immediately recognize that Brie's answer provides evidence of a well-developed number sense for this age level and Leah's answer, a more fragile and less-developed number sense. The knowledge that lies behind this "sense" may be much less apparent. What knowledge does Brie have that enables her to come up with the answer in the first place and to demonstrate good number sense in the process?

1 Knowledge that underlies number sense
Research conducted with the Number Knowledge Test and several other cognitive developmental measures (see Griffin, 2002; Griffin & Case, 1997 for a summary of this research) suggests that the following understandings lie at the heart of the number sense that 5-year-olds like Brie are able to demonstrate on this problem. They know (a) that numbers indicate quantity and therefore, that numbers, themselves, have magnitude; (b) that the word "bigger" or "more" is sensible in this context; (c) that the numbers 7 and 9, like every other number from 1 to 10, occupy fixed positions in the counting sequence; (d) that 7 comes before 9 when you are counting up; (e) that numbers that come later in the sequence— that are higher up— indicate larger quantities and therefore, that 9 is bigger (or more) than 7.

Brie provided evidence of an additional component of number sense in the explanation she provided for her answer. By using the Count-On strategy to show that nine comes two numbers after seven and by suggesting that this means "it has two more than seven," Brie demonstrated that she also knows (f) that each counting number up in the sequence corresponds precisely to an increase of one unit in the size of a set. This understanding, possibly more than any of the others listed above, enables children to use the counting numbers alone, without the need for real objects, to solve quantitative problems involving the joining of two sets. In so doing, it transforms mathematics from something that can only be done out there (e.g., by manipulating real objects) to something that can be done in their own heads, and under their own control.

This set of understandings, the core of *number sense*, forms a knowledge network that Case and Griffin (1990), see also Griffin and Case (1997), have called a *central conceptual structure for number.* Research conducted by these investigators has shown that this structure is central in at least two ways (see Griffin, Case, & Siegler, 1994). First, it enables children to make sense of a broad range of quantitative problems across contexts and to answer questions, for example, about two times on a clock (Which is longer?), two positions on a path (Which is farther?), and two sets of coins (Which is worth more?). Second, it provides the foundation on which children's learning of more complex number concepts, such as those involving double-digit numbers, is built. For this reason, this network of knowledge is an important set of understandings that should be taught in the preschool years, to all children who do not spontaneously acquire them.

2 How can this knowledge be taught?

Number Worlds, a mathematics program for young children (formerly called *Rightstart*), was specifically developed to teach this knowledge and to provide a test for the cognitive developmental theory (i.e., Central Conceptual Structure theory; see Case & Griffin, 1990) on which the program was based. Originally developed for kindergarten, the program (see Griffin & Case, 1995) was expanded to teach a broader range of understandings when research findings provided strong evidence that (a) children who were exposed to the program acquired the knowledge it was designed to teach (i.e., the central conceptual structure for number), and (b) the theoretical postulates on which the program was based were valid (see Griffin & Case, 1996; Griffin, Case, & Capodilupo, 1995; Griffin et al., 1994). Programs for grades one and two were developed to teach the more complex central conceptual structures that underlie base-ten understandings (see Griffin, 1997, 1998) and a program for preschool was developed (see Griffin, 2000) to teach the "precursor" understandings that lay the foundation for the development of the central conceptual structure for number.

Because the four levels of the program are based on a well-developed theory of cognitive development, they provided a finely graded sequence of activities (and associated knowledge objectives) that recapitulates the natural developmental progression for the age range of 3–9 years and that allows each child to enter the program at a point that is appropriate for his or her own development, and to progress through the program to teach 20 or more children at any one time and every effort has been made, in the construction of the **Number Worlds** program, to make it as easy as possible for teachers to accommodate the developmental needs of individual children (or groups of children) in their classroom. Five instructional principles that lie at the heart of the program are described below and are used to illustrate several features of the program that have already been mentioned and several that have not yet been introduced.

2.1 *Principle 1: Build upon children's current knowledge*

Each new idea that is presented to children must connect to their existing knowledge if it is going to make any sense at all. Children must also be allowed to use their existing knowledge to construct new knowledge that is within reach—that is one step beyond where they are now—and a set of bridging contexts and other instructional supports should be in place to enable them to do so.

In the examples of children's thinking presented earlier, three different levels of knowledge are apparent. Brie appears to have acquired the knowledge network that underlies number sense and to be ready, therefore, to move on to the next developmental level: to connect this set of understandings to the written numerals (i.e., the formal symbols) associated with each counting word. Leah appears to have some understanding of some of the components of this network (i.e., that number have magnitude) and to be ready to use this understanding as a base to acquire the remaining understandings (e.g., that a number's magnitude and its position in the counting sequence are directly related). Caitlin demonstrated little understanding of any element of this knowledge network and she might benefit, therefore, from exposure to activities that will help her acquire the "precursor" knowledge needed to build this network, namely knowledge of counting (e.g., the one-to-one correspondence rule) and knowledge of quantity (e.g., an intuitive understanding of relative amount). Although all three children are in kindergarten, each child appears to be at a different point in the developmental trajectory and to require a different set of learning opportunities; ones that will enable each child to use her existing knowledge to construct new knowledge at the next level up.

To meet these individual needs, teachers need (a) a way to assess children's current knowledge, (b) activities that are multi-leveled so children with different entering knowledge can all benefit from exposure to them, and (c) activities that are carefully sequenced and that span several developmental levels so children with different entering knowledge can be exposed to activities that are appropriate for their level of understanding. These are all available in the **Number Worlds** program and are illustrated in various sections of this paper.

Program Research

2.2 Principle 2: Follow the natural developmental progression when selecting new knowledge to be taught

Researchers who have investigated the manner in which children construct number knowledge between the ages of 3 and 9 years have identified a common progression that most, if not all, children follow (see Griffin, 2002; Griffin and Case, 1997 for a summary of this research). As suggested earlier, by the age of 4 years, most children have constructed two "precursor" knowledge networks—knowledge of counting and knowledge of quantity—that are separate in this stage and that provide the base for the next developmental stage. Sometime in kindergarten, children become able to integrate these knowledge networks—to connect the world of counting numbers to the world of quantity— and to construct the central conceptual understandings that were described earlier. Around the age of 6 or 7 years, children connect this integrated knowledge network to the world of formal symbols and, by the age of 8 or 9 years, most children become capable of expanding this knowledge network to deal with double-digit numbers and the base-ten system. A mathematics program that provides opportunities for children to use their current knowledge to construct new knowledge that is a natural next step, and that fits their spontaneous development, will have the best chance of helping children make maximum progress in their mathematics learning and development.

Because there are limits in development on the complexity of information children can handle at any particular age/stage (see Case, 1992), it makes no sense to attempt to speed up the developmental process by accelerating children through the curriculum. However, for children who are at an age when they should have acquired the developmental milestones but for some reason haven't, exposure to a curriculum that will give them ample opportunities to do so makes tremendous sense. It will enable them to catch up to their peers and thus, to benefit from the formal mathematics instruction that is provided in school. Children who are developing normally also benefit from opportunities to broaden and deepen the knowledge networks they are constructing, to strengthen these understandings, and to use them in a variety of contexts.

2.3 Principle 3: Teach computational fluency as well as conceptual understanding

Because computational fluency and conceptual understanding have been found to go hand in hand in children's mathematical development (see Griffin, 2003; Griffin et al., 1994), opportunities to acquire computational fluency, as well as conceptual understanding, are built into every **Number Worlds** activity. This is nicely illustrated in the following activities, drawn from different levels of the program.

In The Mouse and the Cookie Jar Game (created for the preschool program and designed to give 3- to 4-year-olds an intuitive understanding of subtraction), children are given a certain number of counting chips (with each child receiving the same number but a different color) and told to pretend their chips are cookies. They are asked to count their cookies and, making sure they remember how many they have and what their color is, to deposit them in the cookie jar for safe keeping. While the children sleep, a little mouse comes along and takes one (or two) cookies from the jar. The problem that is then posed to the children is "How can we figure out whose cookie(s) the mouse took?"

Although children quickly learn that emptying the jar and counting the set of cookies that bears their own color is a useful strategy to use to solve this problem, it takes considerably longer for many children to realize that, if they now have four cookies (and originally had five), it means that they have one fewer and the mouse has probably taken one of their cookies. Children explore this problem by counting and recounting the remaining sets, comparing them to each other (e.g., by aligning them) to see who has the most or least, and ultimately coming up with a prediction. When a prediction is made, children search the mouse's hole to see whose cookie had been taken and to verify or revise their prediction. As well as providing opportunities to perfect their counting skills, this activity gives children concrete opportunities to experience simple quantity transformations and to discover how the counting numbers can be used to predict and explain differences in amount.

The *Dragon Quest Game* that was developed for the Grade 1 program teaches a much more sophisticated set of understandings. Children are introduced to Phase 1 activity by being told a story about a fire-breathing dragon that has been terrorizing the village where children live. The children playing the game are heroes who have been chosen to seek out the dragon and put out his fire. To extinguish this dragon's fire (as opposed to the other, more powerful dragons they will encounter in later phases) a hero will need at least 10 pails of water. If a hero enters into the dragon's area with less than 10 pails of water, he or she will become the dragon's prisoner and can only be rescued by one of the other players.

To play the game, children take turns rolling a die and moving their playing piece along the colored game board. If they land on a well pile (indicated by a star), they can pick a card from the face-down deck of cards, which illustrate, with images and symbols (e.g., +4) a certain number of pails of water. Children are encouraged to add up their pails of water as they receive them and they are allowed to use a variety of strategies to do so, ranging from mental math (which is encouraged) to the use of tokens to keep track of the quantity accumulated. The first child to reach the dragon's lair with at least 10 pails of water can put out the dragon's fire and free any teammates who have become prisoners.

As children play this game and talk about their progress, they have ample opportunity to connect numbers to several different quantity representations (e.g., dot patterns on the die; distance of their pawn along the path; sets of buckets illustrated on the cards; written numerals also provided on the cards) and to acquire an appreciation of numerical magnitude across these contexts. With repeated play, they also become capable of performing a series of successive addition operations *in their heads* and of expanding the well pile. When they are required to submit formal proof to the mayor of the village that they have amassed sufficient pails of water to put out the dragon's fire before they are allowed to do so, they become capable of writing a series of formal expressions to record the number of pails received and spilled over the course of the game. In contexts such as these children receive ample opportunity to use the formal symbol system in increasingly efficient ways to make sense of quantitative problems they encounter in the course of their own activity.

2.4 Principle 4: provide plenty of opportunity for hands-on exploration, problem-solving, and communication

Like the *Dragon Quest Game* that was just described, many of the activities created for the **Number Worlds** program are set in a game format that provides plenty of opportunity for hands-on exploration of number concepts, for problem-solving and for communication. Communication is explicitly encouraged in a set of question prompts that are included with each small group game (e.g., How far are you now? How many more buckets do you need to put out the dragon's fire? How do you know?) as well as in a more general set of dialogue prompts that are included in the teacher's guide. Opportunities for children to discuss what they learned during game play each day, to share their knowledge with their peers, and to make their reasoning explicit are also provided in a Wrap-Up session that is included at the end of each math lesson.

Finally, in the whole group games and activities that were developed for the Warm-Up portion of each math lesson, children are given ample opportunity to count (e.g., up from 1 and down from 10) and to solve mental math problems, in a variety of contexts. In addition to developing computational fluency, these activities expose children to the language of mathematics and give them practice using it. Although this is valuable for all children, it is especially useful for ESL children, who may know how to count in their native language but not yet in English. Allowing children to take turns in these activities and to perform individually gives teachers opportunities to assess each child's current level of functioning, important for instructional planning, and gives children opportunities to learn from each other.

2.5 Principle 5: Expose children to the major ways number is represented and talked about in developed societies

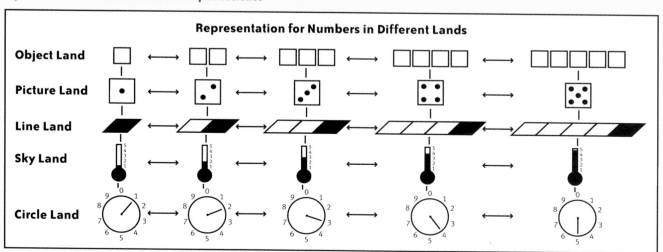

Representation for Numbers in Different Lands

Object Land
Picture Land
Line Land
Sky Land
Circle Land

Program Research

Number is represented in our culture in five major ways: as a group of objects, a dot-set pattern, a position on a line, a position on a scale (e.g., a thermometer), and a point on a dial. In each of these contexts, number is also talked about in different ways, with a larger number (and quantity) described as "more" in the world of dot-sets, as "further along" in the world of paths and lines, as "higher up" in the world of scale measures, and as "further around" in the world of dials. Children who are familiar with these forms of representation and the language used to talk about number in these contexts have a much easier time making sense of the number problems they encounter inside and outside of school.

In the **Number Worlds** program, children are systematically exposed to these forms of representations as they explore five different "lands." Learning activities developed for each land share a particular form of number representation while they simultaneously address specific knowledge goals for each grade level. Many of the games, like *Dragon Quest*, also expose children to multiple representations of number in one activity so children can gradually come to see the ways they are equivalent.

3 Discussion

Children who have been exposed to the **Number Worlds** program do very well on number questions like the one presented in the introduction and on the Number Knowledge Test (Griffin & Case, 1997) from which this question was drawn. In several evaluation studies conducted with children from low-income communities, children who received the **Number Worlds** program made significant gains in conceptual knowledge of number and in number sense, when compared to matched-control groups who received readiness training of a different sort. These gains enabled them to start their formal schooling in grade one on an equal footing with their more advantaged peers, to perform as well as groups of children from China and Japan on a computation test administered at the end of grade one, and to keep pace with their more advantaged peers (and even outperform them on some measures) as they progressed through the first few years of formal schooling (Griffin & Case, 1997).

Teachers also report positive gains from using the **Number Worlds** program and from exposure to the instructional principles on which it is based. Although all teachers acknowledge that implementing the program and putting the principles into action is not an easy task, many claim that their teaching of all subjects has been transformed in the process. They now facilitate discussion rather than dominating it; they pay much more attention to what children say and do; and they now allow children to take more responsibility for their own learning, with positive and surprising results. Above all, they now look forward to teaching math and they and their students are eager to do more of it.

Griffin, S. "Building Number Sense with Number Worlds: A mathematics program for young children." In *Early Childhood Research Quarterly* 19 (2004) 173–180. Elsevier.

References

Case, R. (1992). *The mind's staircase: Exploring the conceptual underpinnings of children's thought and knowledge.* Hillsdale, NJ: Erlbaum.

Case, R., & Griffin, S. (1990). Child cognitive development: The role of central conceptual structures in the development of scientific and social thought. In E.A. Hauert (Ed.), *Developmental psychology: Cognitive, perceptuo-motor, and neurological perspectives* (pp. 193–230). North-Holland: Elsevier.

Griffin, S. (1997). *Number Worlds: Grade one level.* Durham, NH: Number Worlds Alliance Inc.

Griffin, S. (1998). *Number Worlds: Grade two level.* Durham, NH: Number Worlds Alliance Inc.

Griffin, S. (2000). *Number Worlds: Preschool level.* Durham, NH: Number Worlds Alliance Inc.

Griffin, S. (2002). The development of math competence in the preschool and early school years: Cognitive foundations and instructional strategies. In J. Royer (Ed.), *Mathematical cognition* (pp. 1–32). Greenwich, CT: Information Age Publishing.

Griffin, S. (2003). Laying the foundation for computational fluency in early childhood. *Teaching children mathematics, 9,* 306–309.

Griffin, S. (1997). *Number Worlds: Kindergarten level.* Durham, NH: Number Worlds Alliance Inc.

Griffin, S., & Case, R. (1996). Evaluating the breadth and depth of training effects when central conceptual structures are taught. *Society for research in child development monographs, 59,* 90–113.

Griffin, S., & Case, R. (1997). Re-thinking the primary school math curriculum: An approach based on cognitive science. *Issues in Education, 3,* 1–49.

Griffin, S., Case, R., & Siegler, R. (1994). Rightstart: Providing the central conceptual prerequisites for first formal learning of arithmetic to students at-risk for school failure. In K. McGilly (Ed.), *Classroom lessons: Integrating cognitive theory and classroom practice* (pp. 24-49). Cambridge, MA: Bradford Books MIT Press.

Griffin, S., Case, R., & Capodilupo, A. (1995). Teaching for understanding: The importance of central conceptual structures in the elementary mathematics curriculum. In A. McKeough, I. Lupert, & A. Marini (Eds.), *Teaching for transfer: Fostering generalization in learning* (pp. 121–151). Hillsdale, NJ: Erlbaum.

Mathematical proficiency has five strands. These strands are not independent; they represent different aspects of a complex whole . . . they are interwoven and interdependent in the development of proficiency in mathematics.

—Kilpatrick, J., Swafford, J., and Findell, B., eds. *Adding It Up: Helping Children Learn Mathematics.* Washington D.C.: National Research Council/National Academy Press, 2001, pp. 115–133.

Number Worlds develops all five proficiencies in each lesson, so that students build their skills, conceptual understanding, and reasoning powers as well as their abilities to apply mathematics and see it as useful.

Math Proficiencies

1 UNDERSTANDING (Conceptual Understanding): Comprehending mathematical concepts, operations, and relations—knowing what mathematical symbols, diagrams, and procedures mean. *Conceptual Understanding* refers to an integrated and functional grasp of mathematical ideas. Students with conceptual understanding know more than isolated facts and methods. They understand why a mathematical idea is important and the kinds of contexts in which it is useful. They have organized their knowledge into a coherent whole, which enables them to learn new ideas by connecting those ideas to what they already know. Conceptual understanding also supports retention. Because facts and methods learned with understanding are connected, they are easier to remember and use, and they can be reconstructed when forgotten. If students understand a method, they are unlikely to remember it incorrectly.

A significant indicator of conceptual understanding is being able to represent mathematical situations in different ways and knowing how different representations can be useful for different purposes.

Knowledge that has been learned with understanding provides the bases for generating new knowledge and for solving new and unfamiliar problems. When students have acquired conceptual understanding in an area of mathematics, they see the connections among concepts and procedures and can give arguments to explain why some facts are consequences of others. They gain confidence, which then provides a base from which they can move to another level of understanding.

2 COMPUTING (Procedural Fluency): Carrying out mathematical procedures, such as adding, subtracting, multiplying, and dividing numbers flexibly, accurately, efficiently, and appropriately. *Procedural Fluency* refers to knowledge of procedures, knowledge of when and how to use them appropriately, and skill in performing them flexibly, accurately, and efficiently. In the domain of number, procedural fluency is especially needed to support conceptual understanding of place value and the meaning of rational numbers. It also supports the analysis of similarities and differences among methods of calculating. These methods include written procedures, and mental methods for finding certain sums, differences, products, or quotients, and methods that use calculators, computers, or manipulative materials such as blocks, counters, or beads.

Students need to be efficient and accurate in performing basic computations with whole numbers without always having to refer to tables or other aids. They also need to know reasonably efficient and accurate ways to add, subtract, multiply, and divide multidigit numbers, mentally and with pencil and paper. A good conceptual understanding of place value in the base-ten system supports the development of fluency in multidigit computation. Such understanding also supports simplified but accurate mental arithmetic and more flexible ways of dealing with numbers than many students ultimately achieve.

Math Proficiencies

3 **APPLYING** (Strategic Competence): Being able to formulate problems mathematically and to devise strategies for solving them using concepts and procedures appropriately. *Strategic Competence* refers to the ability to formulate mathematical problems, represent them, and solve them. This strand is similar to what has been called problem solving and problem formulation. Although in school, students are often presented with clearly specified problems to solve, outside of school they encounter situations in which part of the difficulty is to figure out exactly what the problem is. Then they need to formulate the problem so that they can use mathematics to solve it. Consequently, they are likely to need experience and practice in problem formulating as well as in problem solving. They should know a variety of solution strategies as well as which strategies might be useful for solving a specific problem.

To represent a problem accurately, students must first understand the situation, including its key features. They then need to generate a mathematical representation of the problem that captures the core mathematical elements and ignores the irrelevant features.

Students develop procedural fluency as they use their strategic competence to choose among effective procedures. They also learn that solving challenging mathematics problems depends on the ability to carry out procedures readily, and conversely, that problem-solving experience helps them acquire new concepts and skills.

4 **REASONING** (Adaptive Reasoning): Using logic to explain and justify a solution to a problem or to extend from something known to something not yet known. *Adaptive Reasoning* refers to the capacity to think logically about the relationships among concepts and situations. Such reasoning is correct and valid, stems from careful consideration of alternatives, and includes knowledge of how to justify the conclusions. In mathematics, adaptive reasoning is the glue that holds everything together and guides learning. One uses it to navigate through the many facts, procedures, concepts, and solution methods and to see that they all fit together in some way that they make sense. In mathematics, deductive reasoning is used to settle disputes and disagreements. Answers are right because they follow from some agreed-upon assumptions through a series of logical steps. Students who disagree about a mathematical answer need not rely on checking with the teacher, collecting opinions from their classmates, or gathering data from outside the classroom. In principle, they need only check that their reasoning is valid.

Research suggests that students are able to display reasoning ability when three conditions are met: They have a sufficient knowledge base, the task is understandable and motivating, and the context is familiar and comfortable.

5 **ENGAGING** (Productive Disposition): Seeing mathematics as sensible, useful, and doable—if you work at it—and being willing to do the work. *Productive disposition* refers to the tendency to see sense in mathematics, to perceive it as both useful and worthwhile, to believe that steady effort in learning mathematics pays off, and to see oneself as an effective learner and doer of mathematics. If students are to develop conceptual understanding, procedural fluency, strategic competence, and adaptive reasoning abilities, they must believe mathematics is understandable, not arbitrary; that with diligent effort, it can be learned and used; and that they are capable of figuring it out. Developing a productive disposition requires frequent opportunities to make sense of mathematics, to recognize the benefits of perseverance, and to experience the rewards of sense making in mathematics.

Students' disposition toward mathematics is a major factor in determining their educational success. Students who have developed a productive disposition are confident in their knowledge and abilities. They see that mathematics is both reasonable and intelligible and believe that, with appropriate effort and experience, they can learn.

Content Strands of Mathematics

Number Sense and Place Value

Understanding of the significance and use of numbers in counting, measuring, comparing, and ordering.

"It is very important for teachers to provide children with opportunities to recognize the meaning of mathematical symbols, mathematical operations, and the patterns or relationships represented in the child's work with numbers. For example, the number sense that a child acquires should be based upon an understanding that inverse operations, such as addition and subtraction, undo the operations of the other. Instructionally, teachers must encourage their students to think beyond simply finding the answer and to actually have them think about the numerical relationships that are being represented or modeled by the symbols, words, or materials being used in the lesson."

Kilpatrick, J., Swafford, J. and Findell, B. eds. Adding It Up: Helping Children Learn Mathematics. Washington, D.C.: National Research Council/National Academy Press, 2001, p. 270–271.

Number Worlds and Number Sense

Goal: Firm understanding of the significance and use of numbers in counting, measuring, comparing and ordering. The ability to think intelligently, using numbers. This basic requirement of numeracy includes the ability to recognize given answers as absurd, without doing a precise calculation, by observing that they violate experience, common sense, elementary logic, or familiar arithmetic patterns. It also includes the use of imagination and insight in using numbers to solve problems. Children should be able to recognize when, for example, a trial-and-error method is likely to be easier to use and more manageable than a standard algorithm.

Developing number sense is a primary goal of **Number Worlds** in every grade. Numbers are presented in a variety of representations and integrated in many contexts so that students develop thorough understanding of numbers.

Algebra

Algebra is the branch of mathematics that uses symbols to represent arithmetic operations. Algebra extends arithmetic through the use of symbols and other notations such as exponents and variables. Algebraic thinking involves understanding patterns, equations, and relationships and includes concepts of functions and inverse operations. Because algebra uses symbols rather than numbers, it can produce general rules that apply to all numbers. What most people commonly think of as algebra involves the manipulation of equations and the solving of equations. Exposure to algebraic ideas can and should occur well before students first study algebra in middle school or high school. Even primary students are capable of understanding many algebraic concepts. Developing algebraic thinking in the early grades smoothes the transition to algebra in middle school and high school and ensures success in future math and science courses as well as in the workplace.

"Algebra begins with a search for patterns. Identifying patterns helps bring order, cohesion, and predictability to seemingly unorganized situations and allows one to make generalizations beyond the information directly available. The recognition and analysis of patterns are important components of the young child's intellectual development because they provide a foundation for the development of algebraic thinking."

Clements, Douglas and Sarama, J., eds. *Engaging Young Children in Mathematics: Standards for Early Childhood Mathematics Education. Mahwah*, New Jersey: Lawrence Erlbaum Associates, Publishers, 2004., p. 52.

Number Worlds and Algebra

Goal: Understanding of functional relationships between variables that represent real world phenomena that are in a constant state of change. Children should be able to draw the graphs of functions and to derive information about functions from their graphs. They should understand the special importance of linear functions and the connection between the study of functions and the solutions of equations and inequalities.

The algebra readiness instruction that begins in the Pre-K level is designed to prepare students for future work in algebra by exposing them to algebraic thinking, including looking for patterns, using variables, working with functions, using integers and exponents, and being aware that mathematics is far more than just arithmetic.

Content Strands

Arithmetic

Arithmetic, one of the oldest branches of mathematics, arises from the most fundamental of mathematical operations: counting. The arithmetic operations—addition, subtraction, multiplication, division, and place holding—form from the basis of the mathematics we use regularly. Mastery of the basic operations with whole numbers (addition, subtraction, multiplication, and division)

> "Although some educators once believed that children memorized their "basic facts" as conditioned responses, research now shows that children do not move from knowing nothing about sums and differences of numbers to having the basic number combinations memorized. Instead, they move through a series of progressively more advance and abstract methods for working out the answers to simple arithmetic problems. Furthermore, as children get older, they use the procedures more and more efficiently."
>
> Kilpatrick, J., Swafford, J., and Findell, B., eds. *Adding It Up: Helping Children Learn Mathematics*. Washington, D.C.: National Research Council/National Academy Press, 2001, pp. 182–183.

Number Worlds and Arithmetic

Goal: Mastery of the basic operations with whole numbers (addition, subtraction, multiplication, and division). Whatever other skills and understandings children acquire, they must have the ability to calculate a precise answer when necessary. This fundamental skill includes not only knowledge of the appropriate arithmetic algorithms but also mastery of the basic addition, subtraction, multiplication, and division facts and understanding of the positional notation (base ten) of the whole numbers.

Cumulative assessment occurs throughout the program to indicate when mastery of concepts and skills is expected. Once taught, arithmetic skills are also integrated into other topics such as data analysis.

Fractions, Decimals, and Percents

Understand rational numbers (fractions, decimals, and percents) and their relationships to each other, including the ability to perform calculations and to use rational numbers in measurement.

> "Children need to learn that rational numbers are numbers in the same way that whole numbers are numbers. For children to use rational numbers to solve problems, they need to learn that the same rational number may be represented in different ways, as a fraction, a decimal, or a percent. Fraction concepts and representations need to be related to those of division, measurement, and ratio. Decimal and fractional representations need to be connected and understood. Building these connections takes extensive experience with rational numbers over a substantial period of time. Researchers have documented that difficulties in working with rational numbers can often be traced to weak conceptual understanding. . . . Instructional sequences in which more time is spent at the outset on developing meaning for the various representations of rational numbers and the concept of unit have been shown to promote mathematical proficiency."
>
> Kilpatrick, J., Swafford, J., and Findell, B., eds. *Adding It Up: Helping Children Learn Mathematics*. Washington, D.C.: National Research Council/National Academy Press, 2001, pp. 415–416.

Number Worlds and Rational Numbers

Goal: Understanding of rational numbers and of the relationship of fractions to decimals. Included here are the ability to do appropriate calculations with fractions or decimals (or both, as in fractions of decimals), the use of decimals in (metric unit) measurements, the multiplication of fractions as a model for the "of" relation and as a model for areas of rectangles.

Goal: Understanding of the meaning of rates and of their relationship to the arithmetic concept of ratio. Children should be able to calculate ratios, proportions, and percentages; understand how to use them intelligently in real-life situations; understand the common units in which rates occur (such as kilometers per hour, cents per gram); understand the meaning of per; and be able to express ratios as fractions.

In **Number Worlds** understanding of rational number begins at the earliest grades with sharing activities and develops understanding of rational numbers with increasing sophistication at each grade.

Geometry

Geometry is the branch of mathematics that deals with the properties of space. Plane geometry is the geometry of flat surfaces and solid geometry is the geometry of three-dimensional solids. Geometry has many more fields, including the study of spaces with four or more dimensions.

"Geometry can be used to understand and to represent the objects, directions, and locations in our world, and the relationships between them. Geometric shapes can be described, analyzed, transformed, and composed and decomposed into other shapes."

Clements, Douglas and Sarama, J., eds. *Engaging Young Children in Mathematics: Standards for Early Childhood Mathematics Education.* Mahwah, New Jersey: Lawrence Erlbaum Associates, Publishers, 2004., p. 39.

Number Worlds and Geometry

Goal: Understanding of an ability to use the geometric concepts of perimeter, area, volume, and congruency as applied to simple figures.

Data Analysis and Applications

Data Analysis encompasses the ability to organize information to make it easier to use and the ability to interpret data and graphs.

"Describing data involves reading displays of data (e.g., tables, lists, graphs); that is, finding information explicitly stated in the display, recognizing graphical conventions, and making direct connections between the original data and the display. The process is essentially what has been called reading the data. . . . The process of organizing and reducing data incorporates mental actions such as ordering, grouping, and summarizing. Data reduction also includes the use of representative measures of center (often termed *measures of central tendency*) such as mean, mode, or media, and measures of spread such as range and standard deviation."

Kilpatrick, J., Swafford, J., and Findell, B., eds. *Adding It Up: Helping Children Learn Mathematics.* Washington, D.C.: National Research Council/National Academy Press, 2001, pp. 289.

Number Worlds and Data Analysis

Goal: Ability to organize and arrange data for greater intelligibility. Children should develop not only the routine skills of tabulating and graphing results but also, at a higher level, the ability to detect patterns and trends in poorly organized data either before or after reorganization. In addition, children need to develop the ability to extrapolate and interpolate from data and from graphic representations. Children should also know when extrapolation or interpolation is justified and when it is not.

In **Number Worlds** students work with graphs beginning in Pre-K. In each grade the program emphasizes understanding what data show.

Problem Solving

Number Worlds and Problem Solving

Goal: Students must develop the critical thinking skills useful for solving problems. Computational skills are, of course, important, but they are not enough. Students also need an arsenal of critical-thinking skills (sometimes called problem-solving strategies and sometimes called heuristics) that they can call upon to solve particular problems. These skills should not be taught in isolation—students should learn to use them in different contexts. By doing so students are more likely to recognize in which situations a particular skill will be useful and when it is not likely to be useful. We can group critical-thinking skills into two categories—those that are useful in virtually all situations and those that are useful in specific contexts.

In **Number Worlds** students solve problems throughout the daily lessons in all levels of the program.

Technology

Technology has changed the world of mathematics. Technological tools have eliminated the need for tedious calculations and have enabled significant advances in applications of mathematics. Technology can also help to make teaching more effective and efficient. Well-designed math software activities have proven effective in advancing children's math achievements. Technology can also help teachers organize planning and instruction and manage record keeping.

Number Worlds integrates two key software programs throughout the lessons to help students develop richer mathematical skills and conceptual understanding: *eMathTools* and *Building Blocks,* which are explained on the following pages.

Building Blocks Software Activities

Building Blocks software provides computer math activities that address specific developmental levels of the math learning trajectories. **Building Blocks** software is critical to **Number Worlds**. The engaging research-based activities provide motivating development and support of concepts.

Some **Building Blocks** activities have different levels of difficulty indicated by ranges in the Activity Names below. The list provides an overview of all of the **Building Blocks** activities along with the domains, descriptions, and appropriate age ranges.

Domain: Trajectory	Activity Name	Description	Age Range (in years)
Geometry: Composition/Decomposition	Create a Scene	Students explore shapes by moving and manipulating them to make pictures.	4–12
Geometry: Composition/Decomposition	Piece Puzzler 1–5, Piece Puzzler Free Explore, and Super Shape 1–7	Students complete pictures using pattern or tangram shapes.	4–12
Geometry: Imagery	Geometry Snapshots 1–8	Students match configurations of a variety of shapes (e.g., line segments in different arrangements, 3-6 tiled shapes, embedded shapes) to corresponding configurations, given only a brief view of the goal shapes.	5–12
Geometry: Shapes (Identifying)	Memory Geometry 1–5	Students match familiar geometric shapes (shapes in same or similar sizes, same orientation) within the framework of a "Concentration" card game.	3–5
Geometry: Shapes (Matching)	Mystery Pictures 1–4 and Mystery Pictures Free Explore	Students construct predefined pictures by selecting shapes that match a series of target shapes.	3–8
Geometry: Shapes (Parts)	Shape Parts 1–7	Students build or fix some real-world object, exploring shape and properties of shapes.	5–12
Geometry: Shapes (Properties)	Legends of the Lost Shape	Students identify target shapes using textual clues provided.	8–12
Geometry: Shapes (Properties)	Shape Shop 1–3	Students identify a wide range of shapes given their names, with more difficult distracters.	8–12
Measurement: Length	Comparisons	Students are shown pictures of two objects and are asked to click on the one that fits the prompt (longer, shorter, heavier, etc.).	4–8
Measurement: Length	Deep Sea Compare	Students compare the length of two objects by representing them with a third object.	5–7
Measurement: Length	Workin' on the Railroad	Students identify the length (in non-standard units) of railroad trestles they built to span a gully.	6–9
Measurement: Length	Reptile Ruler	Students learn about linear measurement by using a ruler to determine the length of various reptiles.	7–10
Multiplication/Division	Arrays in Area	Students build arrays and then determine the length of those arrays.	8–11
Multiplication/Division	Comic Book Shop	Students use skip counting to produce products that are multiples of 10s, 5s, 2s, and 3s. The task is to identify the product, given a number and bundles.	7–9
Multiplication/Division	Egg-stremely Equal	Students divide large sets of eggs into several equal parts.	4–8
Multiplication/Division	Field Trip	Students solve multidigit multiplication problems in a field trip environment (e.g., equal number of students on each bus; the number of tickets needed for all students.)	8–11
Multiplication/Division	Snack Time	Students use direct modeling to solve multiplication problems.	6–8
Multiplication/Division	Word Problems with Tools 5–6, 10	Students use number tools to solve single and multidigit multiplication and division problems.	8–11
Multiplication/Division	Clean the Plates	Students use skip counting to produce products that are multiples of 10s, 5s, 2s, and 3s.	7–9
Numbers: Adding and Subtracting	Barkley's Bones 1–10 and Barkley's Bones 1–20	Students determine the missing addend in X + ___ = Z problems to feed bone treats to a dog. (Z = 10 or less)	5–8
Number: Adding and Subtracting	Double Compare 1–10 and Double Compare 1–20	Students compare sums of cards (to 10 or 20) to determine which sum is greater.	5–8

Building Blocks

Domain: Trajectory	Activity Name	Description	Age Range (in years)
Number: Adding and Subtracting	Word Problems with Tools 1–4, 7–9, 11–12	Students use number tools to solve single and multidigit addition and subtraction problems.	8–12
Number: Adding and Subtracting and Counting	Counting Activities (Road Race Counting Game, Numeral Train Game, etc.)	Students identify numerals or dot amounts (totals to 20) and move forward a corresponding number of spaces on a game board.	3–9
Number: Adding and Subtracting and Multiplication and Division	Function Machine 1–4	Students provide inputs to a function and examine the resulting outputs to determine the definition of that function. Functions include addition, subtraction, multiplication, or division.	6–12
Number: Comparing	Ordinal Construction Company	Students learn ordinal positions (1st through 10th) by moving objects between the floors of a building.	5–7
Number: Comparing	Rocket Blast 1–3	Given a number line with only initial and final endpoints labeled and a location on that line, students determine the number label for that location.	6–12
Number: Comparing and Counting	Party Time 1–3 and Party Time Free Explore	Students use party utensils to practice one-to-one correspondence, identify numerals that represent target amounts, and match object amounts to target numerals.	4–6
Number: Comparing and Multiplication and Division	Number Compare 1–5	Students compare two cards and choose the one with the greater value.	4–11
Number: Comparing, Counting, Adding and Subtracting	Pizza Pizzazz 1–5 and Pizza Pizzazz Free Explore	Students count items, match target amounts, and explore missing addends related to toppings on pizzas.	3–8
Number: Counting (Object)	Countdown Crazy	Students click digits in sequence to count down from 10 to 0.	5–7
Number: Counting (Object)	Memory Number 1–3	Students match displays containing both numerals and collections to matching displays within the framework of a "Concentration" card game.	4–6
Number: Counting (Object) and Adding and Subtracting	Dinosaur Shop 1–4 and Dinosaur Shop Free Explore	Students use toy dinosaurs to identify numerals representing target amounts, match object amounts to target numerals, add groups of objects, and find missing addends.	4–7
Number: Counting (Objects)	Book Stacks	Students "count on" (through at least one decade) from a given number as they load books onto a cart.	6–8
Number: Counting (Objects)	School Supply Shop	Students count school supplies, bundled in groups of ten to reach a target number up to 100.	6–8
Number: Counting (Objects)	Tire Recycling	Students use skip-counting by 2s and 5s to count tires as they are moved.	6–8
Number: Counting (Strategies)	Build Stairs 1–3 and Build Stairs Free Explore	Students practice counting, sequencing, and ordering by building staircases.	4–7
Number: Counting (Strategies)	Math-O-Scope	Students identify the numbers that surround a given number in the context of a 100s chart.	7–9
Number: Counting (Strategies)	Tidal Tally	Students identify missing addends (hidden objects) by counting forward from given addends (visible objects) to reach a numerical total.	6–9
Number: Counting (Verbal)	Count and Race	Students count up to 50 by adding cars to a racetrack one at a time.	3–6
Number: Counting (Verbal)	Before and After Math	Students identify and select numbers that come either just before or right after a target number.	4–7
Number: Counting (Verbal)	Kitchen Counter	Students click on objects one at a time while the numbers from one to ten are counted aloud.	3–6
Number: Subitizing	Number Snapshots 1–10	Students match numerals or dot collections to a corresponding numeral or collection given only a brief view of the goal collections.	3–12
Patterning	Marching Patterns 1–3	Students extend a linear pattern of marchers by one full repetition of an entire unit (AB, AAB, ABB, and ABC patterns).	5–7
Patterning	Pattern Planes 1–3	Students duplicate a linear pattern of flags based on an outline that serves as a guide (AB, AAB, ABB, and ABC patterns).	4–6
Patterning	Pattern Zoo 1–3 and Patterns Free Explore	Students identify a linear pattern of fruit that matches a target pattern to feed zoo animals (AB, AAB, ABB, and ABC patterns). Students explore patterning by creating rhythmic patterns of their own.	3–6

eMathTools

Number Worlds integrates *eMathTools* in appropriate lessons throughout the program. This component provides multimedia formats for demonstrating and exploring concepts and solving problems. The *eMathTools* are described below.

Data Organization and Display Tools

- **Spreadsheet Tool**—allows students to manage, display, sort, and calculate data. Links to the graphing tool for further data display
- **Graphing Tool**—displays data in pie charts or circle graphs, line graphs, bar graphs, or coordinate grids
- **Venn Diagram**—allows students to sort data visually

Measurement and Conversion Tools

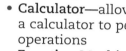

- **Stopwatch**—measures in real time for development of counting and time concepts
- **Calendar**—an electronic calendar to develop concepts of time
- **Metric and Customary Conversion Tool**—converts metric and customary measurements in length, distance, mass and weight, time, temperature, and capacity
- **Estimating Proportion Tool**—allows visual representations of proportions to develop understanding of ratios, fractions, and decimals

Geometric Exploration Tools

- **Tessellations**—allows students to create tessellation patterns by rotating, coloring, and tiling shapes
- **Net Tool**—allows students to manipulate 2-D shapes and then print them to create 3-D shapes
- **Shape Tools**—explores and manipulates shapes to create designs
- **Geometry Sketch Tool**—allows drawing, manipulating, and measuring a wide variety of shapes
- **Pythagorean Theorem Tool**—launches right triangles to explore the Pythagorean Theorem

Calculation and Counting Tools

- **Calculator**—allows students to launch a calculator to perform mathematical operations
- **Function Machine**—an electronic version of a function machine that students use to solve missing variable problems
- **Multiplication and Division Table**—an interactive version of a table that highlights relationships between multipli-cation and division facts
- **Addition and Subtraction Table**—an interactive version of a table that highlights relationships between addition and subtraction facts
- **100s Chart**—an interactive version of a table that highlights patterns and relationships among numbers
- **Number Line**—an electronic number line that allows students to skip count and see the relationships among whole numbers, fractions, decimals, and percents
- **Number Stairs**—a tool to illustrate counting in units
- **Probability Tool**—uses number cubes, spinner, or tumble drum to test scenarios of probability
- **Set Tool**—allows students to visually represent and manipulate different sets of objects for a variety of counting activities
- **Base-Ten Blocks**—allows students to manipulate base-10 block for counting
- **Coins and Money**—uses visual representations of coins and money to represent counting
- **Fraction Tool**—represents fractional units for counting and understanding relationships
- **Array Tool**—presents arrays to represent multiplication and division patterns and relationships

English Learners

English learners enter schools at different grade levels and bring varying levels of English proficiency and academic preparation attained in their primary language. Some students have high levels of academic preparation that allow them to focus more on learning the terminology associated with the math skills they already know. These students have a great advantage as they already understand that mathematics has its own terminology and have a grasp of the basic skills and concepts needed to participate in advancing their math skills.

Other students may have had little schooling or interrupted school experiences, which means they need to catch up to their grade-level peers academically, and they must begin the task of English acquisition at the same time. Many schools use some form of primary language instruction or support to accelerate students' skill base while learning English. This approach contains the added advantage of being able to draw on the student's family as a source of support and inclusion.

Number Worlds and English Learners

Creating Context and Alternate Vocabulary

In addition to the specialized terminology of mathematics, academic language often includes the use of common English idiomatic expressions and content area words that have multiple meanings in English. Previewing these expressions with English learners and helping them keep track of their usage and meaning is excellent practice for extending knowledge of English for academic purposes.

In each *Number Worlds* lesson several words or expressions are noted along with a quick description of their meaning. Some of the words and expressions are math phrases, but many are examples of wider vocabulary used in word problems and direction lines in the lessons. Creating context and alternate vocabulary features highlight some of these key phrases and give teachers a brief explanation.

The inclusion of alternate vocabulary such as the words with double meanings that students encounter in the subject area brings to the attention of teachers those complexities that English speakers may already have mastered but that can make the difference between deep comprehension and confusion for English learners. Alternate vocabulary also highlights some potentially puzzling idiomatic expressions.

Spanish Cognates

One way to rapidly accelerate the acquisition of more sophisticated academic language is to take advantage of words with shared roots across languages. Words that are related by root or are borrowed from another language are called cognates. In English many math and science terms come from Greek, Latin, and Arabic roots. English learners who speak Romance languages such as Spanish have an advantage in these subjects because much of the important concept vocabulary is similar in both languages. By teaching students to look for some basic word parts in English, these students can take advantage of what they know in their primary language to augment their knowledge of English. Pronunciation of these words may be quite different, but the written words often look very similar and mean the same thing.

Children follow natural developmental progressions in learning, developing mathematical ideas in their own way. Curriculum research has revealed sequences of activities that are effective in guiding children through these levels of thinking. These developmental paths are the basis for Building Blocks learning trajectories. Learning trajectories have three parts: a mathematical goal, a developmental path through which children develop to reach that goal, and a set of activities matched to each of those levels that help children develop the next level. Thus, each learning trajectory has levels of understanding, each more sophisticated than the last, with tasks that promote growth from one level to the next. The Building Blocks Learning Trajectories give simple labels, descriptions, and examples of each level. Complete learning trajectories describe the goals of learning, the thinking and learning processes of children at various levels, and the learning activities in which they might engage. This document provides only the developmental levels.

Learning Trajectories for Primary Grades Mathematics
Developmental Levels

Frequently Asked Questions (FAQ)

1. **Why use learning trajectories?** Learning trajectories allow teachers to build the *mathematics of children—the thinking of children as it develops naturally.* So, we know that all the goals and activities are within the developmental capacities of children. We know that each level provides a natural *developmental building block* to the next level. Finally, we know that the activities provide the *mathematical building blocks* for school success, because the research on which they are based typically involves higher-income children.

2. **When are children "at" a level?** Children are at a certain level when most of their behaviors reflect the thinking—ideas and skills—of that level. Often, they show a few behaviors from the next (and previous) levels as they learn.

3. **Can children work at more than one level at the same time?** Yes, although most children work mainly at one level or in transition between two levels (naturally, if they are tired or distracted, they may operate at a much lower level). Levels are not "absolute stages." They are "benchmarks" of complex growth that represent distinct ways of thinking. So, another way to think of them is as a sequence of different patterns of thinking. Children are continually learning, within levels and moving between them.

4. **Can children jump ahead?** Yes, especially if there are separate "sub-topics." For example, we have combined many counting competencies into one "Counting" sequence with sub-topics, such as verbal counting skills. Some children learn to count to 100 at age 6 after learning to count objects to 10 or more, some may learn that verbal skill earlier. The sub-topic of verbal counting skills would still be followed.

5. **How do these developmental levels support teaching and learning?** The levels help teachers, as well as curriculum developers, assess, teach, and sequence activities. *Teachers who understand learning trajectories and the developmental levels that are at their foundation are more effective and efficient.* Through planned teaching and also encouraging informal, incidental mathematics, teachers help children learn *at an appropriate and deep level.*

6. **Should I plan to help children develop just the levels that correspond to my children's ages?** No! The ages in the table are typical ages children develop these ideas. *But these are rough guides only—children differ widely.* Furthermore, the ages below are lower bounds of what children achieve without instruction. So, these are *"starting levels" not goals.* We have found that children who are provided high-quality mathematics experiences are capable of developing to levels one or more years beyond their peers.

Each column in the table below, such as "Counting," represents a main developmental progression that underlies the learning trajectory for that topic.

For some topics, there are "subtrajectories"—strands within the topic. In most cases, the names make this clear. For example, in Comparing and Ordering, some levels are about the "Comparer" levels, and others about building a "Mental Number Line." Similarly, the related subtrajectories of "Composition" and "Decomposition" are easy to distinguish. Sometimes, for clarification, subtrajectories are indicated with a note in italics after the title. For example, in Shapes, Parts and Representing are subtrajectories within the Shapes trajectory.

Clements, D. H., Sarama, J., & DiBiase, A.-M. (Eds.). (2004). *Engaging Young Children in Mathematics: Standards for Early Childhood Mathematics Education.* Mahwah, NJ: Lawrence Erlbaum Associates.

Clements, D. H., & Sarama, J. (in press). "Early Childhood Mathematics Learning." In F. K. Lester, Jr. (Ed.), *Second Handbook of Research on Mathematics Teaching and Learning.* New York: Information Age Publishing.

Developmental Levels for Counting

The ability to count with confidence develops over the course of several years. Beginning in infancy, children show signs of understanding number. With instruction and number experience, most children can count fluently by age 8, with much progress in counting occurring in kindergarten and first grade. Most children follow a natural developmental progression in learning to count with recognizable stages or levels. This developmental path can be described as part of a learning trajectory.

Age Range	Level Name	Level	Description
1–2	Pre-Counter	1	A child at the earliest level of counting may name some numbers meaninglessly. The child may skip numbers and have no sequence.
1–2	Chanter	2	At this level a child may sing-song numbers, but without meaning.
2	Reciter	3	At this level the child verbally counts with separate words, but not necessarily in the correct order.
3	Reciter (10)	4	A child at this level can verbally count to 10 with some correspondence with objects. They may point to objects to count a few items but then lose track.
3	Corresponder	5	At this level a child can keep one-to-one correspondence between counting words and objects—at least for small groups of objects laid in a line. A corresponder may answer "how many" by recounting the objects starting over with one each time.
4	Counter (Small Numbers)	6	*At around 4 years children begin to count meaningfully. They accurately count objects to 5 and answer the "how many" question with the last number counted. When objects are visible, and especially with small numbers, begins to understand cardinality. These children can count verbally to 10 and may write or draw to represent 1–5.*
4	Producer—Counter To (Small Numbers)	7	The next level after counting small numbers is to count out objects up to 5 and produce a group of four objects. When asked to show four of something, for example, this child can give four objects.
4–5	Counter (10)	8	This child can count structured arrangements of objects to 10. He or she may be able to write or draw to represent 10 and can accurately count a line of nine blocks and says there are 9. A child at this level can also find the number just after or just before another number, but only by counting up from 1.
5–6	Counter and Producer—Counter to (10+)	9	Around 5 years of age children begin to count out objects accurately to 10 and then beyond to 30. They can keep track of objects that have and have not been counted, even in different arrangements. They can write or draw to represent 1 to 10 and then 20 and 30, and can give the next number to 20 or 30. These children can recognize errors in others' counting and are able to eliminate most errors in one's own counting.

Age Range	Level Name	Level	Description
5–6	Counter Backward from 10	10	Another milestone at about age 5 is being able to count backwards from 10.
6–7	Counter from N (N+1, N−1)	11	Around 6 years of age children begin to count on, counting verbally and with objects from numbers other than 1. Another noticeable accomplishment is that children can determine immediately the number just before or just after another number without having to start back at 1.
6–7	Skip-Counting by 10s to 100	12	A child at this level can count by tens to 100. They can count through decades knowing that 40 comes after 39, for example.
6–7	Counter to 100	13	A child at this level can count by ones through 100, including the decade transitions from 39 to 40, 49 to 50, and so on, starting at any number.
6–7	Counter On Using Patterns	14	At this level a child keeps track of counting acts by using numerical patterns such as tapping as he or she counts.
6–7	Skip Counter	15	The next level is when children can count by 5s and 2s with understanding.
6–7	Counter of Imagined Items	16	At this level a child can count mental images of hidden objects.
6–7	Counter On Keeping Track	17	A child at this level can keep track of counting acts numerically with the ability to count up one to four more from a given number.
6–7	Counter of Quantitative Units	18	At this level a child can count unusual units such as "wholes" when shown combinations of wholes and parts. For example when shown three whole plastic eggs and four halves, a child at this level will say there are five whole eggs.
6–7	Counter to 200	19	At this level a child counts accurately to 200 and beyond, recognizing the patterns of ones, tens, and hundreds.
7+	Number Conserver	20	A major milestone around age 7 is the ability to conserve number. A child who conserves number understands that a number is unchanged even if a group of objects is rearranged. For example, if there is a row of ten buttons, the child understands there are still ten without recounting, even if they are rearranged in a long row or a circle.
7+	Counter Forward and Back	21	A child at this level counts in either direction and recognizes that sequence of decades sequence mirrors single-digit sequence.

Developmental Levels for Comparing and Ordering Numbers

Comparing and ordering sets is a critical skill for children as they determine whether one set is larger than another to make sure sets are equal and "fair." Prekindergartners can learn to use matching to compare collections or to create equivalent collections. Finding out how many more or fewer in one collection is more demanding than simply comparing two collections. The ability to compare and order sets with fluency develops over the course of several years. With instruction and number experience, most children develop foundational understanding of number relationships and place value at ages 4 and 5. Most children follow a natural developmental progression in learning to compare and order numbers with recognizable stages or levels. This developmental path can be described as part of a learning trajectory.

Age Range	Level Name	Level	Description
2	Object Corresponder	1	At this early level a child puts objects into one-to-one correspondence, but with only intuitive understanding of resulting equivalence. For example, a child may know that each carton has a straw, but doesn't necessarily know there are the same numbers of straws and cartons.
2	Perceptual Comparer	2	At the next level a child can compare collections that are quite different in size (for example, one is at least twice the other) and know that one has more than the other. If the collections are similar, the child can compare very small collections.
2–3	First-Second Ordinal Counter	3	A child at this level can identify the first and often second objects in a sequence.
3	Nonverbal Comparer of Similar Items	4	At this level a child can identify that different organizations of the same number of small groups are equal and different from other sets. (1–4 items).
3	Nonverbal Comparer of Dissimilar Items	5	At the next level a child can match small, equal collections of dissimilar items, such as shells and dots, and show that they are the same number.
4	Matching Comparer	6	As children progress they begin to compare groups of 1–6 by matching. For example, a child gives one toy bone to every dog and says there are the same number of dogs and bones.
4	Knows-to-Count Comparer	7	A significant step occurs when the child begins to count collections to compare. At the early levels children are not always accurate when larger collection's objects are smaller in size than the objects in the smaller collection. For example, a child at this level may accurately count two equal collections, but when asked, says the collection of larger blocks has more.
4	Counting Comparer (Same Size)	8	At the next level children make accurate comparisons via counting, but only when objects are about the same size and groups are small (about 1–5).
5	Counting Comparer (5)	9	As children develop their ability to compare sets, they compare accurately by counting, even when larger collection's objects are smaller. A child at this level can figure out how many more or less.

Age Range	Level Name	Level	Description
5	Ordinal Counter	10	At the next level a child identifies and uses ordinal numbers from "first" to "tenth." For example, the child can identify who is "third in line."
5	Counting Comparer	11	At this level a child can compare by counting, even when the larger collection's objects are smaller. For example, a child can accurately count two collections and say they have the same number even if one has larger objects.
5	Mental Number Line to 10	12	At this level a child uses internal images and knowledge of number relationships to determine relative size and position. For example, the child can determine whether 4 or 9 is closer to 6.
5	Serial Orderer to 6+	13	Children demonstrate development in comparing when they begin to order lengths marked into units (1–6, then beyond). For example, given towers of cubes, this child can put them in order, 1 to 6. Later the child begins to order collections. For example, given cards with one to six dots on them, puts in order.
6	Counting Comparer (10)	14	The next level can be observed when the child compares sets by counting, even when larger collection's objects are smaller, up to 10. A child at this level can accurately count two collections of 9 each, and says they have the same number, even if one collection has larger blocks.
6	Mental Number Line to 10	15	As children move into the next level they begin to use mental rather than physical images and knowledge of number relationships to determine relative size and position. For example, a child at this level can answer which number is closer to 6, 4, or 9 without counting physical objects.
6	Serial Orderer to 6+	16	At this level a child can order lengths marked into units. For example, given towers of cubes the child can put them in order.
7	Place Value Comparer	17	Further development is made when a child begins to compare numbers with place value understandings. For example, a child at this level can explain that "63 is more than 59 because six tens is more than five tens even if there are more than three ones."

Age Range	Level Name	Level	Description
7	Mental Number Line to 100	18	Children demonstrate the next level in comparing and ordering when they can use mental images and knowledge of number relationships, including ones embedded in tens, to determine relative size and position. For example, a child at this level when asked, "Which is closer to 45, 30 or 50?"says "45 is right next to 50, but 30 isn't."

Age Range	Level Name	Level	Description
8+	Mental Number Line to 1000s	19	About age 8 children begin to use mental images of numbers up to 1,000 and knowledge of number relationships, including place value, to determine relative size and position. For example, when asked, "Which is closer to 3,500—2,000 or 7,000?"a child at this level says "70 is double 35, but 20 is only fifteen from 35, so twenty hundreds, 2,000, is closer."

Developmental Levels for Recognizing Number and Subitizing (Instantly Recognizing)

The ability to recognize number values develops over the course of several years and is a foundational part of number sense. Beginning at about age 2, children begin to name groups of objects. The ability to instantly know how many are in a group, called *subitizing*, begins at about age 3. By age 8, with instruction and number experience, most children can identify groups of items and use place values and multiplication skills to count them. Most children follow a natural developmental progression in learning to count with recognizable stages or levels. This developmental path can be described as part of a learning trajectory.

Age Range	Level Name	Level	Description
2	Small Collection Namer	1	The first sign of a child's ability to subitize occurs when the child can name groups of one to two, sometimes three. For example, when shown a pair of shoes, this young child says, "Two shoes."
3	Nonverbal Subitizer	2	The next level occurs when shown a small collection (one to four) only briefly, the child can put out a matching group nonverbally, but cannot necessarily give the number name telling how many. For example, when four objects are shown for only two seconds, then hidden, child makes a set of four objects to "match."
3	Maker of Small Collections	3	At the next level a child can nonverbally make a small collection (no more than five, usually one to three) with the same number as another collection. For example, when shown a collection of three, makes another collection of three.
4	Perceptual Subitizer to 4	4	Progress is made when a child instantly recognizes collections up to four when briefly shown and verbally names the number of items. For example, when shown four objects briefly, says "four."
5	Perceptual Subitizer to 5	5	The next level is the ability to instantly recognize briefly shown collections up to five and verbally name the number of items. For example, when shown five objects briefly, says "five."

Age Range	Level Name	Level	Description
5	Conceptual Subitizer to 5+	6	At the next level the child can verbally label all arrangements to five shown only briefly. For example, a child at this level would say, "I saw 2 and 2 and so I saw 4."
5	Conceptual Subitizer to 10	7	The next step is when the child can verbally label most briefly shown arrangements to six, then up to ten, using groups. For example, a child at this level might say, "In my mind, I made two groups of 3 and one more, so 7."
6	Conceptual Subitizer to 20	8	Next, a child can verbally label structured arrangements up to twenty, shown only briefly, using groups. For example, the child may say, "I saw three 5s, so 5, 10, 15."
7	Conceptual Subitizer with Place Value and Skip Counting	9	At the next level a child is able to use skip counting and place value to verbally label structured arrangements shown only briefly. For example, the child may say, "I saw groups of tens and twos, so 10, 20, 30, 40, 42, 44, 46 . . . 46!"
8+	Conceptual Subitizer with Place Value and Multiplication	10	As children develop their ability to subitize, they use groups, multiplication, and place value to verbally label structured arrangements shown only briefly. At this level a child may say, "I saw groups of tens and threes, so I thought, five tens is 50 and four 3s is 12, so 62 in all."

Developmental Levels for Composing Number
(Knowing Combinations of Numbers)

Composing and decomposing are combining and separating operations that allow children to build concepts of "parts" and "wholes." Most prekindergartners can "see" that two items and one item make three items. Later, children learn to separate a group into parts in various ways and then to count to produce all of the number "partners" of a given number. Eventually children think of a number and know the different addition facts that make that number. Most children follow a natural developmental progression in learning to compose and decompose numbers with recognizable stages or levels. This developmental path can be described as part of a learning trajectory.

Age Range	Level Name	Level	Description
4	Pre-Part-Whole Recognizer	1	At the earliest levels of composing a child only nonverbally recognizes parts and wholes. For example, When shown four red blocks and two blue blocks, a young child may intuitively appreciate that "all the blocks" include the red and blue blocks, but when asked how many there are in all, may name a small number, such as 1.
5	Inexact Part-Whole Recognizer	2	A sign of development in composing is that the child knows that a whole is bigger than parts, but does not accurately quantify. For example, when shown four red blocks and two blue blocks and asked how many there are in all, names a "large number," such as 5 or 10.
5	Composer to 4, then 5	3	The next level is that a child begins to know number combinations. A child at this level quickly names parts of any whole, or the whole given the parts. For example, when shown four, then one is secretly hidden, and then is shown the three remaining, quickly says "1" is hidden.

Age Range	Level Name	Level	Description
6	Composer to 7	4	The next sign of development is when a child knows number combinations to totals of seven. A child at this level quickly names parts of any whole, or the whole given parts and can double numbers to 10. For example, when shown six, then four are secretly hidden, and shown the two remaining, quickly says "4" are hidden.
6	Composer to 10	5	The next level is when a child knows number combinations to totals of 10. A child at this level can quickly name parts of any whole, or the whole given parts and can double numbers to 20. For example, this child would be able to say "9 and 9 is 18."
7	Composer with Tens and Ones	6	At the next level the child understands two-digit numbers as tens and ones; can count with dimes and pennies; and can perform two-digit addition with regrouping. For example, a child at this level can explain, "17 and 36 is like 17 and 3, which is 20, and 33, which is 53."

Developmental Levels for Adding and Subtracting

Learning single-digit addition and subtraction is generally characterized as "learning math facts." It is assumed that children must memorize these facts, yet research has shown that addition and subtraction have their roots in counting, counting on, number sense, the ability to compose and decompose numbers, and place value. Research has shown that learning methods for adding and subtracting with understanding is much more effective than rote memorization of seemingly isolated facts. Most children follow an observable developmental progression in learning to add and subtract numbers with recognizable stages or levels. This developmental path can be described as part of a learning trajectory.

Age Range	Level Name	Level	Description
1	Pre +/−	1	At the earliest level a child shows no sign of being able to add or subtract.
3	Nonverbal +/−	2	The first inkling of development is when a child can add and subtract very small collections nonverbally. For example, when shown two objects, then one object going under a napkin, the child identifies or makes a set of three objects to "match."

Age Range	Level Name	Level	Description
4	Small Number +/−	3	The next level of development is when a child can find sums for joining problems up to 3 + 2 by counting all with objects. For example, when asked, "You have 2 balls and get 1 more. How many in all?" counts out 2, then counts out 1 more, then counts all 3: "1, 2, 3, 3!"

Learning Trajectories

Age Range	Level Name	Level	Description
5	Find Result +/−	4	**Addition** Evidence of the next level in addition is when a child can find sums for joining (you had 3 apples and get 3 more, how many do you have in all?) and part-part-whole (there are 6 girls and 5 boys on the playground, how many children were there in all?) problems by direct modeling, counting all, with objects. For example, when asked, "You have 2 red balls and 3 blue balls. How many in all?" the child counts out 2 red, then counts out 3 blue, then counts all 5. **Subtraction** In subtraction, a child at this level can also solve take-away problems by separating with objects. For example, when asked, "You have 5 balls and give 2 to Tom. How many do you have left?" the child counts out 5 balls, then takes away 2, and then counts the remaining 3.
5	Find Change +/−	5	**Addition** At the next level a child can find the missing addend (5 + __ = 7) by adding on objects. For example, when asked, "You have 5 balls and then get some more. Now you have 7 in all. How many did you get?" the child counts out 5, then counts those 5 again starting at 1, then adds more, counting "6, 7," then counts the balls added to find the answer, 2. **Subtraction** Compares by matching in simple situations. For example, when asked, "Here are 6 dogs and 4 balls. If we give a ball to each dog, how many dogs won't get a ball?" a child at this level counts out 6 dogs, matches 4 balls to 4 of them, then counts the 2 dogs that have no ball.
5	Make It N +/−	6	A significant advancement in addition occurs when a child is able to count on. This child can add on objects to make one number into another, without counting from 1. For example, when asked, "This puppet has 4 balls but she should have 6. Make it 6," puts up 4 fingers on one hand, immediately counts up from 4 while putting up two fingers on the other hand, saying, "5, 6" and then counts or recognizes the two fingers.
6	Counting Strategies +/−	7	The next level occurs when a child can find sums for joining (you had 8 apples and get 3 more . . .) and part-part-whole (6 girls and 5 boys . . .) problems with finger patterns or by adding on objects or counting on. For example, when asked "How much is 4 and 3 more?" the child answers "4 . . . 5, 6, 7 [uses rhythmic or finger pattern]. 7!" Children at this level also can solve missing addend (3 + __ = 7) or compare problems by counting on. When asked, for example, "You have 6 balls. How many more would you need to have 8?" the child says, "6, 7 [puts up first finger], 8 [puts up second finger]. 2!"
6	Part-Whole +/−	8	Further development has occurred when the child has part-whole understanding. This child can solve all problem types using flexible strategies and some derived facts (for example, "5 + 5 is 10, so 5 + 6 is 11"), sometimes can do start unknown (__ + 6 = 11), but only by trial and error. This child when asked, "You had some balls. Then you get 6 more. Now you have 11 balls. How many did you start with?" lays out 6, then 3 more, counts and gets 9. Puts 1 more with the 3, says 10, then puts 1 more. Counts up from 6 to 11, then recounts the group added, and says, "5!"
6	Numbers-in-Numbers +/−	9	Evidence of the next level is when a child recognizes that a number is part of a whole and can solve problems when the start is unknown (__ + 4 = 9) with counting strategies. For example, when asked, "You have some balls, then you get 4 more balls, now you have 9. How many did you have to start with?" this child counts, putting up fingers, "5, 6, 7, 8, 9." Looks at fingers, and says, "5!"
7	Deriver +/−	10	At the next level a child can use flexible strategies and derived combinations (for example, "7 + 7 is 14, so 7 + 8 is 15") to solve all types of problems. For example, when asked, "What's 7 plus 8?" this child thinks: 7 + 8 □ 7 + [7 + 1] □ [7 + 7] + 1 = 14 + 1 = 15. A child at this level can also solve multidigit problems by incrementing or combining tens and ones. For example, when asked "What's 28 + 35?" this child thinks: 20 + 30 = 50; +8 = 58; 2 more is 60, 3 more is 63. Combining tens and ones: 20 + 30 = 50. 8 + 5 is like 8 plus 2 and 3 more, so, it's 13—50 and 13 is 63.
8+	Problem Solver +/−	11	As children develop their addition and subtraction abilities, they can solve all types of problems by using flexible strategies and many known combinations. For example, when asked, "If I have 13 and you have 9, how could we have the same number?" this child says, "9 and 1 is 10, then 3 more to make 13. 1 and 3 is 4. I need 4 more!"
8+	Multidigit +/−	12	Further development is evidenced when children can use composition of tens and all previous strategies to solve multidigit +/− problems. For example, when asked, "What's 37 − 18?" this child says, "I take 1 ten off the 3 tens; that's 2 tens. I take 7 off the 7. That's 2 tens and 0 . . . 20. I have one more to take off. That's 19." Another example would be when asked, "What's 28 + 35?" thinks, 30 + 35 would be 65. But it's 28, so it's 2 less . . . 63.

Developmental Levels for Multiplying and Dividing

Multiplication and division builds on addition and subtraction understandings and is dependent upon counting and place value concepts. As children begin to learn to multiply they make equal groups and count them all. They then learn skip counting and derive related products from products they know. Finding and using patterns aids in learning multiplication and division facts with understanding. Children typically follow an observable developmental progression in learning to multiply and divide numbers with recognizable stages or levels. This developmental path can be described as part of a learning trajectory.

Age Range	Level Name	Level	Description
2	Nonquantitive Sharer "Dumper"	1	Multiplication and division concepts begin very early with the problem of sharing. Early evidence of these concepts can be observed when a child dumps out blocks and gives some (not an equal number) to each person.
3	Beginning Grouper and Distributive Sharer	2	Progression to the next level can be observed when a child is able to make small groups (fewer than 5). This child can share by "dealing out," but often only between two people, although he or she may not appreciate the numerical result. For example, to share four blocks, this child gives each person a block, checks each person has one, and repeats this.
4	Grouper and Distributive Sharer	3	The next level occurs when a child makes small equal groups (fewer than 6). This child can deal out equally between two or more recipients, but may not understand that equal quantities are produced. For example, the child shares 6 blocks by dealing out blocks to herself and a friend 1 at a time.
5	Concrete Modeler ×/÷	4	As children develop, they are able to solve small-number multiplying problems by grouping—making each group and counting all. At this level a child can solve division/sharing problems with informal strategies, using concrete objects—up to twenty objects and two to five people—although the child may not understand equivalence of groups. For example, the child distributes twenty objects by dealing out two blocks to each of five people, then one to each, until blocks are gone.
6	Parts and Wholes ×/÷	5	A new level is evidenced when the child understands the inverse relation between divisor and quotient. For example, this child understands "If you share with more people, each person gets fewer."

Age Range	Level Name	Level	Description
7	Skip Counter ×/÷	6	As children develop understanding in multiplication and division they begin to use skip counting for multiplication and for measurement division (finding out how many groups). For example, given twenty blocks, four to each person, and asked how many people, the child skip counts by 4, holding up one finger for each count of 4. A child at this level also uses trial and error for partitive division (finding out how many in each group). For example, given twenty blocks, five people, and asked how many should each get, this child gives three to each, then one more, then one more.
8+	Deriver ×/÷	7	At the next level children use strategies and derived combinations and solve multidigit problems by operating on tens and ones separately. For example, a child at this level may explain "7 × 6, five 7s is 35, so 7 more is 42."
8+	Array Quantifier	8	Further development can be observed when a child begins to work with arrays. For example, given 7 × 4 with most of 5 × 4 covered, a child at this level may say, "There's eight in these two rows, and five rows of four is 20, so 28 in all."
8+	Partitive Divisor	9	The next level can be observed when a child is able to figure out how many are in each group. For example, given twenty blocks, five people, and asked how many should each get, a child at this level says "four, because 5 groups of 4 is 20."
8+	Multidigit ×/÷	10	As children progress they begin to use multiple strategies for multiplication and division, from compensating to paper-and-pencil procedures. For example, a child becoming fluent in multiplication might explain that "19 times 5 is 95, because twenty 5s is 100, and one less 5 is 95."

Learning Trajectories

Developmental Levels for Measuring

Measurement is one of the main real-world applications of mathematics. Counting is a type of measurement, determining how many items are in a collection. Measurement also involves assigning a number to attributes of length, area, and weight. Prekindergarten children know that mass, weight, and length exist, but they don't know how to reason about these or to accurately measure them.

As children develop their understanding of measurement, they begin to use tools to measure and understand the need for standard units of measure. Children typically follow an observable developmental progression in learning to measure with recognizable stages or levels. This developmental path can be described as part of a learning trajectory.

Age Range	Level Name	Level	Description
3	Length Quantity Recognizer	1	At the earliest level children can identify length as an attribute. For example, they might say, "I'm tall, see?"
4	Length Direct Comparer	2	In the next level children can physically align two objects to determine which is longer or if they are the same length. For example, they can stand two sticks up next to each other on a table and say, "This one's bigger."
5	Indirect Length Comparer	3	A sign of further development is when a child can compare the length of two objects by representing them with a third object. For example, a child might compare length of two objects with a piece of string. Additional evidence of this level is that when asked to measure, the child may assign a length by guessing or moving along a length while counting (without equal length units). The child may also move a finger along a line segment, saying 10, 20, 30, 31, 32.
5	Serial Orderer to 6+	4	At the next level a child can order lengths, marked in one to six units. For example, given towers of cubes, a child at this level puts in order, 1 to 6.
6	End-to-End Length Measurer	5	At the next level the child can lay units end-to-end, although he or she may not see the need for equal-length units. For example, a child might lay 9-inch cubes in a line beside a book to measure how long it is.

Age Range	Level Name	Level	Description
7	Length Unit Iterater	6	A significant change occurs when a child can use a ruler and see the need for identical units.
7	Length Unit Relater	7	At the next level a child can relate size and number of units. For example, the child may explain, "If you measure with centimeters instead of inches, you'll need more of them, because each one is smaller."
8	Length Measurer	8	As children develop measurement ability they begin to measure, knowing the need for identical units, the relationships between different units, partitions of unit, and zero point on rulers. At this level the child also begins to estimate. The child may explain, "I used a meter stick three times, then there was a little left over. So, I lined it up from 0 and found 14 centimeters. So, it's 3 meters, 14 centimeters in all."
8	Conceptual Ruler Measurer	9	Further development in measurement is evidenced when a child possesses an "internal" measurement tool. At this level the child mentally moves along an object, segmenting it, and counting the segments. This child also uses arithmetic to measure and estimates with accuracy. For example, a child at this level may explain, "I imagine one meterstick after another along the edge of the room. That's how I estimated the room's length is 9 meters."

Developalental Levels for Recognizing Geometric Shapes ⭐

Geometric shapes can be used to represent and understand objects. Analyzing, comparing, and classifying shapes helps create new knowledge of shapes and their relationships. Shapes can be decomposed or composed into other shapes. Through their everyday activity, children build both intuitive and explicit knowledge of geometric figures. Most children can recognize and name basic two-dimensional shapes at 4 years of age. However, young children can learn richer concepts about shape if they have varied examples and nonexamples of shape, discussions about shapes and their characteristics, a wide variety of shape classes, and interesting tasks. Children typically follow an observable developmental progression in learning about shapes with recognizable stages or levels. This developmental path can be described as part of a learning trajectory.

Age Range	Level Name	Level	Description
2	Shape Matcher—	1	The earliest sign of understanding shape is when a child can match basic shapes (circle, square, typical triangle) with the same size and orientation. Example: Matches ☐ to ☐. A sign of development is when a child can match basic shapes with different sizes. Example: Matches ☐ to ☐. The next sign of development is when a child can match basic shapes with different orientations. Example: Matches ☐ to ◇.
3	Shape Prototype Recognizer and Identifier	2	A sign of development is when a child can recognize and name prototypical circle, square, and, less often, a typical triangle. For example, the child names this a square ☐. Some children may name different sizes, shapes, and orientations of rectangles, but also accept some shapes that look rectangular but are not rectangles. Children name these shapes "rectangles" (including the non-rectangular parallelogram).
3	Shape Matcher— More Shapes	3	As children develop understanding of shape, they can match a wider variety of shapes with the same size and orientation. —4 Matches wider variety of shapes with different sizes and orientations. Matches these shapes ▬ ╱. —5 Matches combinations of shapes to each other. Matches these shapes ⊙⊙ ⊙⊙.
4	Shape Recognizer— Circles, Squares, and Triangles	4	The next sign of development is when a child can recognize some nonprototypical squares and triangles and may recognize some rectangles, but usually not rhombi (diamonds). Often, the child doesn't differentiate sides/corners. The child at this level may name these as triangles .

Age Range	Level Name	Level	Description
4	Constructor of Shapes from Parts – Looks Like	5	A significant sign of development is when a child represents a shape by making a shape "look like" a goal shape. For example, when asked to make a triangle with sticks, the child creates the following ☐.
5	Shape Recognizer— All Rectangles	6	As children develop understanding of shape, they recognize more rectangle sizes, shapes, and orientations of rectangles. For example, a child at this level correctly names these shapes "rectangles".
5	Side Recognizer	7	A sign of development is when a child recognizes parts of shapes and identifies sides as distinct geometric objects. For example, when asked what this shape is ⋀, the child says it is a quadrilateral (or has four sides) after counting and running a finger along the length of each side.
5	Angle Recognizer	8	At the next level a child can recognize angles as separate geometric objects. For example, when asked, "Why is this a triangle," says, "It has three angles" and counts them, pointing clearly to each vertex (point at the corner).
5	Shape Recognizer	9	As children develop they are able to recognize most basic shapes and prototypical examples of other shapes, such as hexagon, rhombus (diamond), and trapezoid. For example, a child can correctly identify and name all the following shapes.
6	Shape Identifier	10	At the next level the child can name most common shapes, including rhombi, "ellipses-is-not-circle." A child at this level implicitly recognizes right angles, so distinguishes between a rectangle and a parallelogram without right angles. Correctly names all the following shapes:
6	Angle Matcher	11	A sign of development is when the child can match angles concretely. For example, given several triangles, finds two with the same angles by laying the angles on top of one another.

Learning Trajectories

Age Range	Level Name	Level	Description
7	Parts of Shapes Identifier	12	At the next level the child can identify shapes in terms of their components. For example, the child may say, "No matter how skinny it looks, that's a triangle because it has three sides and three angles."
7	Constructor of Shapes from Parts Exact	13	A significant step is when the child can represent a shape with completely correct construction, based on knowledge of components and relationships. For example, asked to make a triangle with sticks, creates the following:
8	Shape Class Identifier	14	As children develop, they begin to use class membership (for example, to sort), not explicitly based on properties. For example, a child at this level may say, "I put the triangles over here, and the quadrilaterals, including squares, rectangles, rhombi, and trapezoids, over there."
8	Shape Property Identifier	15	At the next level a child can use properties explicitly. For example, a child may say, "I put the shapes with opposite sides parallel over here, and those with four sides but not both pairs of sides parallel over there."

Age Range	Level Name	Level	Description
8	Angle Size Comparer	16	The next sign of development is when a child can separate and compare angle sizes. For example, the child may say, "I put all the shapes that have right angles here, and all the ones that have bigger or smaller angles over there."
8	Angle Measurer	17	A significant step in development is when a child can use a protractor to measure angles.
8	Property Class Identifier	18	The next sign of development is when a child can use class membership for shapes (for example, to sort or consider shapes "similar") explicitly based on properties, including angle measure. For example, the child may say, "I put the equilateral triangles over here, and the right triangles over here."
8	Angle Synthesizer	19	As children develop understanding of shape, they can combine various meanings of angle (turn, corner, slant). For example, a child at this level could explain, "This ramp is at a 45° angle to the ground."

Developmental Levels for Composing Geometric Shapes

Children move through levels in the composition and decomposition of two-dimensional figures. Very young children cannot compose shapes but then gain ability to combine shapes into pictures, synthesize combinations of shapes into new shapes, and eventually substitute and build different kinds of shapes. Children typically follow an observable developmental progression in learning to compose shapes with recognizable stages or levels. This developmental path can be described as part of a learning trajectory.

Age Range	Level Name	Level	Description
2	Pre-Composer	1	The earliest sign of development is when a child can manipulate shapes as individuals, but is unable to combine them to compose a larger shape.
3	Pre-DeComposer	2	At the next level a child can decompose shapes, but only by trial and error. For example, given only a hexagon, the child can break it apart to make this simple picture by trial and error:

Age Range	Level Name	Level	Description
4	Piece Assembler	3	Around age 4 a child can begin to make pictures in which each shape represents a unique role (for example, one shape for each body part) and shapes touch. A child at this level can fill simple outline puzzles using trial and error.

Age Range	Level Name	Level	Description
5	Picture Maker	4	As children develop they are able to put several shapes together to make one part of a picture (for example, two shapes for one arm). A child at this level uses trial and error and does not anticipate creation of the new geometric shape. The child can choose shapes using "general shape" or side length and fill "easy" outline puzzles that suggest the placement of each shape (but note below that the child is trying to put a square in the puzzle where its right angles will not fit). Make a Picture Outline Puzzle
5	Simple Decomposer	5	A significant step occurs when the child is able to decompose ("take apart" into smaller shapes) simple shapes that have obvious clues as to their decomposition.
5	Shape Composer	6	A sign of development is when a child composes shapes with anticipation ("I know what will fit!"). A child at this level chooses shapes using angles as well as side lengths. Rotation and flipping are used intentionally to select and place shapes. For example, in the outline puzzle below, all angles are correct, and patterning is evident. Make a Picture Outline Puzzle
6	Substitution Composer	7	A sign of development is when a child is able to make new shapes out of smaller shapes and uses trial and error to substitute groups of shapes for other shapes to create new shapes in different ways. For example, the child can substitute shapes to fill outline puzzles in different ways.

Age Range	Level Name	Level	Description
6	Shape Decomposer (with Help)	8	As children develop they can decompose shapes by using imagery that is suggested and supported by the task or environment. For example, given hexagons, the child at this level can break it apart to make this shape:
7	Shape Composite Repeater	9	The next level is demonstrated when the child can construct and duplicate units of units (shapes made from other shapes) intentionally, and understands each as being both multiple small shapes and one larger shape. For example, the child may continue a pattern of shapes that leads to tiling.
7	Shape Decomposer with Imagery	10	A significant sign of development is when a child is able to decompose shapes flexibly by using independently generated imagery. For example, given hexagons, the child can break it apart to make shapes such as these:
8	Shape Composer— Units of Units	11	Children demonstrate further understanding when they are able to build and apply units of units (shapes made from other shapes). For example, in constructing spatial patterns the child can extend patterning activity to create a tiling with a new unit shape—a unit of unit shapes that he or she recognizes and consciously constructs. For example, the child builds Ts out of four squares, uses four Ts to build squares, and uses squares to tile a rectangle.
8	Shape DeComposer with Units of Units	12	As children develop understanding of shape they can decompose shapes flexibly by using independently generated imagery and planned decompositions of shapes that themselves are decompositions. For example, given only squares, a child at this level can break them apart—and then break the resulting shapes apart again—to make shapes such as these:

Developmental Levels for Comparing Geometric Shapes

As early as 4 years of age children can create and use strategies, such as moving shapes to compare their parts or to place one on top of the other for judging whether two figures are the same shape. From Pre-K to Grade 2 they can develop sophisticated and accurate mathematical procedures for comparing geometric shapes. Children typically follow an observable developmental progression in learning about how shapes are the same and different with recognizable stages or levels. This developmental path can be described as part of a learning trajectory.

Age Range	Level Name	Level	Description
3	"Same Thing" Comparer	1	The first sign of understanding is when the child can compare real-world objects. For example, the child says two pictures of houses are the same or different.
4	"Similar" Comparer	2	The next sign of development occurs when the child judges two shapes the same if they are more visually similar than different. For example, the child may say, "These are the same. They are pointy at the top."
4	Part Comparer	3	At the next level a child can say that two shapes are the same after matching one side on each. For example, "These are the same" (matching the two sides).
4	Some Attributes Comparer	4	As children develop they look for differences in attributes, but may examine only part of a shape. For example, a child at this level may say, "These are the same" (indicating the top halves of the shapes are similar by laying them on top of each other).

Age Range	Level Name	Level	Description
5	Most Attributes Comparer	5	At the next level the child looks for differences in attributes, examining full shapes, but may ignore some spatial relationships. For example, a child may say, "These are the same."
7	Congruence Determiner	6	A sign of development is when a child determines congruence by comparing all attributes and all spatial relationships. For example, a child at this level says that two shapes are the same shape and the same size after comparing every one of their sides and angles.
7	Congruence Superposer	7	As children develop understanding they can move and place objects on top of each other to determine congruence. For example, a child at this level says that two shapes are the same shape and the same size after laying them on top of each other.
8	Congruence Representer	8	Continued development is evidenced as children refer to geometric properties and explain transformations. For example, a child at this level may say, "These must be congruent, because they have equal sides, all square corners, and I can move them on top of each other exactly."

Developmental Levels for Spatial Sense and Motions

Infants and toddlers spend a great deal of time exploring space and learning about the properties and relations of objects in space. Very young children know and use the shape of their environment in navigation activities. With guidance they can learn to "mathematize" this knowledge. They can learn about direction, perspective, distance, symbolization, location, and coordinates. Children typically follow an observable developmental progression in developing spatial sense with recognizable stages or levels. This developmental path can be described as part of a learning trajectory.

Age Range	Level Name	Level	Description
4	Simple Turner	1	An early sign of spatial sense is when a child mentally turns an object to perform easy tasks. For example, given a shape with the top marked with color, correctly identifies which of three shapes it would look like if it were turned "like this" (90 degree turn demonstrated) before physically moving the shape.

Age Range	Level Name	Level	Description
5	Beginning Slider, Flipper, Turner	2	The next sign of development is when a child can use the correct motions, but is not always accurate in direction and amount. For example, a child at this level may know a shape has to be flipped to match another shape, but flips it in the wrong direction.

Age Range	Level Name	Level	Description
6	Slider, Flipper, Turner	3	As children develop spatial sense they can perform slides and flips, often only horizontal and vertical, by using manipulatives. For example, a child at this level can perform turns of 45, 90, and 180 degrees and knows a shape must be turned 90 degrees to the right to fit into a puzzle.

Age Range	Level Name	Level	Description
7	Diagonal Mover	4	A sign of development is when a child can perform diagonal slides and flips. For example, a child at this level knows a shape must be turned or flipped over an oblique line (45 degree orientation) to fit into a puzzle.
8	Mental Mover	5	Further signs of development occur when a child can predict results of moving shapes using mental images. A child at this level may say, "If you turned this 120 degrees, it would be just like this one."

Developmental Levels for Patterning and Early Algebra

Algebra begins with a search for patterns. Identifying patterns helps bring order, cohesion, and predictability to seemingly unorganized situations and allows one to make generalizations beyond the information directly available. The recognition and analysis of patterns are important components of the young child's intellectual development because they provide a foundation for the development of algebraic thinking. Although prekindergarten children engage in pattern-related activities and recognize patterns in their everyday environment, research has revealed that an abstract understanding of patterns develops gradually during the early childhood years. Children typically follow an observable developmental progression in learning about patterns with recognizable stages or levels. This developmental path can be described as part of a learning trajectory.

Age Range	Level Name	Level	Description
2	Pre-Patterner	1	A child at the earliest level does not recognize patterns. For example, a child may name a striped shirt with no repeating unit a "pattern."
3	Pattern Recognizer	2	At the next level the child can recognize a simple pattern. For example, a child at this level may say, "I'm wearing a pattern" about a shirt with black, white, black, white stripes.
3–4	Pattern Fixer	3	A sign of development is when the child fills in a missing element of a pattern. For example, given objects in a row with one missing, the child can identify and fill in the missing element.
4	Pattern Duplicator AB	3	A sign of development is when the child can duplicate an ABABAB pattern, although the child may have to work close to the model pattern. For example, given objects in a row, ABABAB, makes their own ABBABBABB row in a different location.

Age Range	Level Name	Level	Description
4	Pattern Extender AB	4	At the next level the child is able to extend AB repeating patterns.
4	Pattern Duplicator	4	At this level the child can duplicate simple patterns (not just alongside the model pattern). For example, given objects in a row, ABBABBABB, makes their own ABBABBABB row in a different location.
5	Pattern Extender	5	A sign of development is when the child can extend simple patterns. For example, given objects in a row, ABBABBABB, adds ABBABB to the end of the row.
7	Pattern Unit Recognizer	7	At this level a child can identify the smallest unit of a pattern. For example, given objects in a ABBAB_ BABB patterns, identifies the core unit of the pattern as ABB.

Learning Trajectories

Developmental Levels for Classifying and Analyzing Data

Data analysis contains one big idea: classifying, organizing, representing, and using information to ask and answer questions. The developmental continuum for data analysis includes growth in classifying and counting to sort objects and quantify their groups.... Children eventually become capable of simultaneously classifying and counting, for example, counting the number of colors in a group of objects.

Children typically follow an observable developmental progression in learning about patterns with recognizable stages or levels. This developmental path can be described as part of a learning trajectory.

Age Range	Level Name	Level	Description
2	Similarity Recognizer	1	The first sign that a child can classify is when he or she recognizes, intuitively, two or more objects as "similar" in some way. For example, "that's another doggie."
2	Informal Sorter	2	A sign of development is when a child places objects that are alike on some attribute together, but switches criteria and may use functional relationships are the basis for sorting. A child at this level might stack blocks of the same shape or put a cup with its saucer.
3	Attribute Identifier	3	The next level is when the child names attributes of objects and places objects together with a given attribute, but cannot then move to sorting by a new rule. For example, the child may say, "These are both red."
4	Attribute Sorter	4	At the next level the child sorts objects according to a given attributes, forming categories, but may switch attributes during the sorting. A child at this stage can switch rules for sorting if guided. For example, the child might start putting red beads on a string, but switches to the spheres of different colors.
5	Consistent Sorter	5	A sign of development is when the child can sort consistently by a given attribute. For example, the child might put several identical blocks together.
6	Exhaustive Sorter	6	At the next level, the child can sort consistently and exhaustively by an attribute, given or created. This child can use terms "some" and "all" meaningfully. For example, a child at this stage would be able to find all the attribute blocks of a certain size and color.
6	Multiple Attribute Sorter	7	A sign of development is when the child can sort consistently and exhaustively by more than one attribute, sequentially. For example, a child at this level, can put all the attribute blocks together by color, then by shape.
7	Classifier and Counter	8	At the next level, the child is capable of simultaneously classifying and counting. For example, the child counts the number of colors in a group of objects.

Age Range	Level Name	Level	Description
7	List Grapher	9	In the early stage of graphing, the child graphs by simply listing all cases. For example, the child may list each child in the class and each child's response to a question.
8+	Multiple Attribute Classifier	10	A sign of development is when the child can intentionally sort according to multiple attributes, naming and relating the attributes. This child understands that objects could belong to more than one group. For example, the child can complete a two-dimensional classification matrix or forming subgroups within groups.
8+	Classifying Grapher	11	At the next level the child can graph by classifying data (e.g., responses) and represent it according to categories. For example, the child can take a survey, classify the responses, and graph the result.
8+	Classifier	12	At sign of development is when the child creates complete, conscious classifications logically connected to a specific property. For example, a child at this level gives definition of a class in terms of a more general class and one or more specific differences and begins to understand the inclusion relationship.
8+	Hierarchical Classifier	13	At the next level, the child can perform hierarchical classifications. For example, the child recognizes that all squares are rectangles, but not all rectangles are squares.
8+	Data Representer	14	Signs of development are when the child organizes and displays data through both simple numerical summaries such as counts, tables, and tallies, and graphical displays, including picture graphs, line plots, and bar graphs. At this level the child creates graphs and tables, compares parts of the data, makes statements about the data as a whole, and determines whether the graphs answer the questions posed initially.

Student's Name _____

Number _____

Age Range	Counting	Comparing and Ordering Number	Recognizing Number and Subitizing (instantly recognizing)	Composing Number (knowing combinations of numbers)	Adding and Subtracting	Multiplying and Dividing (sharing)
1 year	___ Pre-Counter ___ Chanter				___ Pre +/−	
2	___ Reciter	___ Object Corresponder ___ Perceptual Comparer	___ Small Collection Namer			___ Nonquantitative Sharer
3	___ Reciter (10) ___ Corresponder	___ First-Second Ordinal Counter ___ Nonverbal Comparer of Similar Items (1–4 items)	___ Nonverbal Subitizer ___ Maker of Small Collections		___ Nonverbal +/−	___ Beginning Grouper and Distributive Sharer
4	___ Counter (small numbers) ___ Producer (small numbers) ___ Counter (10)	___ Nonverbal Comparer of Dissimilar Items ___ Matching Comparer ___ Knows-to-Count Comparer ___ Counting Comparer (same size)	___ Perceptual Subitizer to 4	___ Pre-Part-Whole Recognizer	___ Small Number +/−	___ Grouper and Distributive Sharer
5	___ Counter and Producer (10+) ___ Counter Backward from 10	___ Counting Comparer (5) ___ Ordinal Counter	___ Perceptual Subitizer to 5 ___ Conceptual Subitizer to 5+ ___ Conceptual Subitizer to 10	___ Inexact Part-Whole Recognizer ___ Composer to 4, then 5	___ Find Result +/− ___ Find Change +/− ___ Make It N +/−	___ Concrete Modeler ×/÷
6	___ Counter from N (N+1, N−1) ___ Skip Counter by tens to 100 ___ Counter to 100 ___ Counter On Using Patterns ___ Skip Counter ___ Counter of Imagined Items ___ Counter On Keeping Track ___ Counter of Quantitative Units ___ Counter to 200	___ Counting Comparer (10) ___ Mental Number Line to 10 ___ Serial Orderer to 6+	___ Conceptual Subitizer to 20	___ Composer to 7 ___ Composer to 10	___ Counting Strategies +/− ___ Part-Whole +/−	___ Parts and Wholes ×/÷
7	___ Number Conserver ___ Counter Forward and Back	___ Place Value Comparer ___ Mental Number Line to 100	___ Conceptual Subitizer with Place Value and Skip Counting	___ Composer with Tens and Ones	___ Numbers-in-Numbers +/− ___ Deriver +/−	___ Skip Counter ×/÷
8+		___ Mental Number Line to 1,000s	___ Conceptual Subitizer with Place Value and Multiplication		___ Problem Solver +/− ___ Multidigit +/−	___ Deriver ×/÷ ___ Array Quantifier ___ Partitive Divisor ___ Multidigit ×/÷

Student's Name _____

Geometry

Age Range	Shapes	Composing Shapes	Comparing Shapes	Motions and Spatial Sense	Measuring	Patterning	Classifying and Analyzing Data
2 years	Shape Matcher—Identical —Sizes —Orientations					Pre-Patterner	Similarity Recognizer; Informal Sorter
3	Shape Recognizer—Typical; Shape Matcher—More Shapes —Sizes and Orientations —Combinations	Pre-Composer; Pre-Decomposer	"Same Thing" Comparer		Length Quantity Recognizer	Pattern Recognizer	Attribute Identifier
4	Shape Recognizer—Circles, Squares, and Triangles +; Constructor of Shapes from Parts—Looks Like Representing	Piece Assembler	"Similar" Comparer; Part Comparer; Some Attributes Comparer	Simple Turner	Length Direct Comparer	Pattern Fixer; Pattern Duplicator AB; Pattern Extender AB; Pattern Duplicator	Attribute Sorter
5	Shape Recognizer—All Rectangles; Side Recognizer; Angle Recognizer; Shape Recognizer—More Shapes	Picture Maker; Simple Decomposer; Shape Composer	Most Attributes Comparer	Beginning Slider, Flipper, Turner	Indirect Length Comparer	Pattern Extender	Consistent Sorter
6	Shape Identifier; Angle Matcher Parts	Substitution Composer; Shape Decomposer (with help)		Slider, Flipper, Turner	Serial Orderer to 6+; End-to-End Length Measurer		Exhaustive Sorter; Multiple Attribute Sorter
7	Parts of Shapes Identifier; Constructor of Shapes from Parts—Exact Representing	Shape Composite Repeater; Shape Decomposer with Imagery	Congruence Determiner; Congruence Superposer	Diagonal Mover	Length Unit Iterator; Length Unit Relater	Pattern Unit Recognizer	Classifier and Counter; List Grapher
8+	Shape Class Identifier; Shape Property Identifier; Angle Size Comparer; Angle Measurer; Property Class Identifier; Angle Synthesizer	Shape Composer—Units of Units; Shape Decomposer with Units of Units	Congruence Representer	Mental Mover	Length Measurer; Conceptual Ruler Measurer		Multiple Attribute Classifier; Classifying Grapher; Classifier; Hierarchical Classifier; Data Representer

Glossary

A

acute angle An angle with a measure greater than 0 degrees and less than 90 degrees.

addend One of the numbers being added in an addition sentence. In the sentence 41 + 27 = 68, the numbers 41 and 27 are addends.

addition A mathematical operation based on "putting things together." Numbers being added are called *addends*. The result of addition is called a *sum*. In the number sentence 15 + 63 = 78, the numbers 15 and 63 are addends.

additive inverses Two numbers whose sum is 0. For example, 9 + −9 = 0. The additive inverse of 9 is −9, and the additive inverse of −9 is 9.

adjacent angles Two angles with a common side that do not otherwise overlap. In the diagram, angles 1 and 2 are adjacent angles; so are angles 2 and 3, angles 3 and 4, and angles 4 and 1.

algorithm A step-by-step procedure for carrying out a computation or solving a problem.

angle Two rays with a common endpoint. The common endpoint is called the vertex of the angle.

area A measure of the surface inside a closed boundary. The formula for the area of a rectangle or parallelogram is $A = b \times h$, where A represents the area, b represents the length of the base, and h is the height of the figure.

array A rectangular arrangement of objects in rows and columns in which each row has the same number of elements, and each column has the same number of elements.

attribute A feature such as size, shape, or color.

average See **mean**. The **median** and **mode** are also sometimes called the *average*.

axis (plural **axes**) A number line used in a coordinate grid.

B

bar graph A graph in which the lengths of horizontal or vertical bars represent the magnitude of the data represented.

base ten The commonly used numeration system, in which the ten digits 0, 1, 2,..., 9 have values that depend on the place in which they appear in a numeral (ones, tens, hundreds, and so on, to the left of the decimal point; tenths, hundredths, and so on, to the right of the decimal point).

bisect To divide a segment, angle, or figure into two parts of equal measure.

C

capacity A measure of how much liquid or substance a container can hold. See also **volume.**

centi- A prefix for units in the metric system meaning one hundredth.

centimeter (cm) In the metric system, a unit of length defined as 1/100 of a meter; equal to 10 millimeters or 1/10 of a decimeter.

circle The set of all points in a plane that are a given distance (the radius) from a given point (the center of the circle).

circle graph A graph in which a circular region is divided into sectors to represent the categories in a set of data. The circle represents the whole set of data.

circumference The distance around a circle or sphere.

closed figure A figure that divides the plane into two regions, inside and outside the figure. A closed space figure divides space into two regions in the same way.

common denominator Any nonzero number that is a multiple of the denominators of two or more fractions.

common factor Any number that is a factor of two or more numbers.

complementary angles Two angles whose measures total 90 degrees.

composite function A function with two or more operations. For example, this function multiplies the input number by 5 then adds 3.

composite number A whole number that has

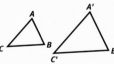

more than two whole number factors. For example, 14 is a composite number because it has more than two whole number factors.

cone A space figure having a circular base, curved surface, and one vertex.

congruent Having identical sizes and shapes. Congruent figures are said to be congruent to each other.

coordinate One of two numbers used to locate a point on a coordinate grid. See also **ordered pair.**

coordinate grid A device for locating points in a plane by means of ordered pairs or coordinates. A coordinate grid is formed by two number lines that intersect at their 0-points.

corresponding angles Two angles in the same relative position in two figures, or in similar locations in relation to a transversal intersecting two lines. In the diagram above, angles 1 and 5, 3 and 7, 2 and 6, and 4 and 8 are corresponding angles. If the lines are parallel, then the corresponding angles are congruent.

corresponding sides Two sides in the same relative position in two figures. In the diagram AB and A'B', BC and B'C', and AC and A'C' are corresponding sides.

cube A space figure whose six faces are congruent squares that meet at right angles.

cubic centimeter (cm³) A metric unit of volume; the volume of a cube 1 centimeter on an edge. 1 cubic centimeter is equal to 1 milliliter.

cubic unit A unit used in a volume and capacity measurement.

customary system of measurement The measuring system used most often in the United States. Units for linear measure (length, distance) include inch, foot, yard, and mile; units for weight include ounce and pound; units for capacity (amount of liquid or other substance a container can hold) include fluid ounce, cup, pint, quart, and gallon.

cylinder A space figure having a curved surface and parallel circular or elliptical bases that are congruent.

D

decimal A number written in standard notation, usually one containing a decimal point, as in 3.78.

decimal approximation A decimal that is close to the value of a rational number. By extending the decimal approximation to additional digits, it is possible to come as close as desired to the value of the rational number. For example, decimal approximations of $\frac{1}{12}$ are 0.083, 0.0833, 0.08333, and so on.

decimal equivalent A decimal that names the same number as a fraction. For example, the decimal equivalent of $\frac{3}{4}$ is 0.75. The only rational numbers with decimal equivalents are those that can be written as fractions whose denominators have prime factors only of 2 and 5. For example, $\frac{1}{2}, \frac{1}{4}$, and $\frac{1}{20}$ have decimal equivalents, but $\frac{5}{6}, \frac{1}{7}$, and $\frac{1}{9}$ have only decimal approximations.

degree (°) A unit of measure for angles; based on dividing a circle into 360 equal parts. Also, a unit of measure for temperature.

degree Celsius (°C) In the metric system, the unit for measuring temperature. Water freezes at 0°C and boils at 100°C.

degree Fahrenheit (°F) In the U.S. customary system, the unit for measuring temperature. Water freezes at 32°F and boils at 212°F.

denominator The number of equal parts into which a whole is divided. In the fraction $\frac{a}{b}$, b is the denominator. **See also numerator.**

diameter A line segment, going through the center of a circle, that starts at one point on the circle and ends at the opposite point on the circle; also, the length of such a line segment. The diameter of a circle is twice its radius. AB is a diameter of this circle. See also **circle.**

difference The result of subtraction. In the subtraction sentence 40 − 10 = 30, the difference is 30.

digit In the base-ten numeration system, one of the symbols 0, 1, 2, 3, 4, 5, 6, 7, 8, 9. Digits can be used to write a numeral for any whole number in the base-ten numbering system. For example, the numeral 145 is made of the digits 1, 4, and 5.

distributive law A law that relates two operations on numbers, usually multiplication and addition, or multiplication and subtraction. Distributive law of multiplication over addition: $a \times (b + c) = (a \times b) + (a \times c)$

dividend See **division.**

division A mathematical operation based on "equal sharing" or "separating into equal parts." The *dividend* is the total before sharing. The divisor is the number of equal parts or the number in each equal part. The *quotient* is the result of division. For example, in 35 ÷ 5 = 7, 35 is the dividend, 5 is the divisor, and 7 is the quotient. If 35 objects are separated into 5 equal parts, there are 7 objects in each part. If 35 objects are separated into parts with 5 in each part, there are 7 equal parts. The number left over when a set of objects is shared equally or separated into equal groups is called the *remainder*. For 35 ÷ 5, the quotient is 7 and the remainder is 0. For 36 ÷ 5, the quotient is 7 and the remainder is 1.

Glossary

divisor See **division.**

E

edge The line segment where two faces of a polyhedron meet.

endpoint The point at either end of a line segment; also, the point at the end of a ray. Line segments are named after their endpoints; a line segment between and including points A and B is called segment AB or segment BA.

equation A mathematical sentence that states the equality of two expressions. For example, $3 + 7 = 10$, $y = x + 7$, and $4 + 7 = 8 + 3$ are equations.

equilateral polygon A polygon in which all sides are the same length.

equivalent Equal in value, but in a different form. For example, $\frac{1}{2}$, $\frac{2}{4}$, 0.5, and 50% are equivalent forms of the same number.

equivalent fractions Fractions that have different numerators and denominators but name the same number. For example, $\frac{2}{3}$ and $\frac{6}{9}$ are equivalent fractions.

estimate A judgment of time, measurement, number, or other quantity that may not be exactly right.

evaluate an algebraic expression To replace each variable in an algebraic expression with a particular number and then to calculate the value of the expression.

evaluate a numerical expression To carry out the operations in a numerical expression to find the value of the expression.

even number A whole number such as 0, 2, 4, 6, and so on, that can be divided by 2 with no remainder. See also **odd number.**

event A happening or occurrence. The tossing of a coin is an event.

expression A group of mathematical symbols (numbers, operation signs, variables, grouping symbols) that represents a number (or can represent a number if values are assigned to any variables it contains).

F

face A flat surface on a space figure.

fact family A group of addition or multiplication facts grouped together with the related subtraction or division facts. For example, $4 + 8 = 12$, $8 + 4 = 12$, $12 - 4 = 8$, and $12 - 8 = 4$ form an addition fact family. The facts $4 \times 3 = 12$, $3 \times 4 = 12$, $12 \div 3 = 4$, and $12 \div 4 = 3$ form a multiplication fact family.

factor (noun) One of the numbers that is multiplied in a multiplication expression. For example, in $4 \times 1.5 = 6$, the factors are 4 and 1.5. See also **multiplication.**

factor (verb) To represent a quantity as a product of factors. For example, 20 factors to 4×5, 2×10, or $2 \times 2 \times 5$.

factor of a whole number n A whole number, which, when multiplied by another whole number, results in the number n. The whole number n is divisible by its factors. For example, 3 and 5 are factors of 15 because $3 \times 5 = 15$, and 15 is divisible by 3 and 5.

factor tree A method used to obtain the prime factorization of a number. The original number is represented as a product of factors, and each of those factors is represented as a product of factors, and so on, until the factor string consists of prime numbers.

formula A general rule for finding the value of something. A formula is usually written as an equation with variables representing unknown quantities. For example, a formula for distance traveled at a constant rate of speed is $d = r \times t$, where d stands for distance, r is for rate, and t is for time.

fraction A number in the form $\frac{a}{b}$, where a and b are integers and b is not 0. Fractions are used to name part of a whole object or part of a whole collection of objects, or to compare two quantities. A fraction can represent division; for example, $\frac{2}{5}$ can be thought of as 2 divided by 5.

frequency The number of times an event or value occurs in a set of data.

function machine An imaginary machine that processes numbers according to a certain rule. A number (input) is put into the machine and is transformed into a second number (output) by application of the rule.

G

greatest common factor The largest factor that two or more numbers have in common. For example, the common factors of 24 and 30 are 1, 2, 3, and 6. The greatest common factor of 24 and 30 is 6.

H

height (of a parallelogram) The length of the line segment between the base of the parallelogram and the opposite side (or an extension of the opposite side), running perpendicular to the base.

height (of a polyhedron) The perpendicular distance between the bases of the polyhedron or between a base and the opposite vertex.

height (of a rectangle) The length of the side perpendicular to the side considered the base of the rectangle. (Base and height of a rectangle are interchangeable.)

height (of a triangle) The length of the line segment perpendicular to the base of the triangle (or an extension of the base) from the opposite vertex.

hexagon A polygon with six sides.

histogram A bar graph in which the labels for the bars are numerical intervals.

hypotenuse In a right triangle, the side opposite the right angle.

I

improper fraction A fraction that names a number greater than or equal to 1; a fraction whose numerator is equal to or greater than its denominator. Examples of improper fractions are $\frac{4}{3}$, $\frac{10}{8}$, and $\frac{4}{4}$.

inch (in.) In the U. S. customary system, a unit of length equal to $\frac{1}{12}$ of a foot.

indirect measurement Methods for determining heights, distances, and other quantities that cannot be measured or are not measured directly.

inequality A number sentence stating that two quantities are not equal. Relation symbols for inequalities include < (is less than), > (is greater than), and ≠ (is not equal to).

integers The set of integers is $\{..., -4, -3, -2, -1, 0, 1, 2, 3, 4, ...\}$. The set of integers consists of whole numbers and their opposites.

intersect To meet (at a point, a line, and so on), sharing a common point or points.

interior The set of all points in a plane "inside" a closed plane figure, such as a polygon or circle. Also, the set of all points in space "inside" a closed space figure, such as a polyhedron or sphere.

isosceles Having two sides of the same length; commonly used to refer to triangles and trapezoids.

K

kilo- A prefix for units in the metric system meaning one thousand.

L

least common denominator The least common multiple of the denominators of every fraction in a given set of fractions. For example, 12 is the least common denominator of $\frac{2}{3}$, $\frac{1}{4}$, and $\frac{5}{6}$. See also **least common multiple.**

least common multiple The smallest number that is a multiple of two or more numbers. For example, some common multiples of 6 and 8 are 24, 48, and 72. 24 is the least common multiple of 6 and 8.

leg of a right triangle A side of a right triangle that is not the hypotenuse.

line A straight path that extends infinitely in opposite directions.

line graph (broken-line graph) A graph in which points are connected by line segments to represent data.

line of symmetry A line that separates a figure into halves. The figure can be folded along this line into two parts which exactly fit on top of each other.

line segment A straight path joining two points, called *endpoints* of the line segment. A straight path can be described as the shortest distance between two points.

line symmetry A figure has line symmetry (also called *bilateral symmetry*) if a line of symmetry can be drawn through the figure.

liter (L) A metric unit of capacity, equal to the volume of a cube 10 centimeters on an edge. $1 \text{ L} = 1{,}000 \text{ mL} = 1{,}000 \text{ cm}^3$. A liter is slightly larger than a quart. See also **milliliter (mL).**

M

map scale A ratio that compares the distance between two locations shown on a map with the actual distance between them.

mean A typical or central value that may be used to describe a set of numbers. It can be found by adding the numbers in the set and dividing the sum by the number of numbers. The mean is often referred to as the *average.*

median The middle value in a set of data when the data are listed in order from least to greatest (or greatest to least). If the number of values in the set is even (so that there is no "middle" value), the median is the mean of the two middle values.

meter (m) The basic unit of length in the metric system, equal to 10 decimeters, 100 centimeters, and 1,000 millimeters.

metric system of measurement A measurement system based on the base-ten numeration system and used in most countries in the world. Units for linear measure (length, distance) include millimeter, centimeter, meter, kilometer; units for mass (weight) include gram and kilogram; units for capacity (amount of liquid or other substance a container can hold) include milliliter and liter.

Glossary

midpoint A point halfway between two points.

milli- A prefix for units in the metric system meaning one thousandth.

milliliter (mL) A metric unit of capacity, equal to 1/1,000 of a liter and 1 cubic centimeter.

millimeter (mm) In the metric system, a unit of length equal to 1/10 of a centimeter and 1/1,000 of a meter.

minuend See **subtraction**.

mixed number A number greater than 1, written as a whole number and a fraction less than 1. For example, $5\frac{1}{2}$ is equal to $5 + \frac{1}{2}$.

mode The value or values that occur most often in a set of data.

multiple of a number *n* The product of a whole number and the number *n*. For example, the numbers 0, 4, 8, 12, and 16 are all multiples of 4 because $4 \times 0 = 0$, $4 \times 1 = 4$, $4 \times 2 = 8$, $4 \times 3 = 12$, and $4 \times 4 = 16$.

multiplication A mathematical operation used to find the total number of things in several equal groups, or to find a quantity that is a certain number of times as much or as many as another number. Numbers being multiplied are called *factors*. The result of multiplication is called the *product*. In $8 \times 12 = 96$, 8 and 12 are the factors and 96 is the product.

multiplicative inverses Two numbers whose product is 1. For example, the multiplicative inverse of $\frac{2}{5}$ is $\frac{5}{2}$, and the multiplicative inverse of 8 is $\frac{1}{8}$. Multiplicative inverses are also called *reciprocals* of each other.

N

negative number A number less than 0; a number to the left of 0 on a horizontal number line.

number line A line on which equidistant points correspond to integers in order.

number sentence A sentence that is made up of numerals and a relation symbol ($<$, $>$, or $=$). Most number sentences also contain at least one operation symbol. Number sentences may also have grouping symbols, such as parentheses.

numeral The written name of a number.

numerator In a whole divided into a number of equal parts, the number of equal parts being considered. In the fraction $\frac{a}{b}$, *a* is the numerator.

O

obtuse angle An angle with a measure greater than 90 degrees and less than 180 degrees.

octagon An eight-sided polygon.

odd number A whole number that is not divisible by 2, such as 1, 3, 5, and so on. When an odd number is divided by 2, the remainder is 1. A whole number is either an odd number or an even number.

opposite of a number A number that is the same distance from 0 on the number line as the given number, but on the opposite side of 0. If *a* is a negative number, the opposite of *a* will be a positive number. For example, if $a = -5$, then $-a$ is 5. See also **additive inverses**.

ordered pair Two numbers or objects for which order is important. Often, two numbers in a specific order used to locate a point on a coordinate grid. They are usually written inside parentheses; for example, (2, 3). See also **coordinate.**

ordinal number A number used to express position or order in a series, such as first, third, tenth. People generally use ordinal numbers to name dates; for example, "May fifth" rather than "May five."

origin The point where the *x*- and *y*-axes intersect on a coordinate grid. The coordinates of the origin are (0, 0).

outcome The result of an event. Heads and tails are the two outcomes of the event of tossing a coin.

P

parallel lines (segments, rays) Lines (segments, rays) going in the same direction that are the same distance apart and never meet.

parallelogram A quadrilateral that has two pairs of parallel sides. Pairs of opposite sides and opposite angles of a parallelogram are congruent.

parentheses A pair of symbols, (and), used to show in which order operations should be done. For example, the expression $(3 \times 5) + 7$ says to multiply 5 by 3 then add 7. The expression $3 \times (5 + 7)$ says to add 5 and 7 and then multiply by 3.

pattern A model, plan, or rule that uses words or variables to describe a set of shapes or numbers that repeat in a predictable way.

pentagon A polygon with five sides.

percent A rational number that can be written as a fraction with a denominator of 100. The symbol % is used to represent percent. 1% means 1/100 or 0.01. For example, "53% of the students in the school are girls" means that of every 100 students in the school, 53 are girls.

perimeter The distance along a path around a plane figure. A formula for the perimeter of a rectangle is $P = 2 \times (B + H)$, where *B* represents the base and *H* is the height of the rectangle. Perimeter may also refer to the path itself.

perpendicular Two rays, lines, line segments, or other figures that form right angles are said to be perpendicular to each other.

pi The ratio of the circumference of a circle to its diameter. Pi is the same for every circle, approximately 3.14 or $\frac{22}{7}$. Also written as the Greek letter π.

pictograph A graph constructed with pictures or icons, in which each picture stands for a certain number. Pictographs make it easier to visually compare quantities.

place value A way of determining the value of a digit in a numeral, written in standard notation, according to its position, or place, in the numeral. In base-ten numbers, each place has a value ten times that of the place to its right and one-tenth the value of the place to its left.

plane A flat surface that extends forever.

plane figure A figure that can be contained in a plane (that is, having length and width but no height).

point A basic concept of geometry; usually thought of as a location in space, without size.

polygon A closed plane figure consisting of line segments (sides) connected endpoint to endpoint. The interior of a polygon consists of all the points of the plane "inside" the polygon. An *n*-gon is a polygon with *n* sides; for example, an 8-gon has 8 sides.

polyhedron A closed space figure, all of whose surfaces (faces) are flat. Each face consists of a polygon and the interior of the polygon.

power A product of factors that are all the same. For example, $6 \times 6 \times 6$ (or 216) is called 6 to the third power, or the third power of 6, because 6 is a factor three times. The expression $6 \times 6 \times 6$ can also be written as 6^3.

power of 10 A whole number that can be written as a product using only 10 as a factor. For example, 100 is equal to 10×10 or 10^2, so 100 is called 10 squared, the second power of 10, or 10 to the second power. Other powers of 10 include 10^1, or 10, and 10^3, or 1,000.

prime factorization A whole number expressed as a product of prime factors. For example, the prime factorization of 18 is $2 \times 3 \times 3$. A number has only one prime factorization (except for the order in which the factors are written).

prime number A whole number greater than 1 that has exactly two whole number factors, 1 and itself. For example, 13 is a prime number because its only factors are 1 and 13. A prime number is divisible only by 1 and itself. The first five prime numbers are 2, 3, 5, 7, and 11. See also **composite number.**

prism A polyhedron with two parallel faces (bases) that are the same size and shape. Prisms are classified according to the shape of the two parallel bases. The bases of a prism are connected by parallelograms that are often rectangular.

probability A number between 0 and 1 that indicates the likelihood that something (an event) will happen. The closer a probability is to 1, the more likely it is that an event will happen.

product See **multiplication.**

protractor A tool for measuring or drawing angles. When measuring an angle, the vertex of the angle should be at the center of the protractor and one side should be aligned with the 0 mark.

pyramid A polyhedron in which one face (the base) is a polygon and the other faces are formed by triangles with a common vertex (the apex). A pyramid is classified according to the shape of its base, as a triangular pyramid, square pyramid, pentagonal pyramid, and so on.

Pythagorean Theorem A mathematical theorem, proven by the Greek mathematician Pythagoras and known to many others before and since, that states that if the legs of a right triangle have lengths *a* and *b*, and the hypotenuse has length *c*, then $a^2 + b^2 = c^2$.

Q

quadrilateral A polygon with four sides.

quotient See **division.**

Glossary

R

radius A line segment that goes from the center of a circle to any point on the circle; also, the length of such a line segment.

random sample A sample taken from a population in a way that gives all members of the population the same chance of being selected.

range The difference between the maximum and minimum values in a set of data.

rate A ratio comparing two quantities with unlike units. For example, a measure such as 23 miles per gallon of gas compares mileage with gas usage.

ratio A comparison of two quantities using division. Ratios can be expressed with fractions, decimals, percents, or words. For example, if a team wins 4 games out of 5 games played, the ratio of wins to total games is $\frac{4}{5}$, 0.8, or 80%.

rational number Any number that can be represented in the form $a \div b$ or $\frac{a}{b}$, where a and b are integers and b is positive. Some, but not all, rational numbers have exact decimal equivalents.

ray A straight path that extends infinitely in one direction from a point, which is called its *endpoint*.

reciprocal See **multiplicative inverses.**

rectangle A parallelogram with four right angles.

reduced form A fraction in which the numerator and denominator have no common factors except 1.

reflection A transformation in which a figure "flips" so that its image is the reverse of the original.

regular polygon A convex polygon in which all the sides are the same length and all the angles have the same measure.

relation symbol A symbol used to express the relationship between two numbers or expressions. Among the symbols used in number sentences are = for "is equal to," < for "is less than," > for "is greater than," and ≠ for "is not equal to."

remainder See **division.**

rhombus A parallelogram whose sides are all the same length.

right angle An angle with a measure of 90 degrees, representing a quarter of a full turn.

right triangle A triangle that has a right angle.

rotation A transformation in which a figure "turns" around a center point or axis.

rotational symmetry Property of a figure that can be rotated around a point (less than a full, 360-degree turn) in such a way that the resulting figure exactly matches the original figure. If a figure has rotational symmetry, its order of rotational symmetry is the number of different ways it can be rotated to match itself exactly. "No rotation" is counted as one of the ways.

rounding Changing a number to another number that is easier to work with and is close enough for the purpose. For example, 12,924 rounded to the nearest thousand is 13,000 and rounded to the nearest hundred is 12,900.

S

sample A subset of a group used to represent the whole group.

scale The ratio of the distance on a map or drawing to the actual distance.

scalene triangle A triangle in which all three sides have different lengths.

scale drawing An accurate picture of an object in which all parts are drawn to the same scale. If an actual object measures 32 by 48 meters, a scale drawing of it might measure 32 by 48 millimeters.

scale model A model that represents an object or display in proportions based on a determined scale.

similar figures Figures that are exactly the same shape but not necessarily the same size.

space figure A figure which cannot be contained in a plane. Common space figures include the rectangular prism, square pyramid, cylinder, cone, and sphere.

sphere The set of all points in space that are a given distance (the radius) from a given point (the center). A ball is shaped like a sphere.

square number A number that is the product of a whole number and itself. The number 36 is a square number, because 36 = 6 × 6.

square of a number The product of a number multiplied by itself. For example, 2.5 squared is $(2.5)^2$.

square root The square root of a number n is a number which, when multiplied by itself, results in the number n. For example, 8 is a square root of 64, because 8 × 8 = 64.

square unit A unit used to measure area—usually a square that is 1 inch, 1 centimeter, 1 yard, or other standard unit of length on each side.

standard notation The most familiar way of representing whole numbers, integers, and decimals by writing digits in specified places; the way numbers are usually written in everyday situations.

statistics The science of collecting, classifying, and interpreting numerical data as it is related to a particular subject.

stem-and-leaf plot A display of data in which digits with larger place values are named as stems, and digits with smaller place values are named as leaves.

straight angle An angle of 180 degrees; a line with one point identified as the vertex of the angle.

subtraction A mathematical operation based on "taking away" or comparing ("How much more?"). The number being subtracted is called the *subtrahend*; the number it is subtracted from is called the *minuend*; the result of subtraction is called the *difference*. In the number sentence 63 − 45 = 18, 63 is the minuend, 45 is the subtrahend, and 18 is the difference.

subtrahend See **subtraction.**

supplementary angles Two angles whose measures total 180 degrees.

surface area The sum of the areas of the faces of a space figure.

symmetrical Having the same size and shape across a dividing line or around a point.

T

tessellation An arrangement of closed shapes that covers a surface completely without overlaps or gaps.

tetrahedron A space figure with four faces, each formed by an equilateral triangle.

transformation An operation that moves or changes a geometric figure in a specified way. Rotations, reflections, and translations are types of transformations.

translation A transformation in which a figure "slides" along a line.

transversal A line which intersects two or more other lines.

trapezoid A quadrilateral with exactly one pair of parallel sides.

tree diagram A tool used to solve probability problems in which there is a series of events. This tree diagram represents a situation where the first event has three possible outcomes and the second event has two possible outcomes.

triangle A polygon with three sides. An *equilateral* triangle has three sides of the same length. An *isosceles* triangle has two sides of the same length. A *scalene* triangle has no sides of the same length.

U

unit (of measure) An agreed-upon standard with which measurements are compared.

unit fraction A fraction whose numerator is 1. For example, $\frac{1}{2}$, $\frac{1}{3}$, and $\frac{1}{10}$ are unit fractions.

unit cost The cost of one item or one specified amount of an item. If 20 pencils cost 60¢, then the unit cost is 3¢ per pencil.

unlike denominators Unequal denominators, as in $\frac{3}{4}$ and $\frac{5}{6}$.

V

variable A letter or other symbol that represents a number, one specific number, or many different values.

Venn diagram A picture that uses circles to show relationships between sets. Elements that belong to more than one set are placed in the overlap between the circles.

vertex The point at which the rays of an angle, two sides of a polygon, or the edges of a polyhedron meet.

vertical angles Two intersecting lines form four adjacent angles. In the diagram, angles 2 and 4 are vertical angles. They have no sides in common. Their measures are equal. Similarly, angles 1 and 3 are vertical angles.

volume A measure of the amount of space occupied.

W

whole number Any of the numbers 0, 1, 2, 3, 4, and so on. Whole numbers are the numbers used for counting and zero.

Scope and Sequence

The topics addressed at each grade level were determined after extensive analysis of national and state mathematics teaching expectations, standardized assessments, and topics covered in mathematics basal programs. Across all levels, the program follows the optimal sequence outlined by the learning trajectories of primary mathematics.

	A	B	C	D	E	F	G	H
Addition (whole numbers)								
Basic facts	•	•	•	•	•	•	•	•
Three or more addends			•	•	•	•	•	•
Two-digit numbers				•	•	•	•	•
Three-digit numbers					•	•	•	•
Greater numbers							•	•
Estimating sums						•	•	•
Algebra								
Properties of whole numbers	•	•	•	•	•	•	•	•
Integers (negative numbers)							•	•
Operations with integers							•	•
Make and solve number sentences and equations		•	•	•	•	•	•	•
Variables						•	•	•
Order of operations						•	•	•
Writing variable expressions						•	•	•
Evaluating expressions						•	•	•
Solving one-step equations	•	•	•		•	•	•	•
Solving two-step equations				•	•	•	•	•
Combining like terms						•	•	•
Solving inequalities						•	•	•
Function machines/tables					•	•	•	•
Coordinate graphing					•	•	•	•
Graphing linear functions						•	•	•
Using formulas						•	•	•
Decimals and Money								
Place value						•	•	•
Adding						•	•	•
Subtracting						•	•	•
Multiplying by a whole number							•	•
Multiplying by a decimal								•
Multiplying by powers of 10						•	•	•

Scope and Sequence

	A	B	C	D	E	F	G	H
Decimals and Money								
Dividing by a whole number							●	●
Dividing by a decimal								●
Identifying and counting currency				●	●	●	●	●
Exchanging money				●	●	●	●	●
Computing with money				●	●	●	●	●
Division								
Basic facts							●	●
Remainders							●	●
One-digit divisors							●	●
Two-digit divisors							●	●
Greater divisors							●	●
Estimating quotients							●	●
Fractions								
Fractions of a whole				●	●	●	●	●
Comparing/ordering				●	●	●	●	●
Equivalent fractions						●	●	●
Reduced form						●	●	●
Mixed numbers/improper fractions								●
Adding—like denominators								●
Adding—unlike denominators								●
Geometry								
Plane figures	●	●	●	●	●	●	●	●
Classifying figures	●	●	●	●	●	●	●	●
Solid figures	●	●	●	●	●	●	●	●
Congruence				●	●	●	●	●
Symmetry				●	●	●	●	●
Symmetry (rotational)					●	●	●	●
Slides/Flips/Turns					●	●	●	●
Angles					●	●	●	●
Classifying triangles				●	●	●	●	●
Classifying quadrilaterals						●	●	●
Parallel and perpendicular lines						●	●	●
Perimeter				●	●	●	●	●
Radius and diameter							●	●
Circumference							●	●
Surface Area				●	●	●	●	●
Volume				●	●	●	●	●

Scope and Sequence

	A	B	C	D	E	F	G	H
Measurement								
Length								
Use customary units				•	•	•	•	•
Use metric units				•	•	•	•	•
Mass/Weight								
Use customary units				•	•	•	•	•
Use metric units				•	•	•	•	•
Capacity								
Use customary units				•	•	•	•	•
Use metric units				•	•	•	•	•
Temperature								
Use degrees Fahrenheit				•	•	•	•	
Use degrees Celsius				•	•	•	•	
Converting within customary system						•	•	•
Converting within metric system						•	•	•
Telling Time								
to the hour	•	•			•			
to the half hour					•			
to the quarter hour					•			
to the minute					•			
Converting units of time					•	•		
Reading a calendar					•			
Multiplication								
Basic facts					•	•	•	•
One-digit multipliers						•	•	•
Two-digit multipliers						•	•	•
Greater multipliers							•	•
Estimating products							•	•
Number and Numeration								
Reading and writing numbers	•	•	•	•	•	•		
Counting	•	•	•	•	•	•		
Skip counting			•	•	•	•		
Ordinal numbers				•	•	•		
Place value			•	•	•	•		
Even/odd numbers				•	•			
Comparing and ordering numbers	•	•	•	•	•	•	•	•
Rounding						•	•	•
Estimation/Approximation				•	•	•	•	•

Scope and Sequence

	A	B	C	D	E	F	G	H
Comparing and ordering integers							•	•
Integers (negative numbers)							•	•
Prime and composite numbers							•	•
Factors and prime factorization							•	•
Common factors							•	•
Common multiples						•	•	•

Subtraction (whole numbers)

	A	B	C	D	E	F	G	H
Basic facts	•	•	•	•	•	•	•	•
Two-digit numbers			•	•	•	•	•	•
Three-digit numbers					•	•	•	•
Greater numbers						•	•	•
Estimating differences	•	•	•	•	•	•	•	•

Patterns, Relations, and Functions

	A	B	C	D	E	F	G	H
Number patterns	•	•	•	•	•	•	•	•
Geometric patterns	•	•	•	•	•	•	•	•
Inequalities	•	•	•	•	•	•	•	•

Ratio and Proportion

	A	B	C	D	E	F	G	H
Meaning/Use							•	•
Similar Figures							•	•
Meaning of Percent							•	•
Percent of a Number							•	•

Statistics and Graphing

	A	B	C	D	E	F	G	H
Real and picture graphs	•	•	•	•	•	•	•	•
Bar graphs	•	•	•	•	•	•	•	•
Line graphs			•	•	•	•		
Circle graphs					•	•	•	•
Analyzing graphs	•	•	•	•	•	•	•	•
Finding the mean							•	•
Finding the median							•	•
Finding the mode							•	•
Finding the range							•	•